THE CAMBRIDGE GUIDE TO PREHISTORIC MAN

THE CAMBRIDGE GUIDE TO PREHISTORIC MAN

David Lambert
and the Diagram Group

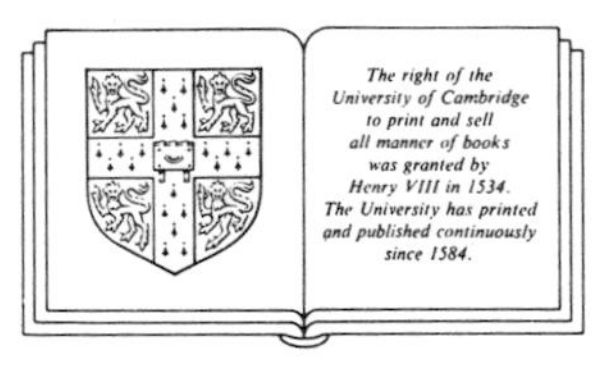

CAMBRIDGE UNIVERSITY PRESS
Cambridge
New York Port Chester Melbourne Sydney

The Diagram Group

Editor	Denis Kennedy
Editorial assistant	Annabel Else
Indexer	David Harding
Designer	Arthur Lockwood
Art editor	Richard Hummerstone
Artists	Graham Rosewarne, and Joe Bonello, Ray Burrows, Richard Czapnik, Brian Hewson, Philip Patenall

Published by the Press Syndicate of the University of Cambridge
The Pitt Building, Trumpington Street, Cambridge, CB2 1RP
40 West 20th Street, New York, NY 10011, USA
10 Stamford Road, Oakleigh, Melbourne 3166, Australia

First published 1987
Reprinted 1989

British Library cataloguing in publication data:

Lambert, David, *1932–*
The Cambridge guide to prehistoric man
1. Human evolution
I. Title
573.2 GN 281

ISBN 0 521 33364 4 hard covers
ISBN 0 521 33644 9 paperback

Printed in Portugal by Resopal Indústria Gráfica, Ida. – Sintra

Consultants

Dr Leslie Aiello, University College, University of London
Dr Peter Andrews, British Museum (Natural History), London
Dr Peter L. Drewett, Institute of Archaeology, University of London
Dr Christopher Stringer, British Museum (Natural History), London
Professor Bernard Wood, The University of Liverpool, England

Acknowledgment
In the course of preparing this book for publication we have consulted many reference sources, the majority of which are listed below. We should like to thank the respective authors/publishers of these books and apologize for any omissions from the list.

Aiello, L. *Discovering the Origins of Mankind* Longman, 1982
Archer, M. and Clayton, G. (editors) *Vertebrate Zoogeography & Evolution in Australasia* Hesperian Press, Australia 1984
Bordaz, J. *Tools of the Old and New Stone Age* David and Charles, 1971
Brace, C.E. and Montagu, A. *Human Evolution* Maçmillan, 1977
Clapham, F.M. (editor) *The Rise of Man* Sampson Low, 1976
Cole, S. *The Neolithic Revolution* British Museum (Natural History), 1970
Coles, J.M. and Higgs, E.S. *The Archaeology of Early Man* Faber and Faber, 1969
Day, M.H. *Fossil Man* Hamlyn, 1969
Gribbin, J. and Cherfas, J. *The Monkey Puzzle* Bodley Head, 1982
Jurmain, R., Nelson, H., Kurashina, H., and Turnbaugh, W.A. *Understanding Physical Anthropology and Archeology* West Publishing Co., 1981
Kennedy, G.E. *Paleoanthropology* McGraw Hill, 1980
Leakey, M. *Olduvai Gorge* Collins, 1979
Leakey, R.E. *The Making of Mankind* Michael Joseph, 1981
Lewin, R. *Human Evolution* Blackwell Scientific Publications, 1984
McEvedy, C. and Jones, R. *Atlas of World Population History* Allen Lane, 1978
Napier, P. *Monkeys and Apes* Hamlyn, 1970
Oakley, K.P. *Man the Tool-Maker* British Museum (Natural History), 1972
Palmer, S. *Mesolithic Cultures of Britain* Dolphin Press, 1977
Phillips, P. *The Prehistory of Europe* Penguin Books, 1981
Poirier, F.E. *Fossil Evidence* C.V. Mosby Co., 1981
Reader, J. *Missing Links* Collins, 1981
Romer, A.S. *Man and the Vertebrates* Penguin Books, 1954
Romer, A.S. *The Vertebrate Body* W.B. Saunders Co., 1970
Semenov, S.A. *Prehistoric Technology* Adams & Dart, 1964
Shackley, M. *Neanderthal Man* Duckworth, 1980
Szalay, F.S. and Delson, E. *Evolutionary History of the Primates* Academic Press, 1979
Tomkins, S. *The Origins of Mankind* Cambridge University Press, 1984
Weiner, J.S. *Man's Natural History* Weidenfeld and Nicolson, 1971
Wood, B. *The Evolution of Early Man* Peter Lowe, 1976

Various: *The Emergence of Man* series published by Time-Life Books, 1972–73; *Encyclopaedia Britannica; National Geographic.*

FOREWORD

This book offers a concise, up-to-date key to the origins of humankind. Large, labeled, life-like restorations and reconstructed skeletons, marginal "field-guide" illustrations, diagrams, and family trees are integrated with a text combining popular and scientific terms. The result is a guide accessible to anyone from the inquiring eleven-year-old to the budding scientist.

There are 11 chapters. Each has a brief explanatory introduction, followed by topics arranged under bold headings.

Chapter 1 (What is Man?) outlines key features that distinguish us from other living things, and shows how evolution produced human organs and body systems.

Chapter 2 (Primitive Primates) deals with the origins of primates – the mammal group including man – and describes extinct and living prosimians: the lower primates.

Chapter 3 (Evolving Anthropoids) traces evolutionary trends among the higher primates: the group including monkeys, apes, and man. The chapter takes us up to prehistoric apes apparently ancestral to modern apes and man.

Chapter 4 (Apes and Man) points up the many similarities that underline the close relationship between the living great apes and ourselves.

Chapter 5 ("Men-apes" and Early Man) looks at evolutionary trends that transformed prehistoric apes into early man. These pages describe the australopiths and their possible descendant *Homo habilis*, the first reputed member of our genus.

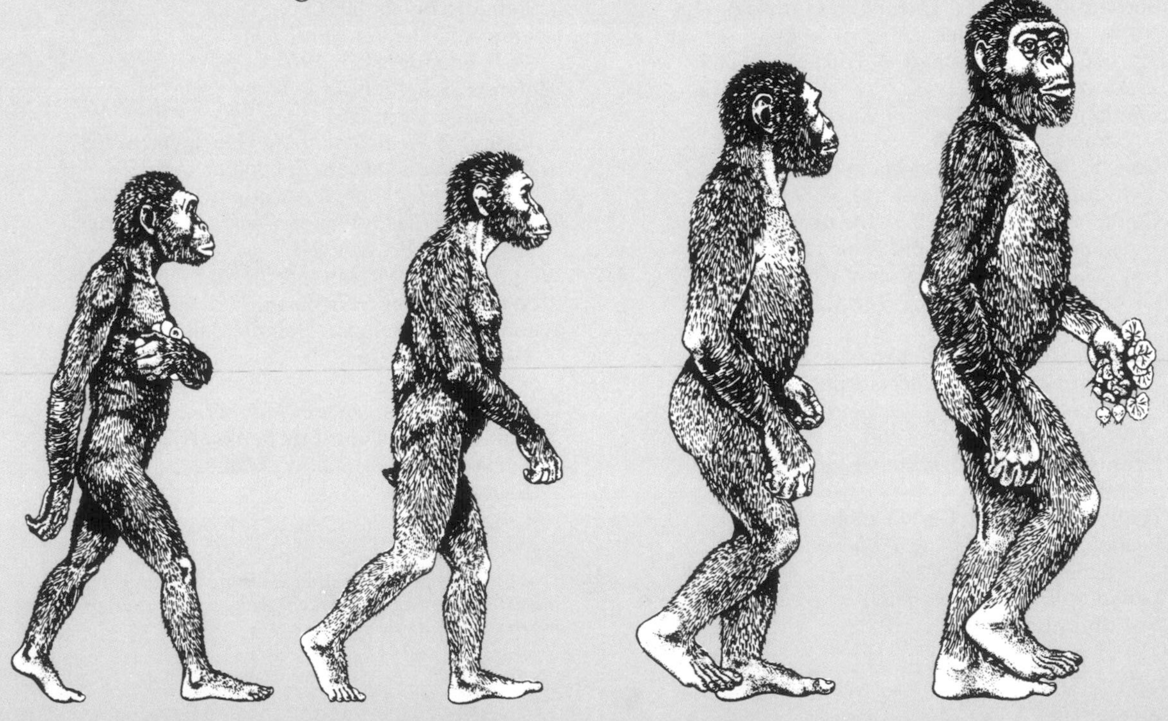

Chapter 6 (Upright Man) describes the form and lifestyle of *Homo erectus*, a widespread human species apparently ancestral to our own.

Chapter 7 (Neandertal Man) covers archaic members of our species *Homo sapiens*.

Chapter 8 (Modern Man in Europe) deals with early European members of fully modern man, our own subspecies: *Homo sapiens sapiens*.

Chapter 9 (Modern Man Worldwide) embraces the global spread of fully modern man.

Chapter 10 (Since the Ice Age) shows how cultural evolution has transformed human life in the last 10,000 years.

Chapter 11 (Discovering Man's Past) explains how scientists trace our prehistoric origins, includes achievements of major paleontologists and archeologists, and gives brief details of museum displays around the world.

Lastly, there is a list of books for further reading, and an index.

The producers of this guide have slightly modified Szalay and Delson's 1979 classification of primates, adopted a 1982 Cambridge revision of some time units, otherwise indicating major points where experts disagree. The author is answerable for all facts here presented, but thanks those named and unnamed experts whose advice has helped to make this book more accurate and up to date.

CONTENTS

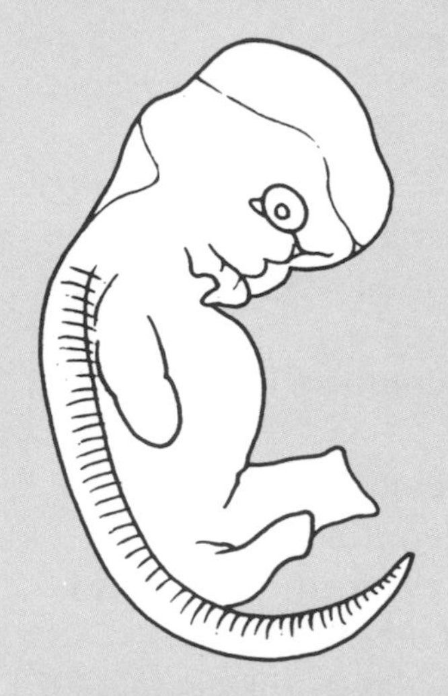

Chapter 1
WHAT IS MAN?

Chapter 2
PRIMITIVE PRIMATES

Chapter 3
EVOLVING ANTHROPOIDS

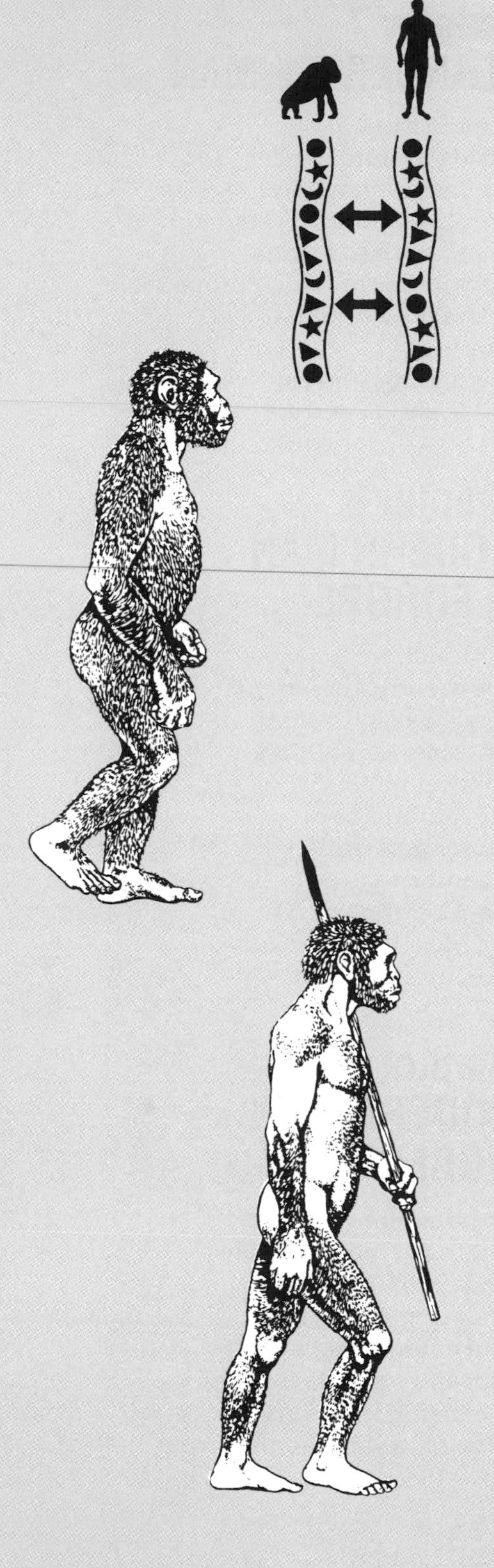

Chapter 4 APES AND MAN

Chapter 5 "MEN-APES" AND EARLY MAN

Chapter 6 UPRIGHT MAN

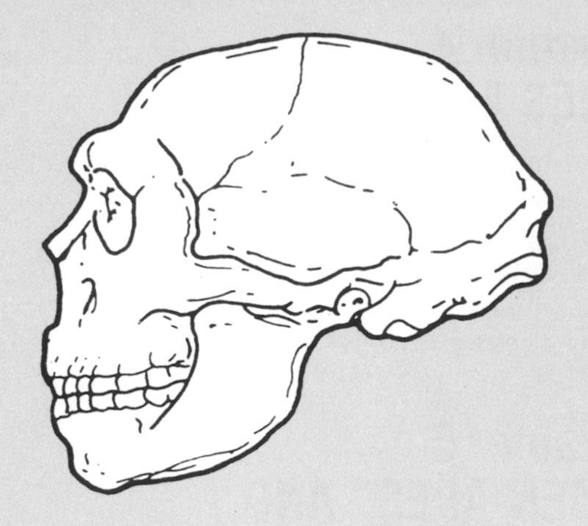

Chapter 7
NEANDERTAL MAN

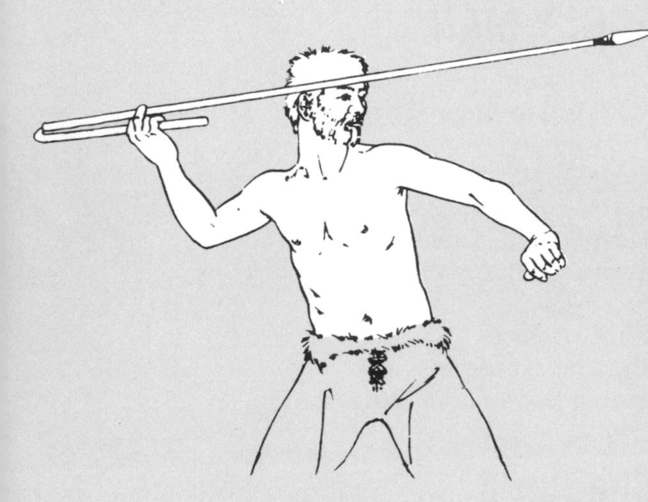

Chapter 8
MODERN MAN IN EUROPE

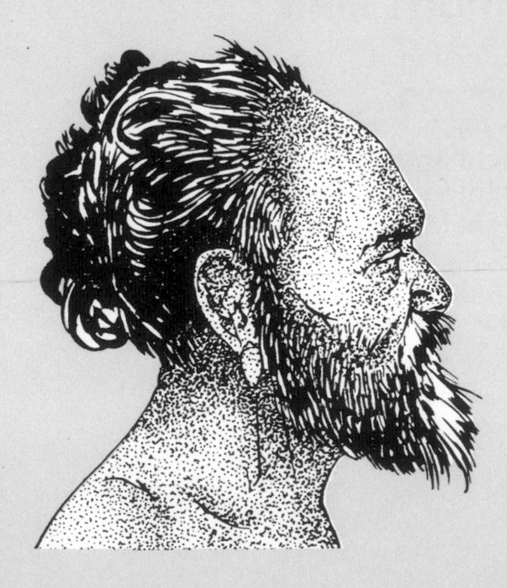

Chapter 9
MODERN MAN WORLDWIDE

Chapter 10
SINCE THE ICE AGE

Chapter 11
DISCOVERING MAN'S PAST

Chapter 1 WHAT IS MAN?

This chapter outlines the combined physical and mental attributes that make our species, *Homo sapiens*, unique.

As evolutionary products, though, our bodies comprise systems inherited from other creatures. Comparisons with various backboned animals – some living, some extinct – reveal major changes that created human bones, lungs, blood supply, and other features.

The chapter ends by placing man in context, as a member of the group of mammals known as primates.

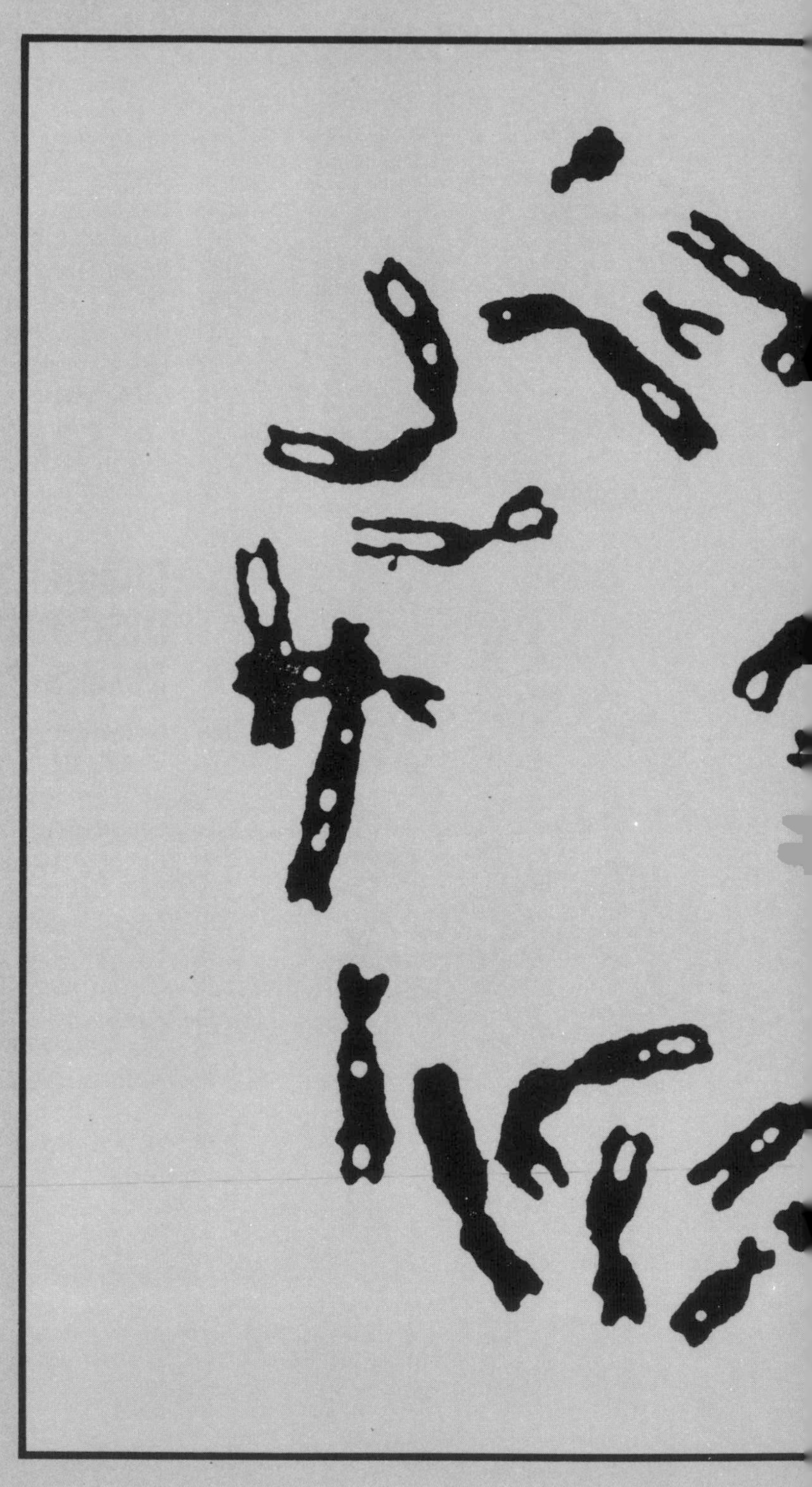

This illustration shows the 46 chromosomes in a dividing human body cell, enlarged 5000 times. The genes in human chromosomes bear hereditary information that makes our species different from all others. (Adapted from *Scientific American*)

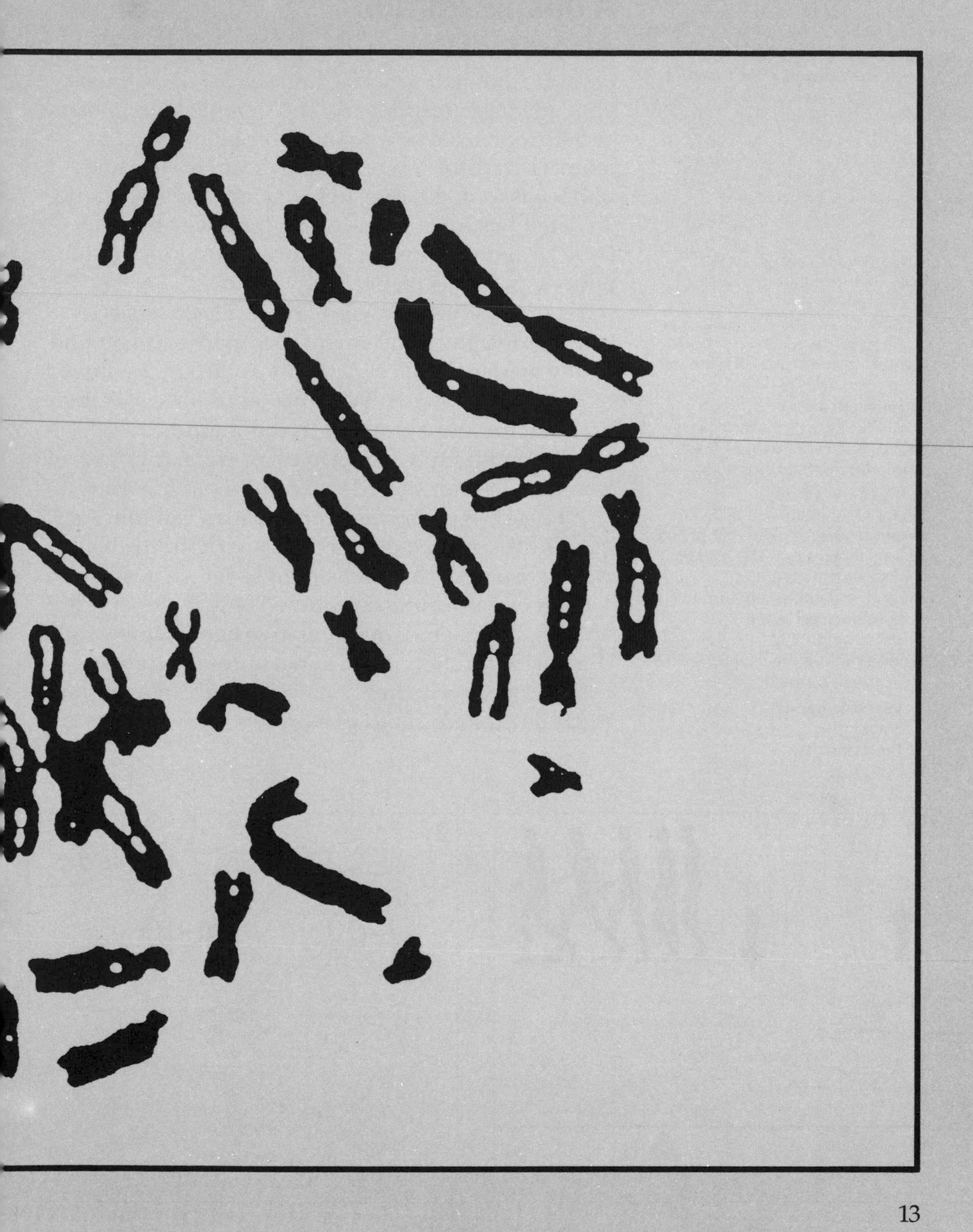

A unique animal

Our species has been defined variously as intelligent, political, tool-using, social, and self aware. Armed with advanced technology, and organized in massive social groups, we now manipulate plants and animals, transforming the Earth's surface to satisfy our needs for food, fuel, living space, and transport. To defend us from others of our kind we now also have weaponry capable of wiping most kinds of life entirely from this planet.

Yet the uniquely powerful being *Homo sapiens* contains just the same chemical ingredients you find in any mammal – such elements as carbon, hydrogen, oxygen, and nitrogen. What makes us special is the way that these are grouped in compounds, cells, tissues, and body systems to create a creature with all round ability not shared by any other living thing.

A horse can gallop faster than a man can run. An eagle's eyes are keener than our own. Our teeth and claws are no match for a leopard's. Yet we possess four crucial attributes, combined only in our species: an upright skeleton, manipulative hands, three-dimensional color vision, and a uniquely complex brain. Collectively these four confer advantages that made us masters of our planet.

Human attributes
Four illustrations show aspects that collectively distinguish us from other living things.

1 Bipedal walking
Weight is transmitted from heel through outer edge of foot to ball of foot and big toe. The leg and foot act as a lever with:
a Load, transmitted by tibia
b Fulcrum (ball of foot)
c Effort, applied by the Achilles tendon pulling the heel up as the calf muscle contracts.

2 Versatile hands
a Power grip
b Precision grip
c Cupping

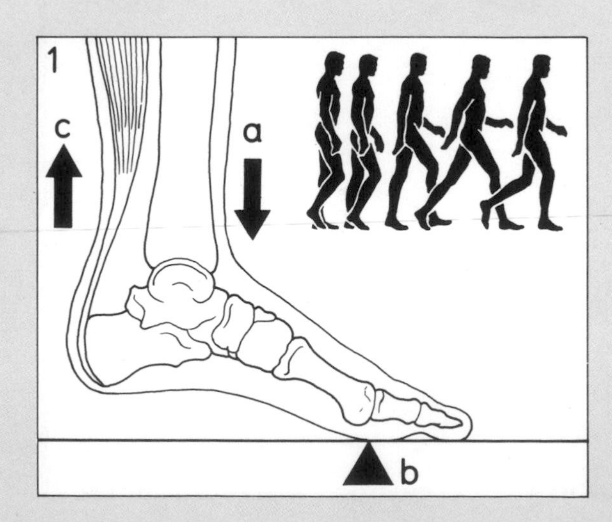

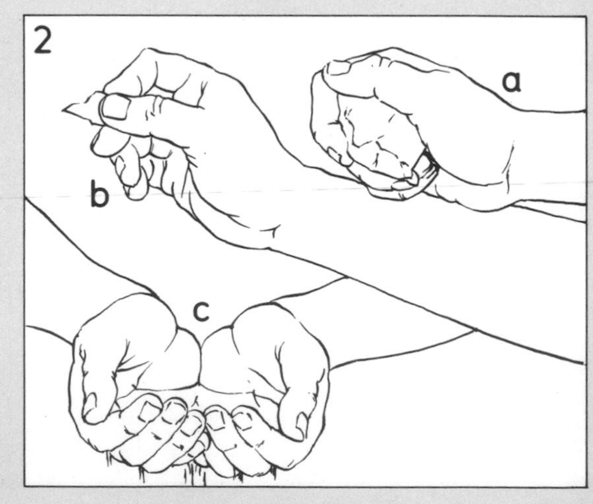

Our upright skeleton enables us to travel on two legs with a unique heel-toe action, each step a balancing act demanding split-second coordination of muscles in the back, hips, legs, and feet. Not only can we walk but run, jump, swim, dive, and climb a cliff or tree. Long-distance runners have more endurance than a deer.

With forelimbs not needed for support, we use the flexible and sensitive thumbs and fingers of our hands to explore surfaces by touch, and to grip objects strongly or precisely. Swapping tool for tool at will, we control our environment far more effectively than any mammal stuck with forelimbs shod with hooves or armed with claws.

Forward-facing eyes sensitive to color enable us to focus sharp images, judge distance accurately, and distinguish hue as well as shape and brightness – abilities few other mammals share. Rotating in their sockets, eyes follow movement without the need to move our head. And because we stand high off the ground we see much farther than other ground-based mammals of our size.

Lastly comes a brain, large relative to body size and superbly capable of learning, reasoning, controlling speech, and precise hand-eye coordination.

3 Binocular vision
Both eyes combine to focus on objects in a wide range of locations:
a Near
b Left
c Right
d Distant

4 Big brain
Seen here from above, the big human brain is deeply convoluted. Its wrinkles produce a huge surface area – the part that largely gives the brain intelligence.
a Normally visible surface
b Surface if spread flat: an area of 324 sq in (2090 sq cm).

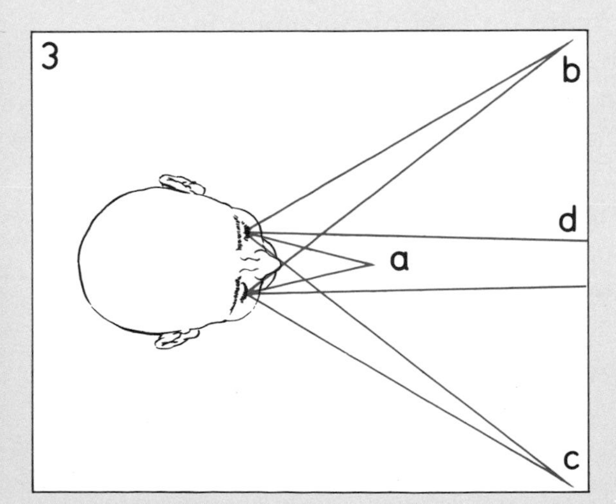

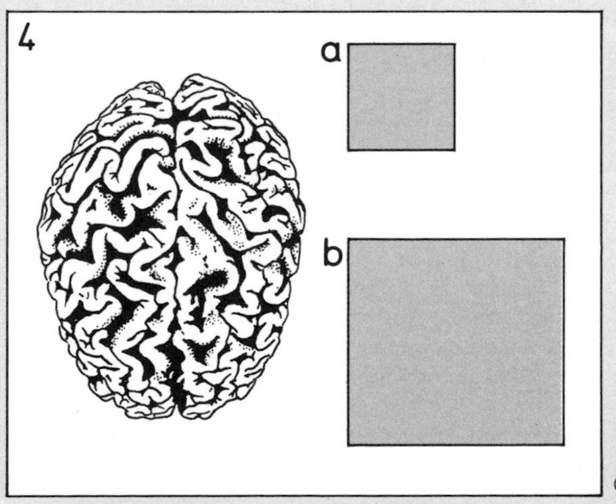

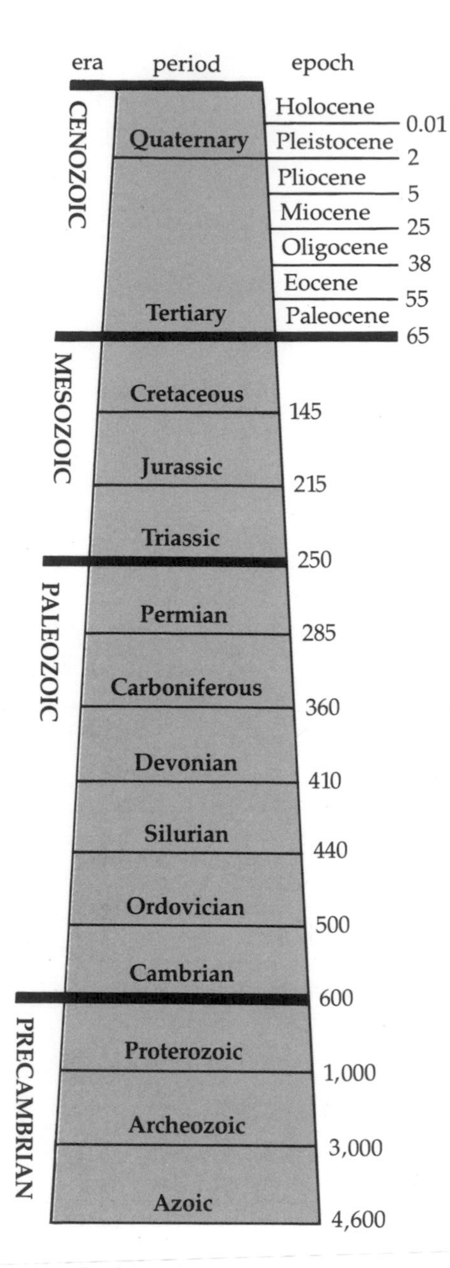

Prehistoric time
This shows major and minor subdivisions of the geological column: prehistoric time, as represented by successive layers of sedimentary rocks. Figures are in millions of years ago. Early and late Carboniferous are North America's Mississippian and Pennsylvanian.

Man's early ancestors

Scientists now recognize that the human body and its special attributes are products of 3500 million years of evolution that gave rise also to the million other kinds of creature now alive, and many millions that died out long ago. Modern knowledge of heredity shows how evolution works, but the chief clues to our early ancestors are their fossilized remains in ancient rocks, laid down throughout prehistory (see also p. 222).

New kinds of animal evolved from old by cumulative changes in their genes – minute hereditary units in every body cell, dictating how each organism grows, and passing similar instructions to its offspring. Only genetically compatible individuals' matings produce fertile offspring, and only genetic changes favoring survival persisted. So arose the many different species, each adapted to life in a particular environment. Environmental change and competition for survival, selecting from random genetic change, steered evolution from microscopic primeval organisms to the complex creature, man.

Life seemingly began in seas, with minute one-celled organisms that split to multiply. Imperfect splitting probably created metazoans: many-celled animals. By 700 million years ago, warm seas held worms and other creatures with specialized cells grouped in organs for feeding, locomotion, reproduction, and so on. Worm-like creatures indirectly led to fishes – creatures with an inner skeleton. From air-breathing fishes came amphibians with limbs. From these came reptiles, the first land-breeding vertebrates. Some evolved into birds, others into mammals – warm-blooded vertebrates possessing hair and different types of teeth, mostly giving birth instead of laying eggs, and suckling their young. This group of animals embraces man.

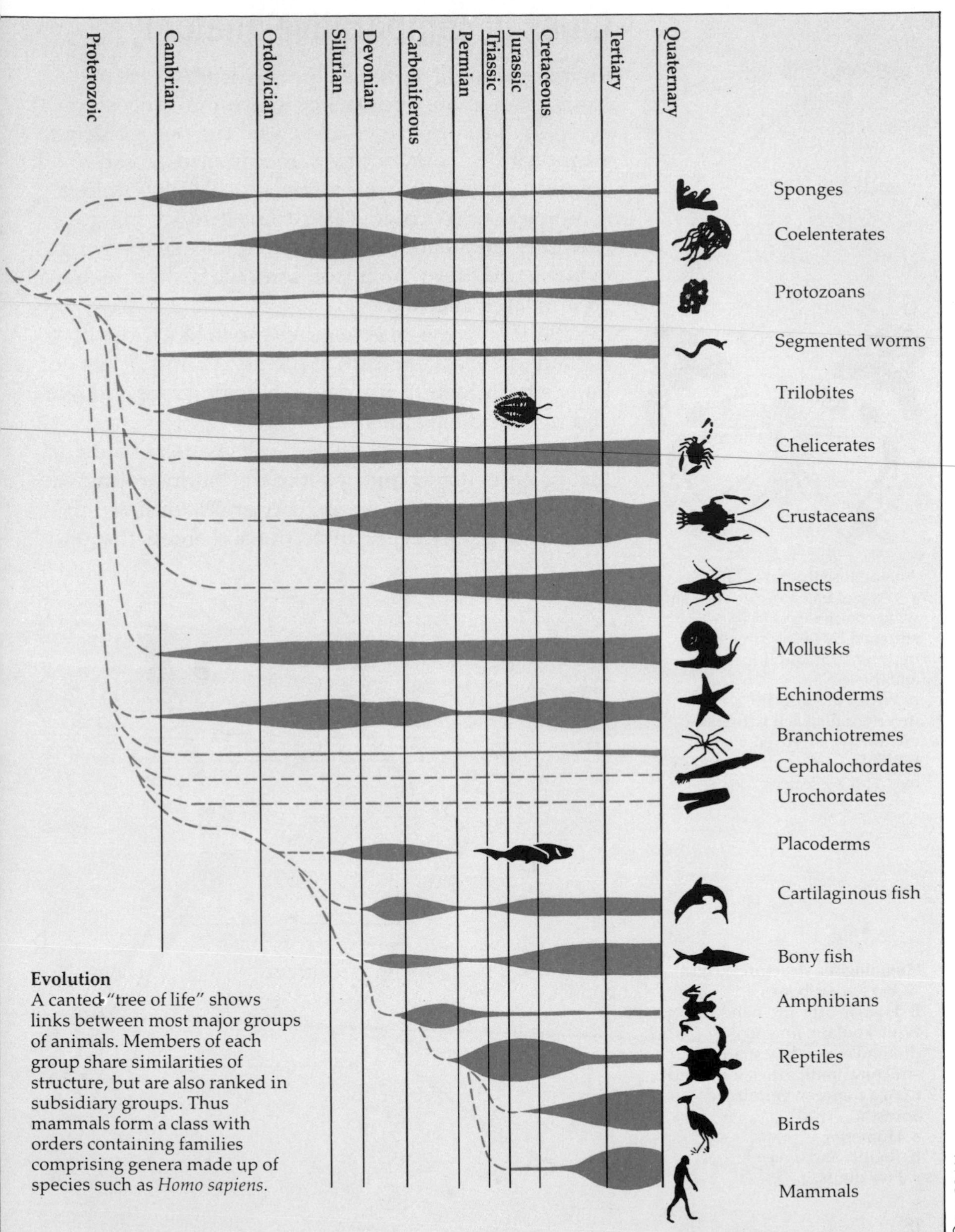

Evolution
A canted "tree of life" shows links between most major groups of animals. Members of each group share similarities of structure, but are also ranked in subsidiary groups. Thus mammals form a class with orders containing families comprising genera made up of species such as *Homo sapiens*.

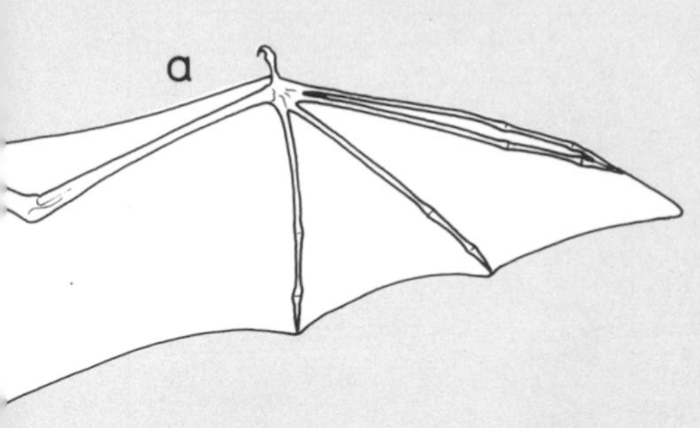

Clues in comparative anatomy

Living and fossil animals with structures that correspond to ours hold clues to common ancestors: the more the similarities, the closer the relationship to us appears. But anatomists carefully distinguish between *homologous* structures, those which share a common origin (a man's arm and a bird's wing, for example), and *analogous* structures such as the wings of birds and flies which look somewhat alike yet have an unrelated ancestry.

Primitive living sea creatures much like those of 700 million years ago hint at the great antiquity of our body's basic layout: its inside, outside, front and rear, and right and left sides.

A sea anemone has only two cell layers: an outside that protects it and informs it of the world around, and a layer lining an inner cavity and handling nourishment and reproduction. Food enters through

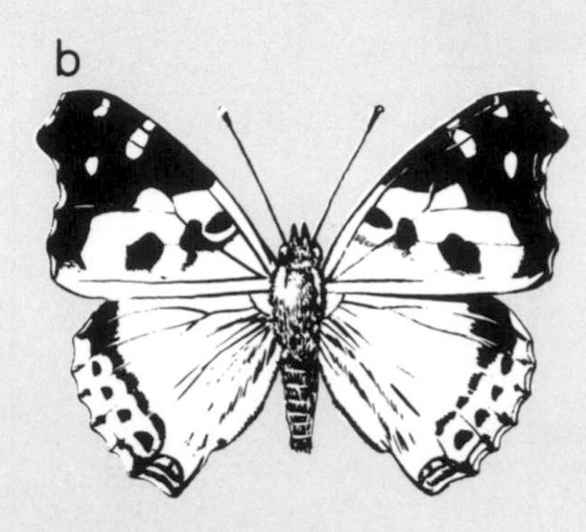

Analogous structures (above)
a Wing of bat, a vertebrate. This wing comprises skin flaps stiffened by being stretched between arm and finger bones and the legs.
b Wings of butterfly, an invertebrate. Each wing is a membrane stretched over stiff, tube-like veins, and covered by overlapping scales.

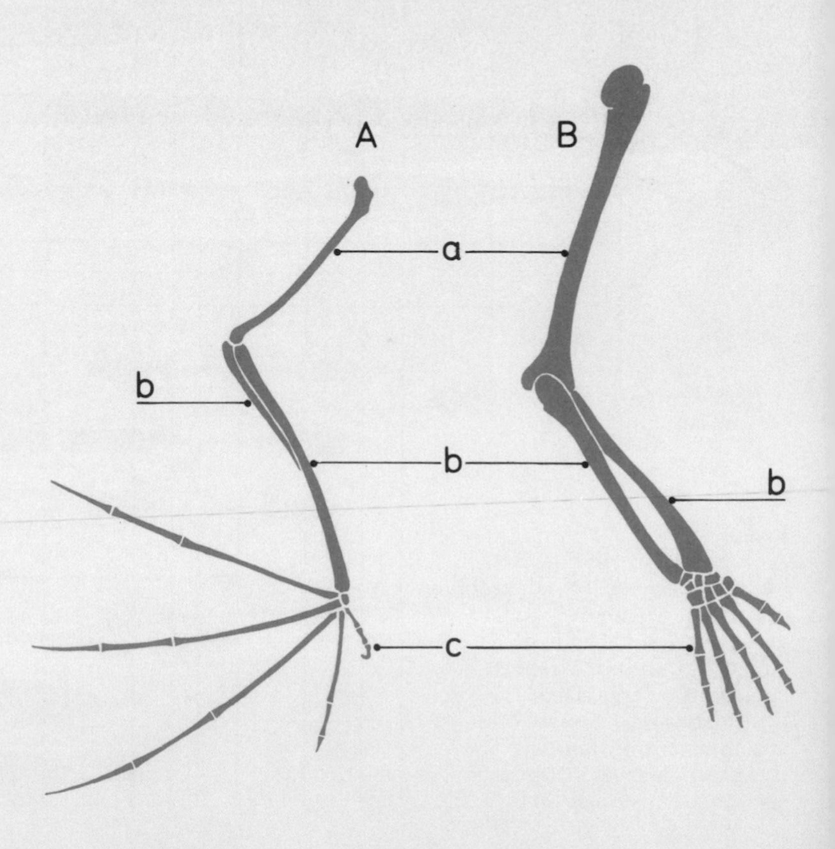

Homologous structures (right)
A Bat's wing bones
B Human arm and hand bones. Wing and arm may look dissimilar, yet they share structural similarities inherited from a common vertebrate ancestor:
a Humerus
b Radius and ulna
c Five digits

a single opening to be digested by cells that line this pouch, while the same opening releases eggs and body waste. In worms the pouch has become a tube, with a mouth at one end and an anus at the other – the basic pattern of our own digestive system.

This layout automatically gave its owner a front and rear. Cells specialized and grouped as muscles helped mobile creatures move forward to find food for the mouth. Tentacles or teeth clustered around the mouth for seizing food. Cells specialized for finding food by scent or sight also became concentrated at the front, which gained a head with a brain coordinating signals for an increasingly complex nervous system.

Meanwhile, in man's early worm-like ancestors, the body had become bilaterally symmetrical: each side a mirror image of the other. This made it easier to travel forward, steer from side to side, and stay the right way up. Our own paired limbs, eyes, ears, and nostrils are legacies of this arrangement.

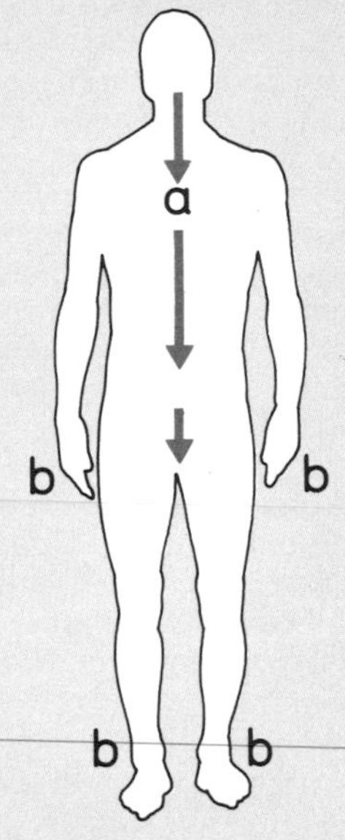

Lowly legacies (above)
We probably have lowly wormlike ancestors to thank for our body's basic layout:
a Central digestive system, with mouth, gut, anus
b Bilateral symmetry: with paired limbs and other organs

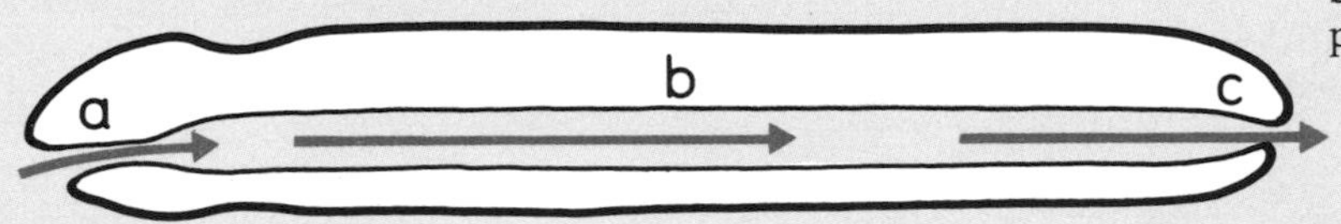

Front and rear (above)
This longitudinal section through a worm shows the one-way flow of food made possible by the evolution of three crucial structures:
a Mouth
b Gut
c Anus

Amphioxus (below)
Amphioxus, the living lancelet, lacks a head, jaws, or vertebrae, but such beasts might have given rise to all backboned animals from fish to man. like a fish, a lancelet has:
a Gills
b Nerve cord
c Notochord (a precursor of the backbone)
d Tail fin

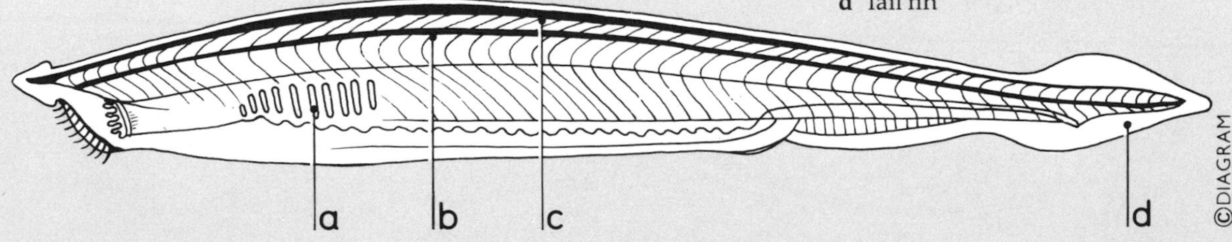

Bones and muscles

Evolving backbones
Six diagrams trace the evolution of the backbone from horizontal rod to cantilevered arch, and vertical column.
1 *Eusthenopteron's* straight backbone gave its swimming muscles leverage. This fish lived 375 million years ago.
2 A strong curved backbone supported early amphibian sprawlers like *Ichthyostega* as they crawled about on land.
3 Complex changes to the spine and ribs improved land mobility in the ancient mammal-like reptile *Thrinaxodon*, perhaps ancestral to the mammals.

Most of the human body is bone and muscle – products of a middle body layer that first appeared between the other two among some early animals without a backbone. Many of these invertebrates evolved an outside shell or skeleton for protection and support, but vertebrates evolved internal bony skeletons: stores of calcium and phosphorus doubling as guards for vital organs and attachment points for muscles, giving leverage for limbs or fins. A crab's shell must be shed from time to time with growth, leaving its owner unprotected until a new shell hardens; but internal skeletons grow at the same rate as the whole body.

In jawless fishes and all later vertebrates the basic bony structure is a backbone of interlocking vertebrae. These shield the vulnerable spinal cord,

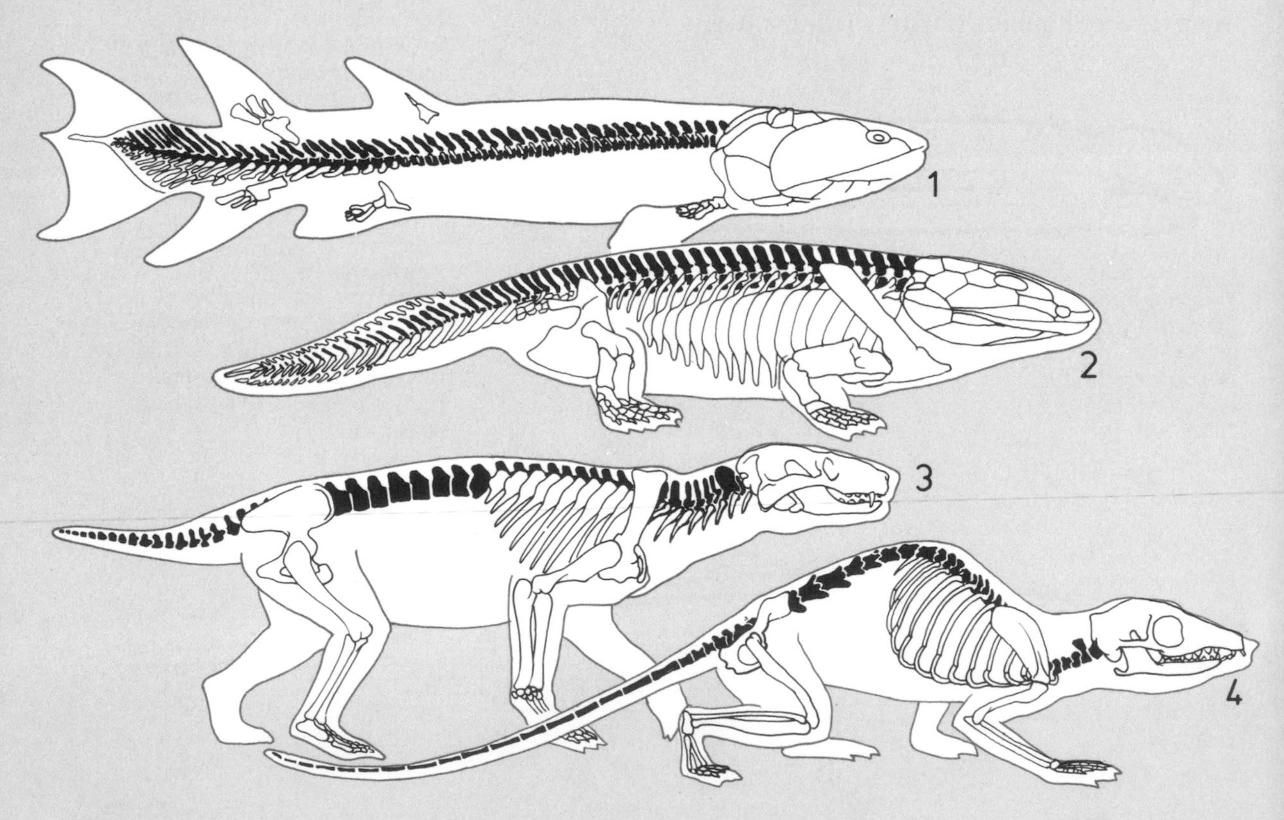

while ribs projecting at intervals from each side of the backbone guard soft internal organs and provide attachment points for intervening muscle segments. With these a fish waggles its body and tail which thrust water back to drive the body forward.

Fishes have an almost straight backbone with relatively weak vertebrae. Vertebrates that walk on land need stronger internal support. Early amphibians like *Ichthyostega* had a sturdy backbone curved in a shallow arch – a suspension bridge with the body's trunk slung beneath – and a large rib cage to support the body when lying on the ground. Ribs also probably expanded and contracted to help operate the lungs.

Some land-based vertebrates showed complex spinal changes. As limbs took over locomotion, the tail dwindled to a balancer or fly whisk, and in apes and man disappeared, while ribs tended to vanish from the waist and neck. In reptiles ancestral to mammals, neck ribs shrank, allowing freer head movements. Also neck vertebrae curved up to lift the head. In certain mammals, changes to the first two neck vertebrae allowed the head to turn more freely. In man, a second spinal curve low in the back pushed the chest up and back, bringing chest and head above the hips for good bipedal balance.

4 Flexible backbones of early insect-eating mammals resembling this living tree shrew adapted them for climbing trees.
5 The semi-upright chimpanzee has a straighter backbone than its quadrupedal monkey-like ancestors, whose spine was arched between supporting hips and shoulders.
6 Man's spinal column forms a sinuous column combining strength and flexibility. Its curves are shaped to keep the head and body above the center of gravity.

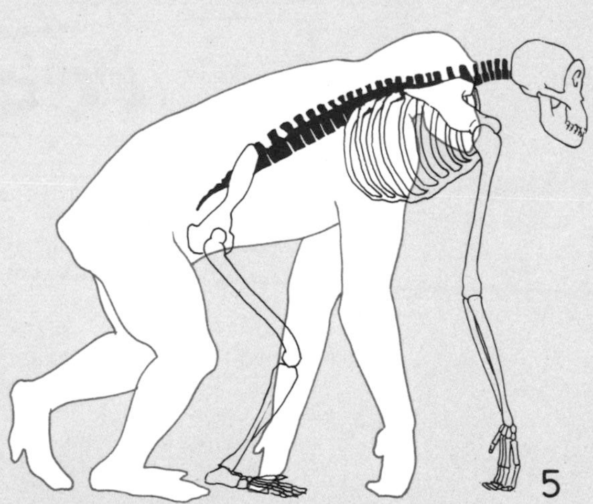

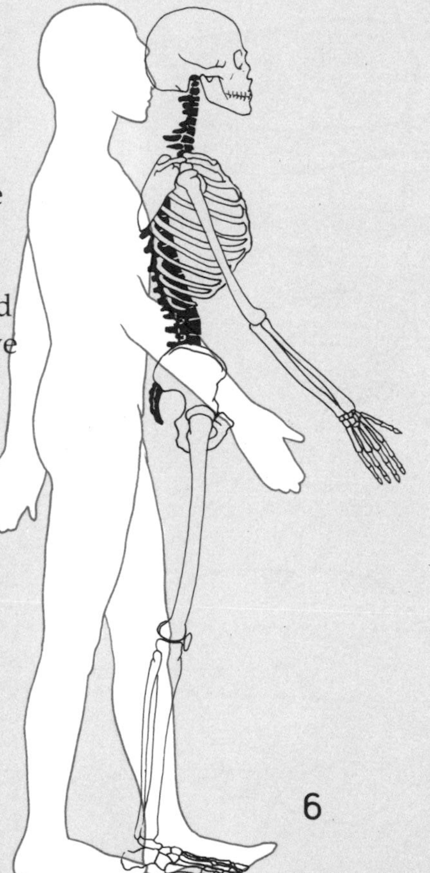

Skull, jaws, and teeth

Our skull, jaws, and teeth have various and complex origins.

Early vertebrates, somewhat resembling those living jawless fishes lampreys, had no true skull or jaws; just a box of gristle or bone guarding brain and internal ear and protecting eyes and nostrils, plus bars of bone or gristle stiffening gill slits in the side of the head. Bony plates set in the skin protected head and neck.

In certain fishes, some gill bars disappeared but one pair enlarged to form hinged jaws, in time equipped with teeth derived from denticles: sharp, enamel-coated points embedded in the skin (as they still are in sharks). Meanwhile some armor plates migrated from the skin to cover most of the head, fusing with, reinforcing, and largely replacing braincase and upper jaw, converting all into a single bony skull and adding extra bones onto the lower jaw.

1

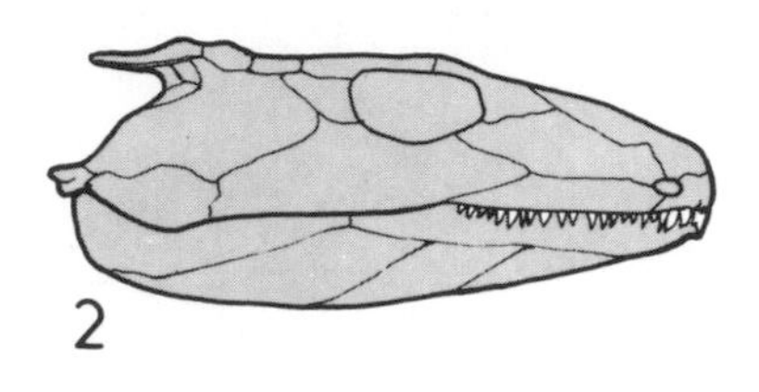

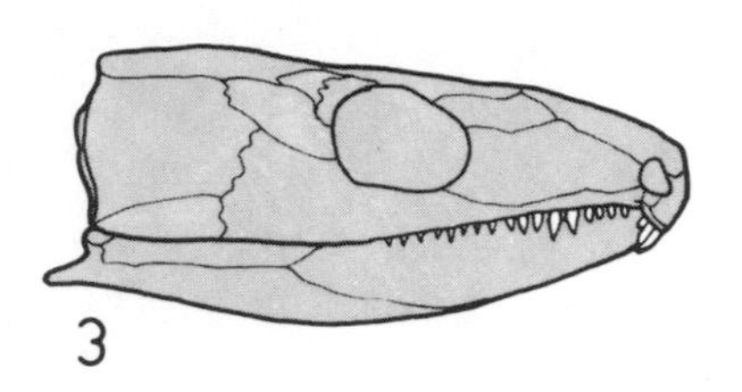

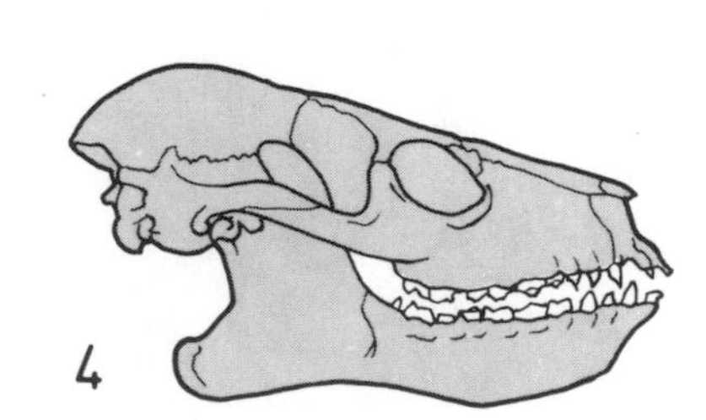

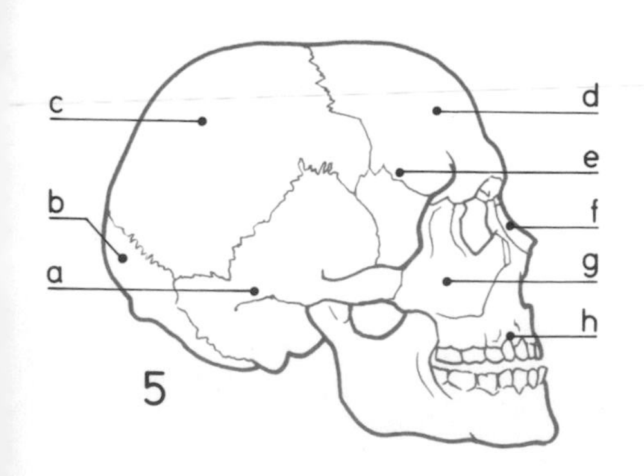

Evolving skulls (left)
Five skulls (four from extinct animals) show an evolutionary trend to fewer bones.
1 Lobe-finned fish
2 Primitive amphibian
3 Primitive reptile
4 Fossil lemur (a mammal)
5 Modern man

Bones of a human skull
a Temporal
b Occipital
c Parietal
d Frontal
e Sphenoid
f Nasal
g Zygomatic
h Maxilla
i Mandible

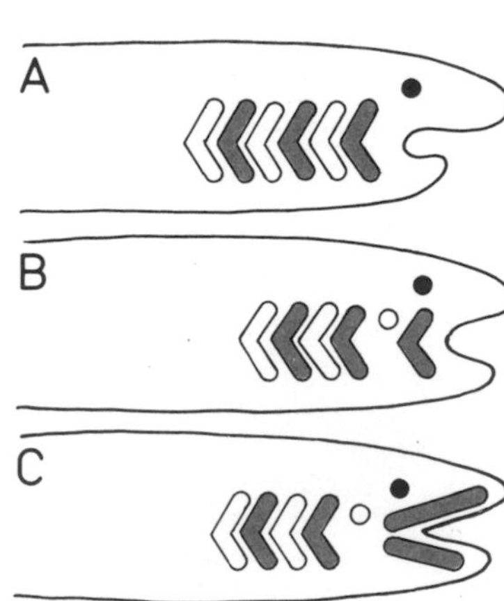

Evolving jaws (above)
Three diagrams trace the evolution of jaws from a fish's gill bars, stiffening gill slits in the sides of the head.
A Jawless fish. Bony gill bars (tinted) alternate with gills.
B The first pair of gills shrinks to a spiracle, a tiny hole for drawing in water free from mud.
C The first pair of gill bars expands to form hinged jaws

In our own skull ancient "skin bones" form the top and front, and most of the sides and mouth roof. Only the back and base of our brain are still protected by primeval braincase bone.

Almost all our skull bones can be traced back to those in lobe-finned fishes of around 350 million years ago. But with redesigning for life on land, bones guarding gills and throat dropped out. Mammal-like reptiles gained differentiated teeth, a skull opening behind the eye to give jaw muscles freer action, and a bony partition above the mouth roof, an aid to breathing while eating. Meanwhile some skull and jaw bones disappeared. Of the early reptiles' seven jaw bones only one – the tooth-bearing dentary – survived in mammals. These gained a new joint where jaw articulates with skull, while tiny bones from the old jaw joint migrated in the skull becoming two middle ear bones whose vibrations help us hear. Meanwhile early mammals' braincases expanded to accommodate expanding brains. Further redesigning of the skull accommodated the early primate brain ancestral to our own (see pp. 30–31).

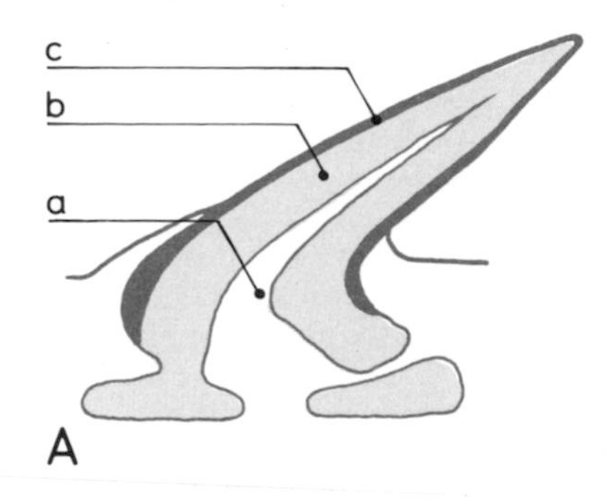

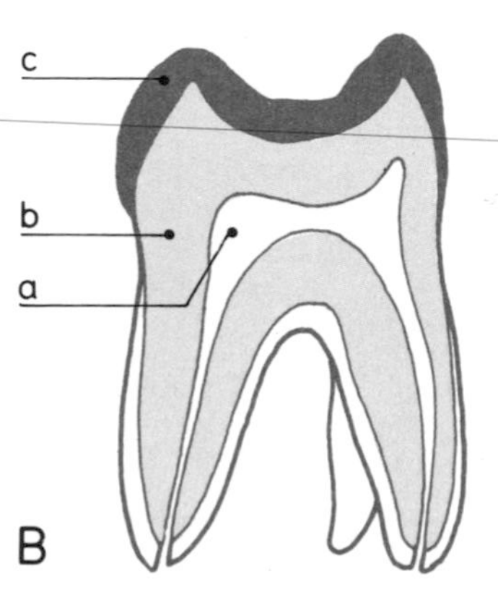

Teeth from denticles (above)
Sections through a shark's denticle and a mammal's tooth show similarities which hint that teeth came from structures set in skin.
A Shark's denticle
B Mammal molar tooth
a Pulp cavity
b Dentine
c Enamel-like vitrodentine (in shark) or (in mammal) enamel

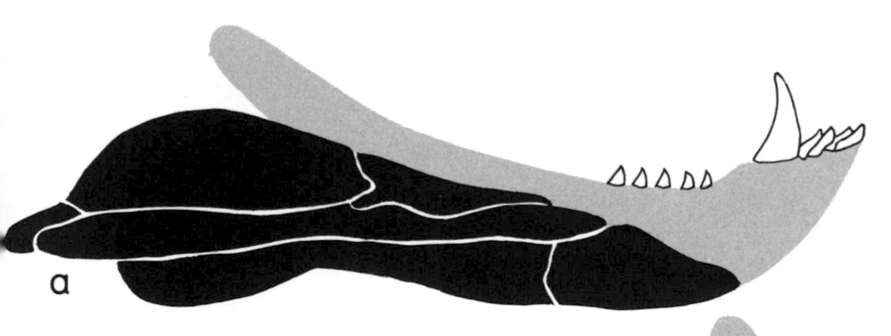

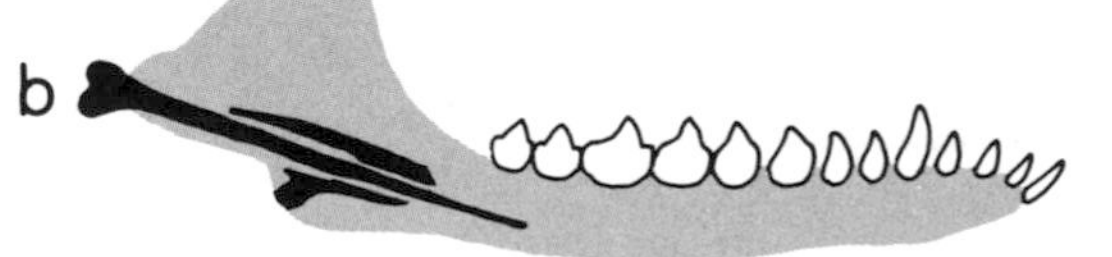

The successful dentary
Inner-jaw views show a trend to fewer bones. The dentary is tinted.
a Mammal-like reptile: seven bones. Mammal-like reptiles gave rise to mammals.
b Early mammal: four bones.
c Another early mammal: one bone as in modern mammals.

Limbs

Evolving limbs
Six diagrams trace pelvic and pectoral (hip and shoulder) bone changes that improved limb strength and leverage.
1 *Eusthenopteron*'s pelvis was unattached to the spine and this rhipidistian fish's shoulder girdle was joined to an almost rigid skull.
2 A pelvic girdle linking hindlimb to spine and a shoulder girdle separated from the skull produced stronger limb leverage and freer movement for early walkers like amphibian *Ichthyostega*.
3 Further changes gave mammal-like reptile *Thrinaxodon* more limb power and mobility.
4 A tree shrew's long, narrow pelvis and its flexible shoulder joint, allowing arm raising, are aids to climbing.

Legs and arms are outgrowths of the body that we can trace back through the fossil record to the fleshy fins of rhipidistians – lobe-finned fishes probably ancestral to all land vertebrates.

These fins were mainly balancers, but strong enough to haul their air-breathing owners a short way overland. Front fins (the arms' precursors) joined shoulder girdles fixed to the skull. Hind fins (precursors of the thighs and legs) joined pelvic bones inside the body.

In early amphibians, fins had given rise to jointed limbs. Knees and elbows jutted sideways from the body, but hind limbs joined a pelvic girdle braced by

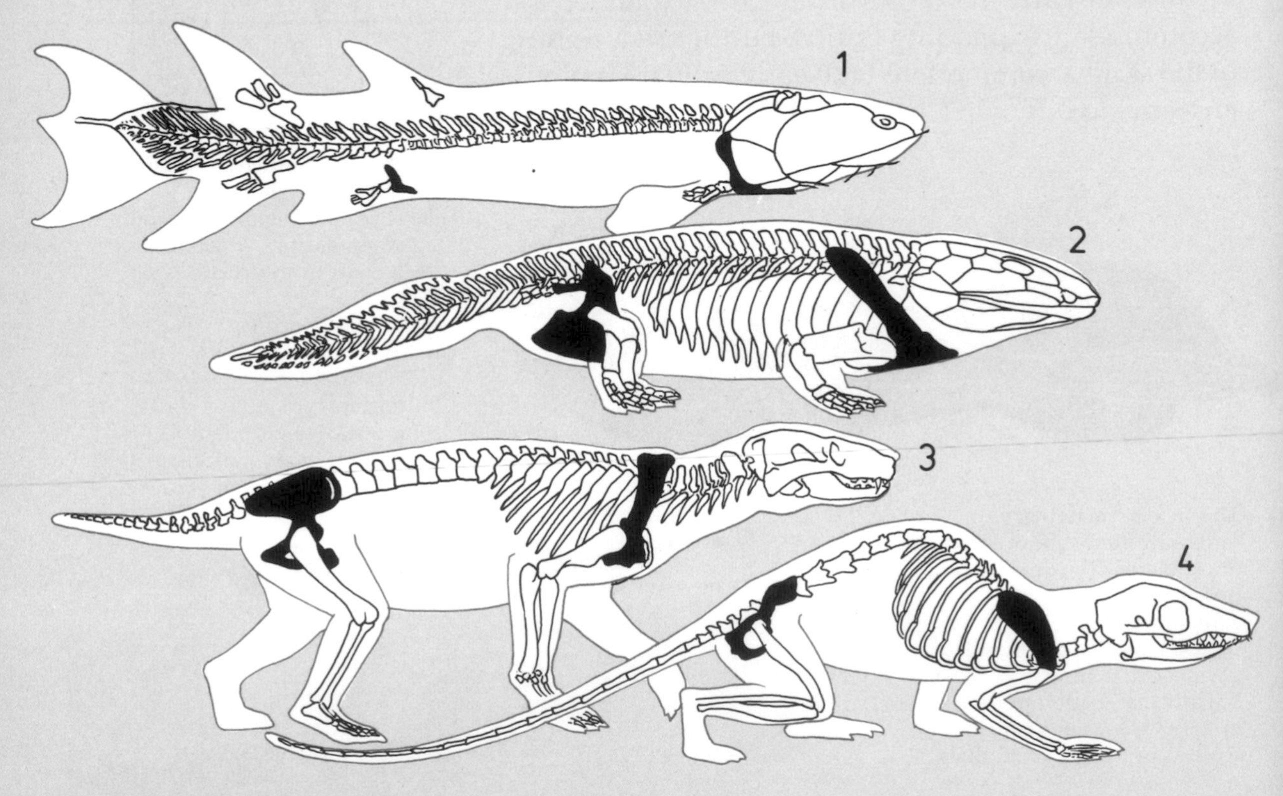

its attachment to the spine, so aiding forward movement. Meanwhile, the shoulder girdle had been freed from the skull so that head and front limbs now moved separately. Almost every bone inside these ancient prehistoric limbs finds its equivalent in man. Change has been mainly in the proportion not the number of bones.

Advanced mammal-like reptiles remotely descended from amphibians evolved limbs with knees and elbows rotated underneath the body, giving a longer stride and faster movement over land. Each limb now ended in five forward-jutting, clawed digits, and the old massive shoulder girdle had grown lighter, making forelimbs much more mobile.

This scheme persisted in the early mammals. Refinements culminating in bipedal walking and manipulative hands figure in the chapters yet to come.

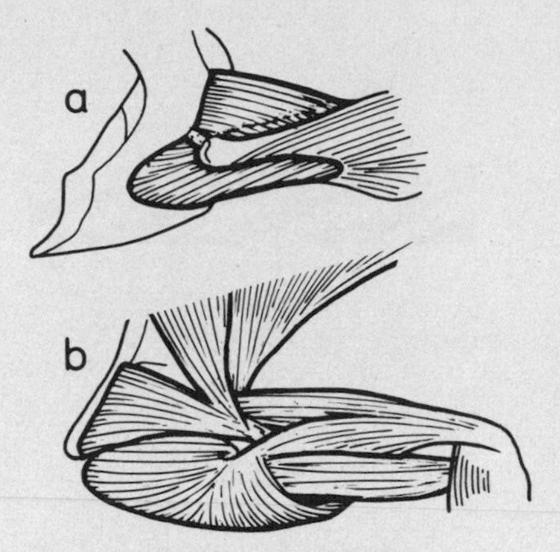

Multiplying muscles (above)
a Two muscles are enough to work a fish's pectoral fin.
b Seven muscles operate the shoulder and upper arm of a lizard, a land reptile.

5 This ape's pelvis and shoulder girdle show adaptations for walking and climbing on all fours and swinging from branches by the arms.
6 Man's short, wide, pelvis and broad, flexible shoulder blade between them allow bipedal walking and free arm rotation.

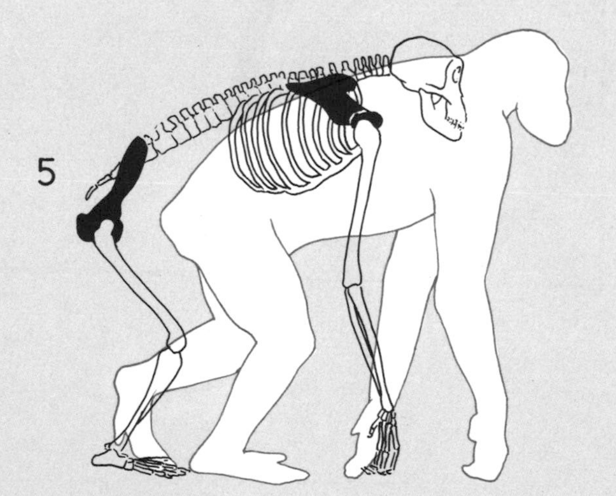

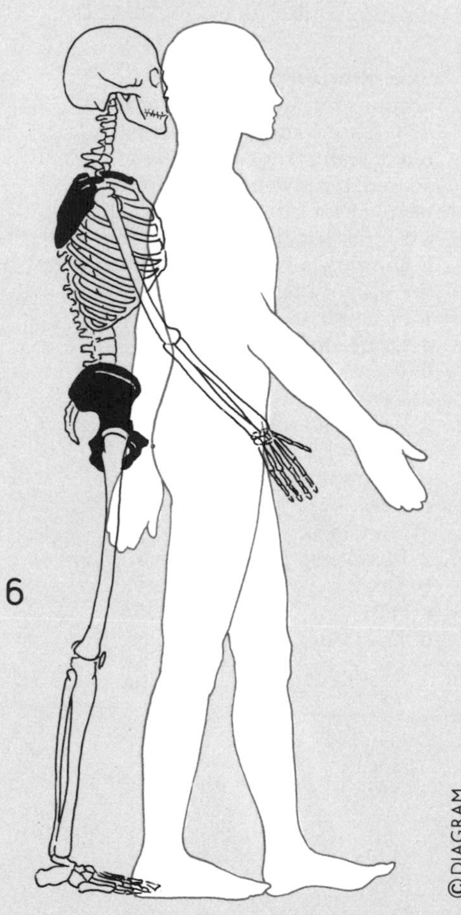

The skin

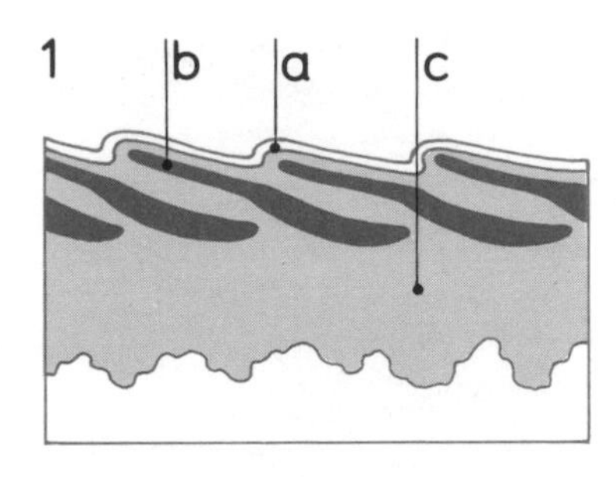

Our skin and nervous system both seem to have originated in the outer layer of our ancestral invertebrates' two-layered bodies. As in their descendants, fishes, our own skin protects the body's soft, moist inside from outside dangers. But our only relics of an early fish's bony armour have disappeared or sunk to form our collarbone and large parts of the skull.

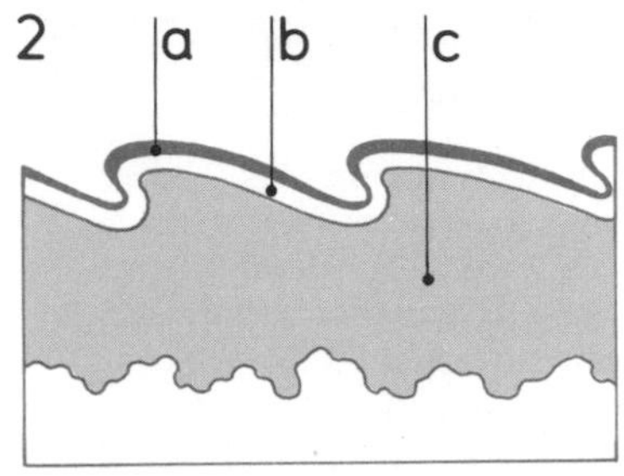

Like a reptile's skin, our own is waterproof, yet of the horny scales covering our reptile ancestors, our only legacy is nails: flattened toe and finger guards derived from claws. Grip-assisting ridges on our toes and fingers are tokens of the tough pads that helped our mammal forebears walk and climb.

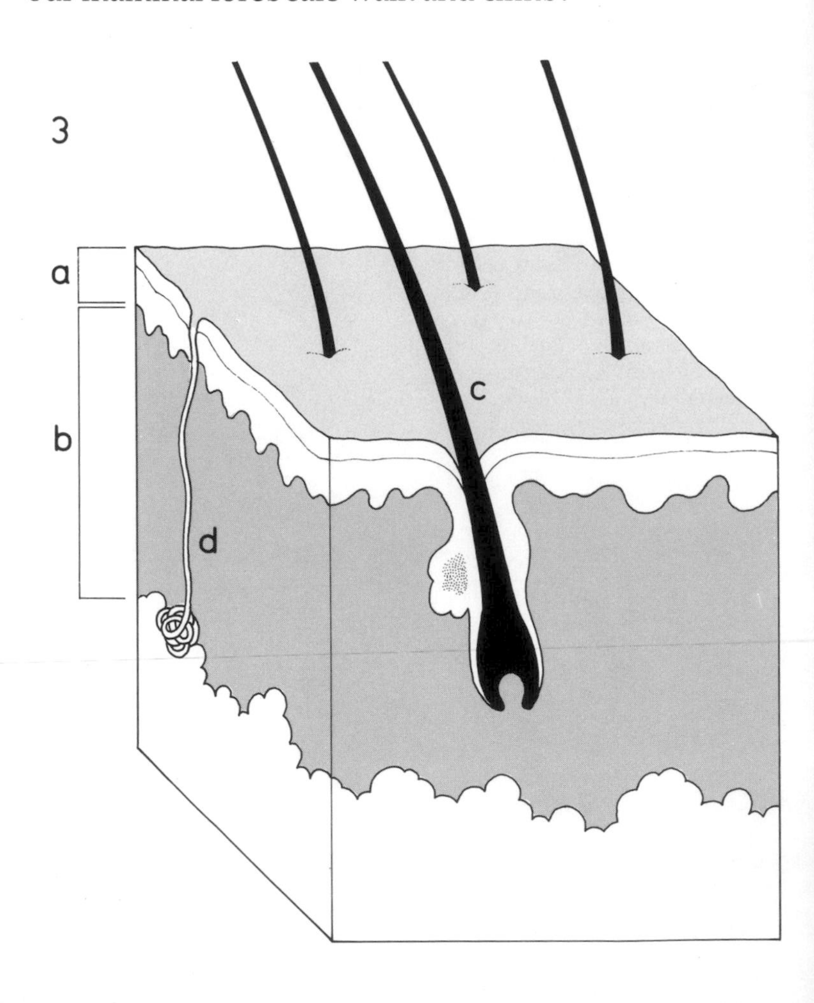

Skin structures
Sections through the skin of a fish, reptile, and human show major features including two, shared, basic structures: a thick, deep, fibrous dermis, underlying a thinner superficial epidermis. Both imply a common origin for the skin of all the vertebrates.
1 Fish skin in section
a Epidermis
b Bony scale
c Dermis
2 Reptile skin in section
a Horny scale
b Epidermis
c Dermis
3 Human skin in section
a Epidermis
b Dermis
c Hair
d Duct from sweat gland

Our insulating body hair probably derives from sensitive outgrowths of the skin first seen in warm-blooded theriodonts, reptiles that led directly to the mammals. The origins of skin glands are less obvious, but milk glands probably arose in early mammals from the sebaceous glands whose oil keeps skin and hair in good condition. Some people, incidentally, have extra nipples, relics of the rows that evidently ran down early mammals' undersides.

Below the facial skin, subcutaneous muscle that evolved with our primate ancestors endows us with a language of facial expressions denied to lower vertebrates like fishes, whose faces stay, inscrutably, a mask.

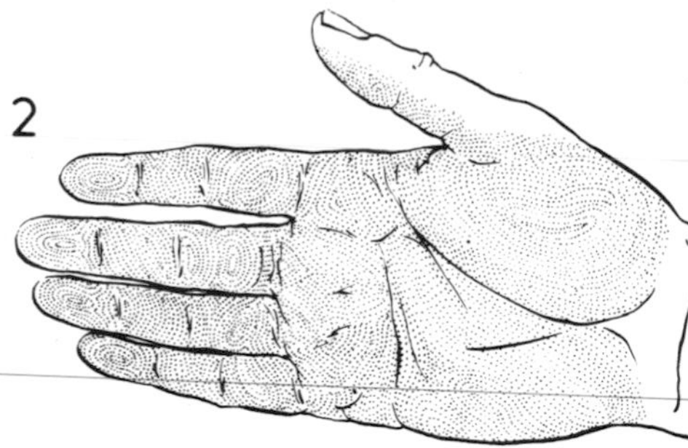

Pads and ridges (above)
1 Insectivore's hand, showing wear-resistant skin pads rich in keratin – a horny epidermal protein
2 Human hand with grip-assisting friction ridges patterned like the pads of insectivore ancestors

Evolving expressions
A A fish's blank stare reflects a lack of facial muscles.
B Facial muscles and facial expressions in an opossum.
C Highly evolved facial muscles permit the wide range of human facial expressions.

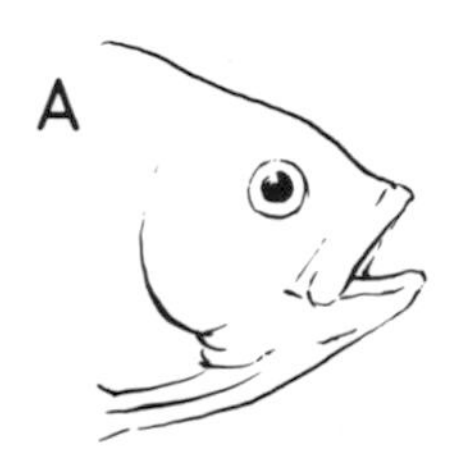

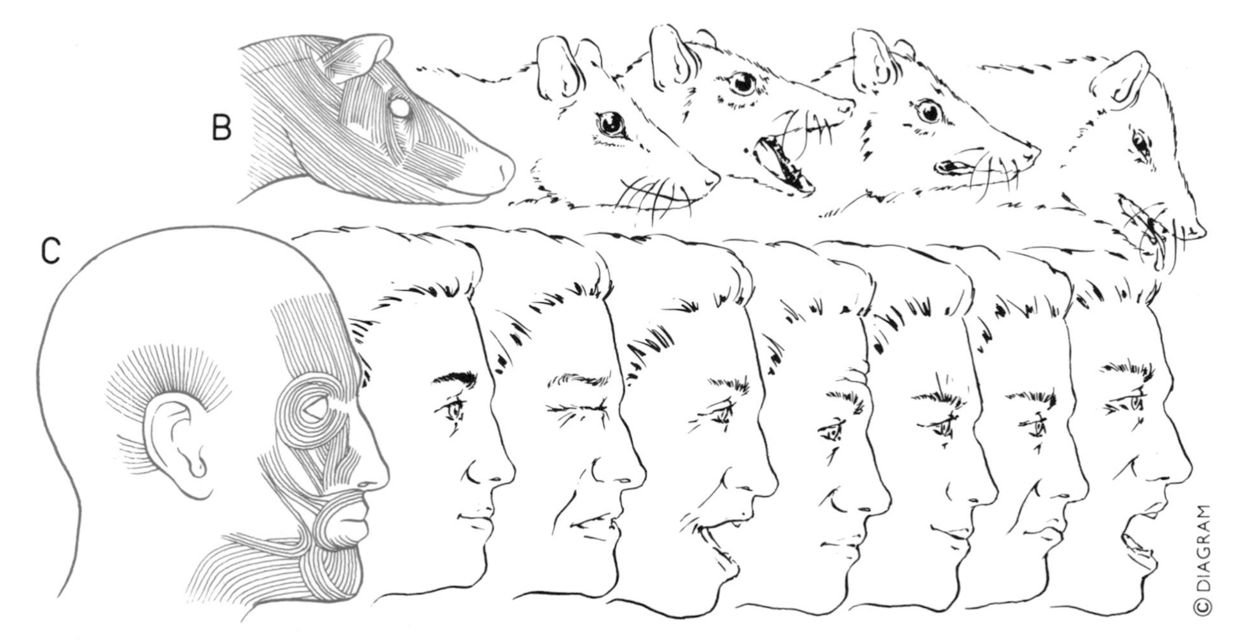

Nerves and senses

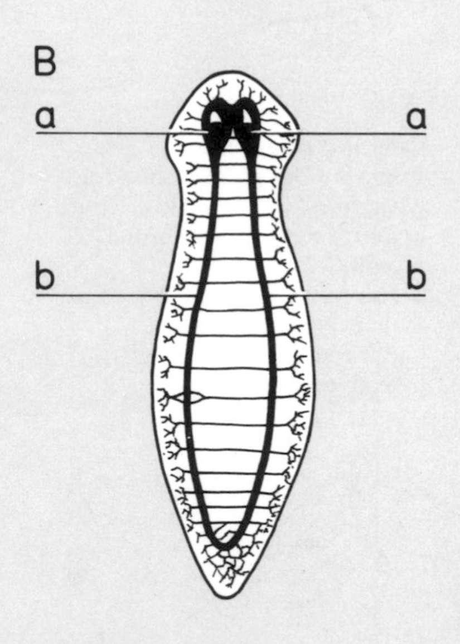

Evolving nervous system
Two invertebrates (shown much enlarged) represent major stages in the evolution of the nervous system.
A *Hydra*, a freshwater relative of jellyfish and sea anemones, has a nerve net – a diffused mesh of nerve cells.
B *Planaria*, a flatworm, has bilateral symmetry and a central nervous system with:
a Brain in the head end
b Nerve cords (foreshadowing our spinal cord).

All animals depend on nerve cells for communication and control. Our own complex nervous system originated in the loose nerve net of lowly creatures like the jellyfish. With fishes came a coherent nervous structure, with a spinal cord handling reflex actions, taking signals to a brain from nerve endings specialized for tasting, smelling, seeing, and maintaining balance, and relaying orders from the brain to specific muscle groups.

In us, as in the early vertebrates, key sense organs cluster in the head – the body's old front end. In man, as in a fish, most work by sensing dissolved substances, or vibrations passing through a fluid. Thus taste buds on the human tongue detect substances dissolved in water. And nasal membranes must be moist to pick up scents. Because our nasal membranes are connected to the mouth, flavors are really tastes and smells combined.

As in a fish, fluid flowing in closed tubes of the inner ear provides our sense of balance. But evolutionary changes – first in amphibians, then reptiles, then mammals – remodeled parts of the old fishes' gills, giving ears external openings and making them efficient hearing organs, sensitive to airborne sounds. Internally, our ears convert these to vibrations in liquid that surrounds nerve endings in the cochlea, a coiled tube of the inner ear.

Eyes evolved even earlier than ears, deriving from light-sensitive spots in primitive invertebrates. Even primeval fishes had paired eyes much like our own, designed as cameras. But tear glands, which moisten eyes exposed to air, came with amphibians; eyelids later, with the reptiles; while our sharp, binocular color vision originated later still, with day-active ancestors of apes.

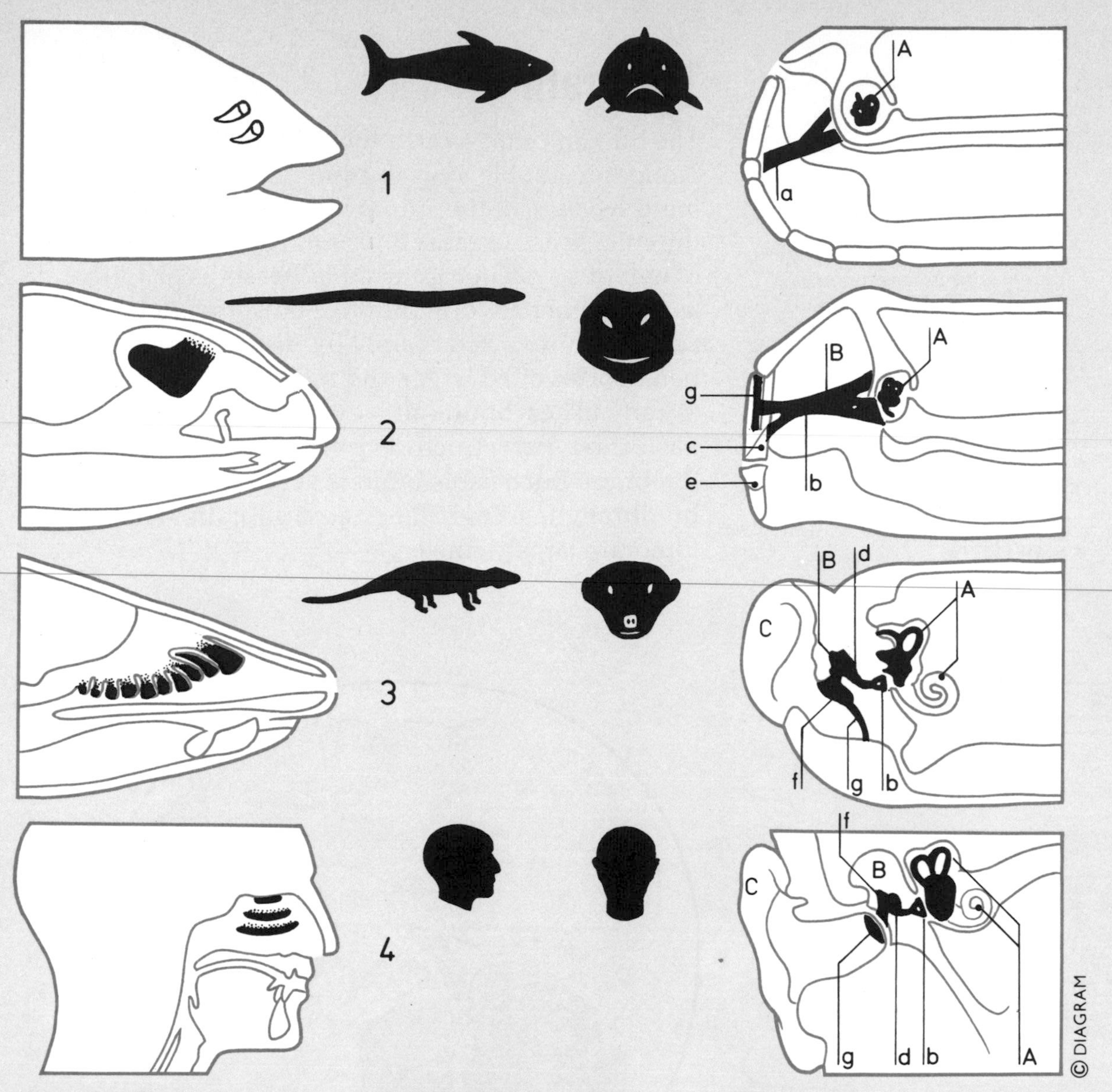

Sense of smell
Longitudinal sections through four heads show evolving organs for detecting scent.
1 Fish: olfactory membranes in pockets separate from mouth
2 Reptile: olfactory membranes in nasal cavity connected to the mouth
3 Typical mammal: enlarged olfactory area giving a keen sense of smell
4 Human: reduced olfactory area reflecting dependence more on sight than smell

Evolving ears
Cross sections through the same four heads show bone migrations that have helped amplify airborne sounds and transmit them to the inner ear.
A Inner Ear
B Middle Ear
C External Ear
a Hyomandibular bone becomes:
b Stapes.
c Quadrate jaw bone becomes:
d Incus.
e Articular bone becomes:
f Malleus.
g Eardrum, transmitting sound waves to the middle ear.

The brain

The human brain, which thought up everything from Stone Age pebble tools to atom bombs, began as a mere swelling of the spinal cord at its head end near the chief sense organs. Brains as coordinating centers of nervous systems occurred in beasts as primitive as the living flatworm. But like most invertebrates, such creatures acted mainly by instinct: intelligent behavior awaited larger and more complex brains.

Early fishes' brains already showed our own brain's basic three-part structure: hindbrain, midbrain, and forebrain. But a fish's brain is very limited. Its hindbrain handles balance, its midbrain vision, its forebrain largely smell.

Evolving brains
Longitudinal sections through four heads retrace progressive changes in proportions of the brain's basic three divisions:
1 Fish's brain
2 Reptile's brain
3 Primitive mammal's brain
4 Human brain
For each we show:
a Hindbrain (including its offshoot, the cerebellum)
b Midbrain
c Forebrain

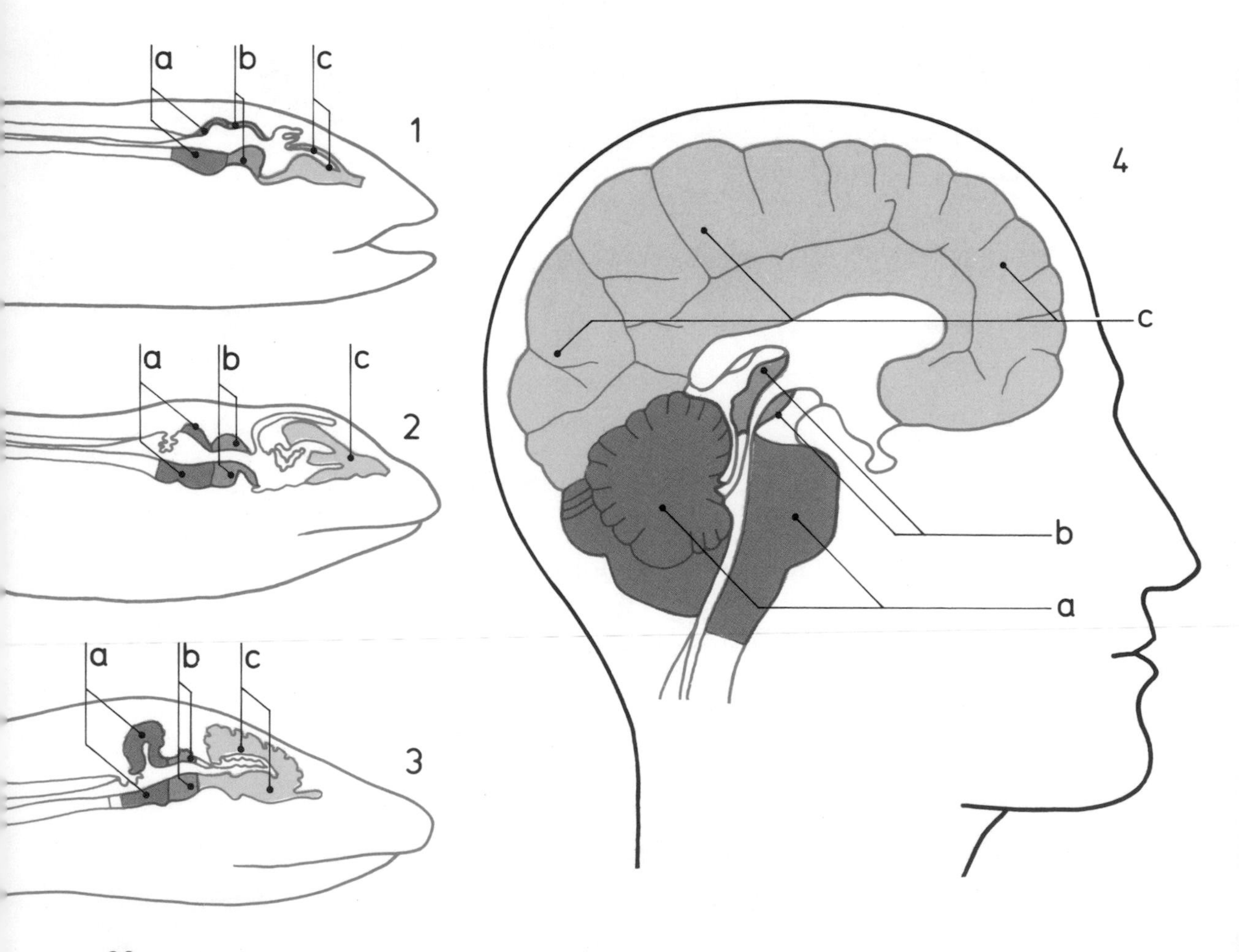

Reptiles' brains have a relatively larger hindbrain and midbrain, giving improved hearing, vision, and sensory coordination – evolutionary changes geared to life on land.

In early mammals, the brain grew much bigger and more complex than a reptile's brain. The hindbrain gained a large, cabbage-shaped cerebellum, co-ordinating complex movements. Sense coordination shifted forward to the forebrain, and this acquired a large, many-wrinkled cerebrum to cope with memory and learning. Even in such small-brained mammals as the hedgehog, the forebrain overgrows the midbrain. In more advanced mammals such as monkeys, the evolving forebrain has grown relatively larger still – a process referred to in the next few chapters. In man, the forebrain's enormously expanded cerebrum, the seat of reason, has overgrown and dwarfed the rest of the brain, remodeling much of the skull.

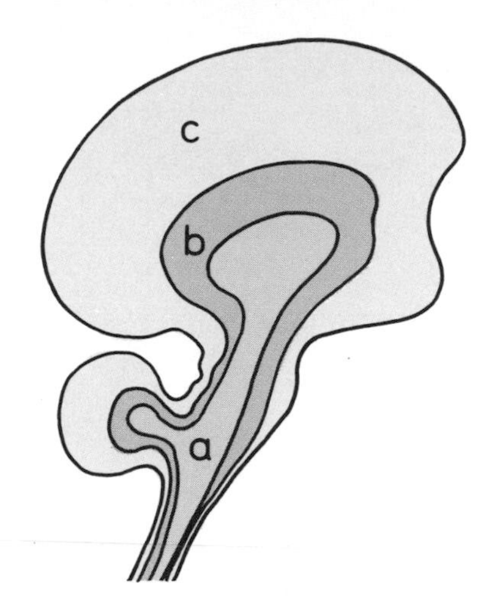

A layered brain (above)
One theory holds that our brain has three layers that betray stages in its evolution:
a Reptilian (hindbrain and midbrain) – slave to precedent
b Paleomammalian (around the upper brainstem) – seat of the emotions
c Neomammalian (cerebral cortex) – seat of reason

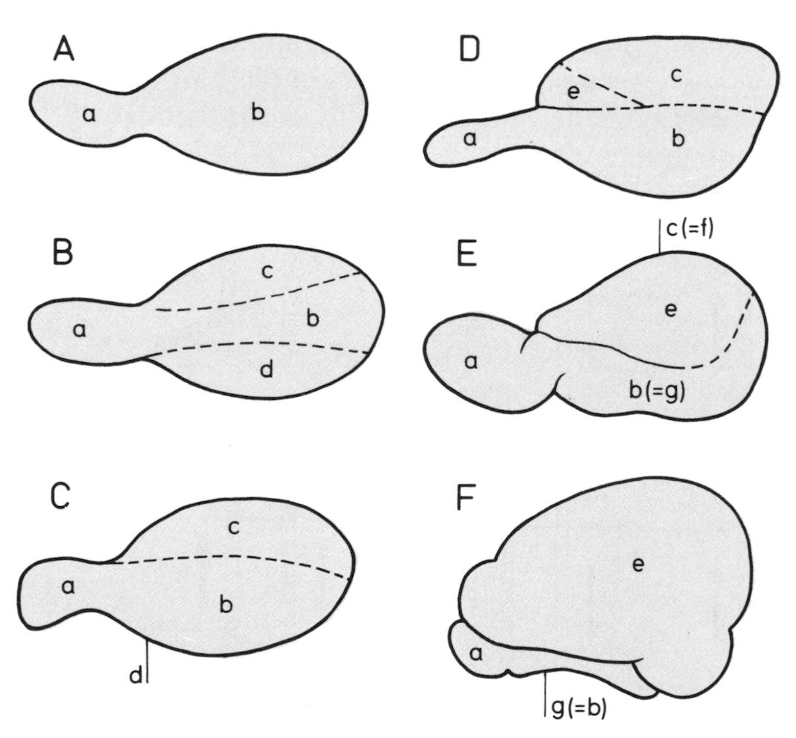

Cerebral evolution (left)
Schematic side views trace trends in evolution of the brain's left cerebral hemisphere. Most gray matter, rich in nerve cells, migrates to the brain surface to form its cortex or *pallium* ("cloak").
A Primitive form
B Amphibian
C Primitive reptile
D Advanced reptile
E Primitive mammal
F Advanced mammal
Structures shown are:
a Olfactory bulb
b Paleopallium ("old covering")
c Archipallium ("primitive covering")
d Basal nuclei
e Neopallium (rich in association centers)
f Hippocampus (the old archipallium)
g Olfactory lobe (the old paleopallium or "smell brain")

Three body systems

Lungs, blood supply, and kidneys serve vital functions. For instance, lungs provide the body with oxygen for "burning" food to fuel life's processes and expel the combustion waste gas, carbon dioxide. Blood brings cells the nutrients and oxygen they need. Kidneys purify the blood, removing harmful body wastes.

Lungs probably began as moist throat pouches that lobe-finned fishes used for gulping atmospheric air in swampy pools low in oxygen. In amphibians and even reptiles lung sacs still had and have small oxygen-absorbing surfaces. In mammals these grew large and complex. Meanwhile the amphibians' throat-muscle pumping action gave way to the mammals' vacuum pump operated by muscles of the ribs and diaphragm.

In lung-breathing fishes, the nose – at first a smelling organ only – had become also a tube to suck air into lungs while keeping water out. Reptiles gained a bony palate above the mouth partly separating food from breathed-in air. Mammals added a soft palate making possible continuous breathing while eating – vital for warm-blooded animals.

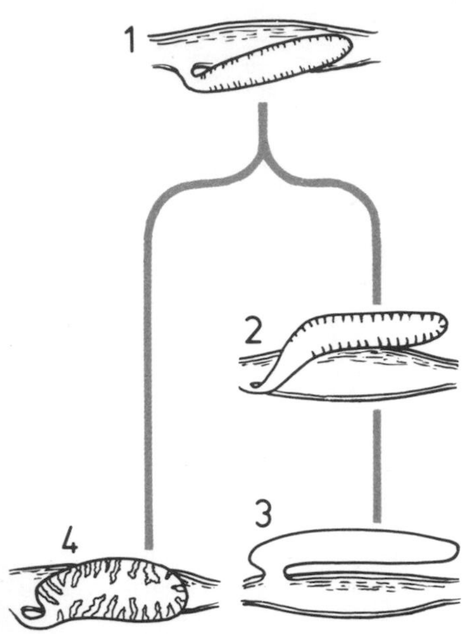

How lungs began (above)
Longitudinal sections show the common origin of land vertebrates' lungs and fishes' swim bladders.
1 Primitive fish lung, evolved from a throat pouch
2 Fish lung modified
3 Air-filled swim bladder, controlling buoyancy in bony fishes
4 Land vertebrate's lung, with a much-folded inner wall giving more efficiency than the simpler fish lung

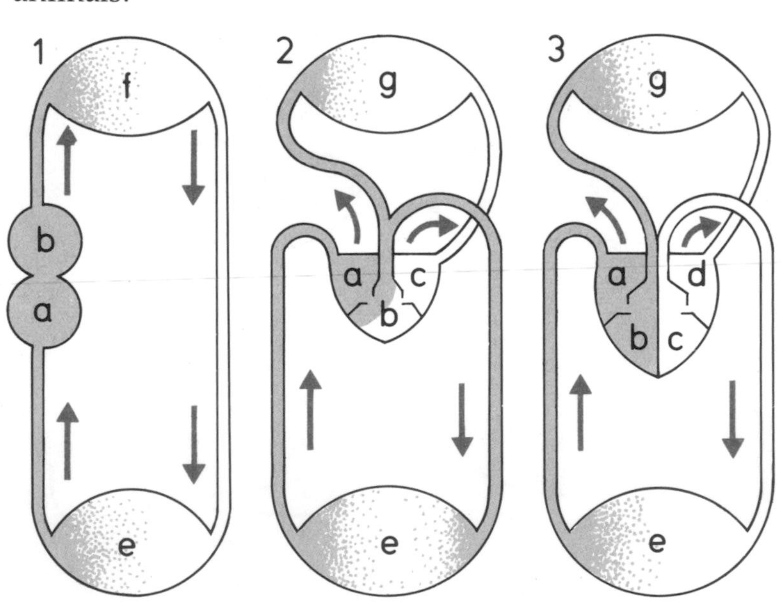

Evolving circulation (right)
1 Fish circulation: A two-chambered heart pumps stale blood from the body to the gills. Fresh blood flows from gills to body.
2 One reptilian system: A three-chambered heart pumps mixed blood to lungs and body.
3 Mammal circulation: A four-chambered heart keeps stale and fresh blood separate.
a–d Heart chambers
e Body
f Gill capillaries
g Lungs

Our blood supply is an ancient legacy indeed, its saltiness hinting at our origin from simple cells designed for life surrounded by the sea. The circulation of the blood, however, has been revolutionized. Since the ancient heart-gill system of ancestral fishes, there has been redesigning of the heart and arteries to take stale blood to lungs and bring fresh blood – enriched with oxygen – from lungs to cells throughout the body. In our cold-blooded forebears, amphibians and reptiles, lung evolution outpaced that of circulation. Only birds and mammals have four-chambered "double-barreled" hearts completely separating fresh blood from stale, so giving an efficient circulation system making possible the warm bloodedness that keeps us capable of rapid and sustained bodily activity by day or night, even in cold weather.

Kidneys seemingly began among invertebrates as tiny isolated tubes pumping body wastes from cavities between organs. Fishes added filters to these pumps and grouped them in two long strips inside the body. Here, kidneys purified the blood and emptied urine from a common exit. In reptiles, kidneys became two solid masses concentrated far back in the body. In man and other mammals, each kidney tubule has acquired a long loop releasing water back into the body to stop this drying up.

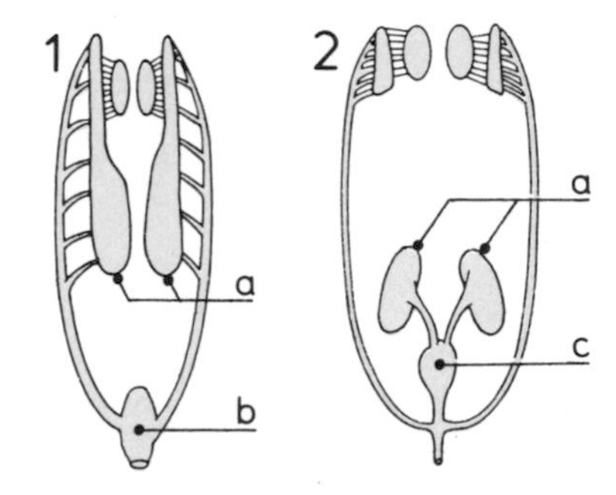

Evolving kidneys (above)
1 In lower vertebrates kidneys form two long strips excreting urine through a cloaca along with solid body waste.
2 In mammals kidneys form two bean-shaped organs excreting urine via a bladder.
a Kidneys
b Cloaca
c Bladder

Vanishing tubes (below)
Five diagrams show reduction in numbers of arterial tubes accompanying the change from gills to lungs.
A Primitive fish
B Lung-bearing fish
C Amphibian
D Reptile
E Mammal
1–6 Aortic arches (originally gill-related)
a Heart chambers
b Lungs

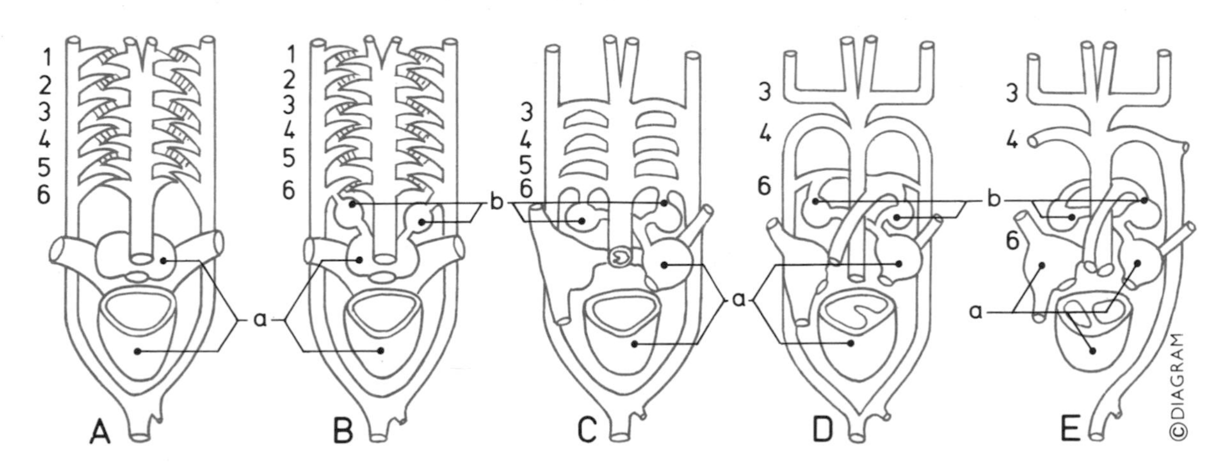

Reproduction

External fertilization
a Female fish releases eggs.
b Male fish sheds a cloud of sperm that fertilize the eggs outside the parents' bodies.

How the human egg is fertilized and grows inside the mother's body is a far cry from the chancy way in which our early ancestors reproduced themselves. Like their living counterparts, these old marine invertebrates most likely shed huge quantities of sperms and eggs haphazardly into the sea. But vertebrates evolved more selective and efficient ways of releasing sex cells and ensuring that egg cells were met by sperms and fertilized.

In primitive vertebrates, sex glands shed sperms and eggs into the body cavity from where they left through pores near openings releasing body wastes. But fishes evolved special tubes to launch their sex glands' products on the outside world.

Like most fishes' eggs, amphibians' lack shells so must be laid in water where they are fertilized outside the female's body. With reptiles, though, reproduction became geared for life on land. Males injected sperms directly into females' bodies where moisture kept sperms alive until these fertilized the

Evolving sex organs
Modifications of male and female sex organs aided internal fertilization.
1 Male lower vertebrate: Sperm from testes exit through the cloaca.
2 Female lower vertebrate: Eggs from ovaries exit through the cloaca.
3 Male mammal: Sperm from testes (usually descended) pass via urethra in penis (not shown) into vagina.
4 Female mammal: After a sperm fertilizes an egg this develops in a uterus. Some mammals have two uteruses.
a Testes
b Cloaca
c Ovaries
d Urethra
e Vagina
f Uterus
g Kidneys

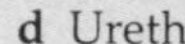

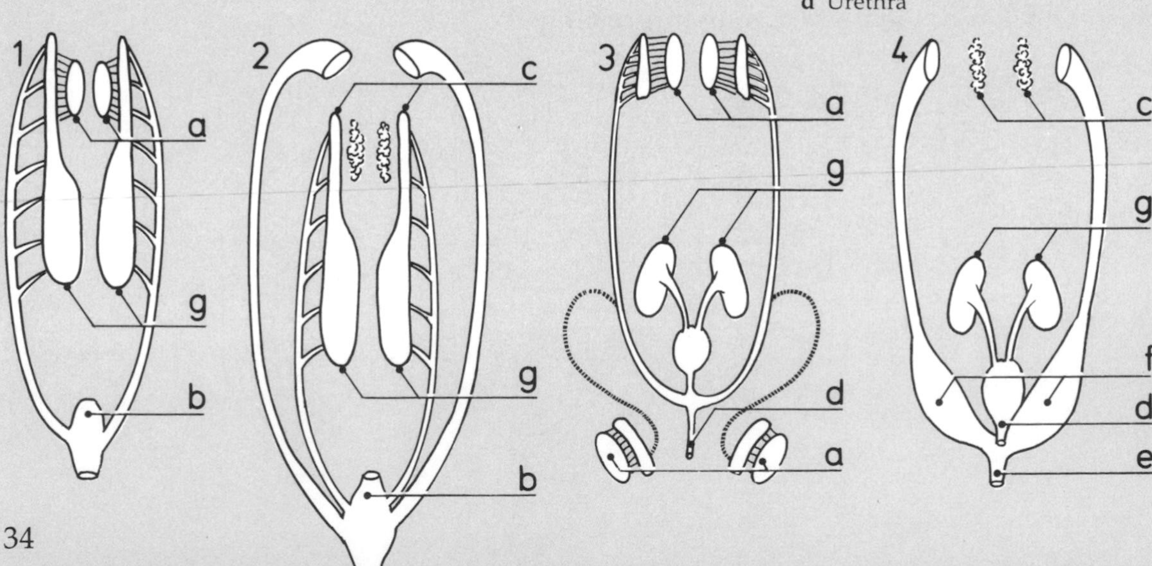

eggs. Internal fertilization meant that a female reptile need lay fewer eggs than, say, a frog or toad to ensure enough would hatch to keep her species going.

Reptiles' eggs are engineered to hatch on land, with a shell, membrane, and fluid-filled sac to guard the growing embryo from injury and drought; a large yolk food supply; a kind of "lung;" and a dump for body wastes.

Man and other mammals have inherited the ground plan of this so-called amniotic egg, but while most reptiles lay eggs and leave them for the sun to hatch, placental mammals' eggs develop warm and safe inside the mother's body. There, a temporary organ, the placenta, supplies her unborn young with more food than could be crammed inside an egg surrounded by a shell, and removes the fetal body wastes. This reproductive innovation has proved so successful that our species flourishes despite its long pregnancies, most culminating in the birth of just a single, helpless, young.

Embryos and evolution
Three sections trace changes in protection and nourishment that led to development of young inside the mother.
A Amphibian embryo
B Reptile embryo
C Human (mammal) embryo
a Embryo
b Yolk (food supply)
c Albumin (food supply)
d Shell (protection)
e Amnion (shock absorbing sac)
f Chorion (air space letting oxygen in, waste gas out)
g Allantois (dump for body wastes)
h Placenta (a temporary organ with chorionic and allantoic structures. Via the mother's blood supply it provides an embryo with nutrients and oxygen, and removes wastes.)
i Uterus (the organ in which an embryo develops inside its mother's body)

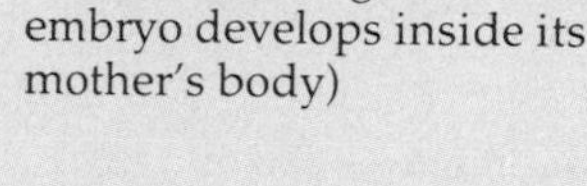

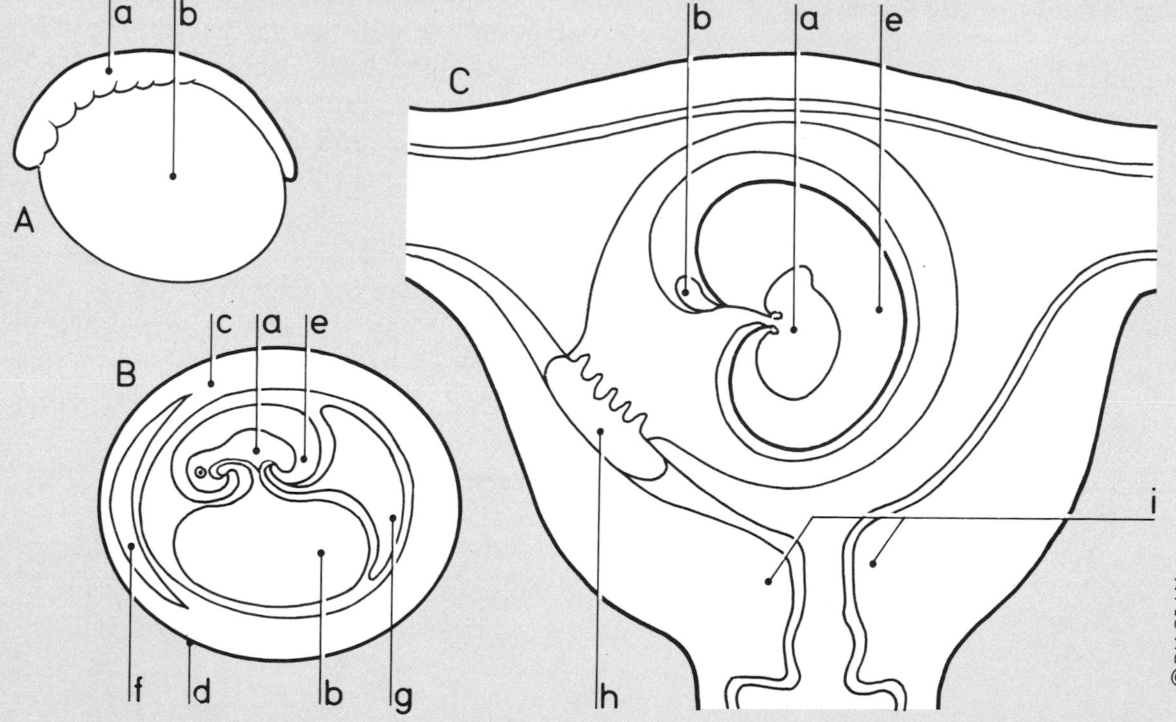

Man's hidden history

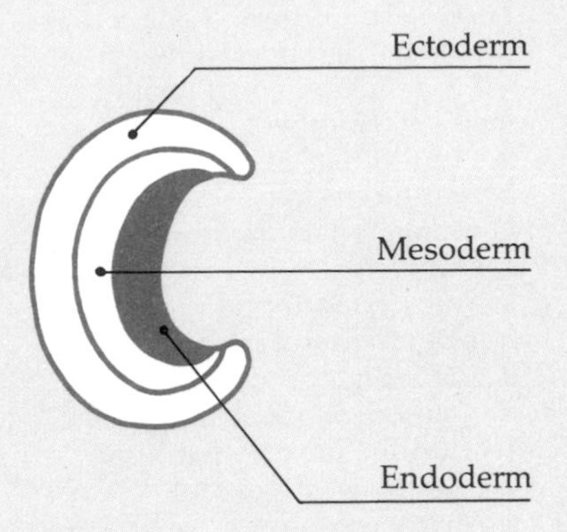

Three cell layers (above)
A human embryo a few days old resembles that of any other vertebrate. Each consists of ectoderm, mesoderm, and endoderm – three layers of cells from which all body structures will arise.

Vestigial features (below)
Here are three evolutionary products that persist in man yet serve no valuable purpose.
a Appendix
b Ear muscles
c Coccyx (vestigial tail bone)

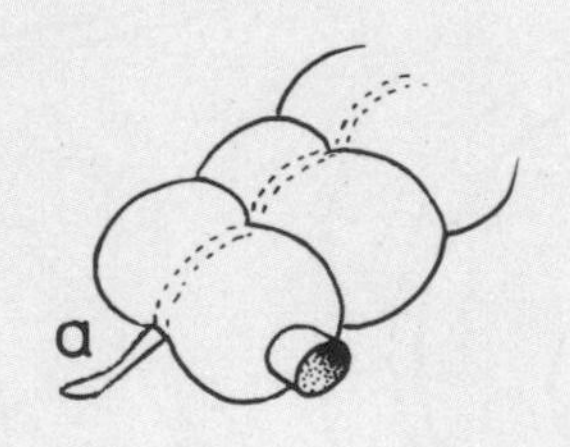

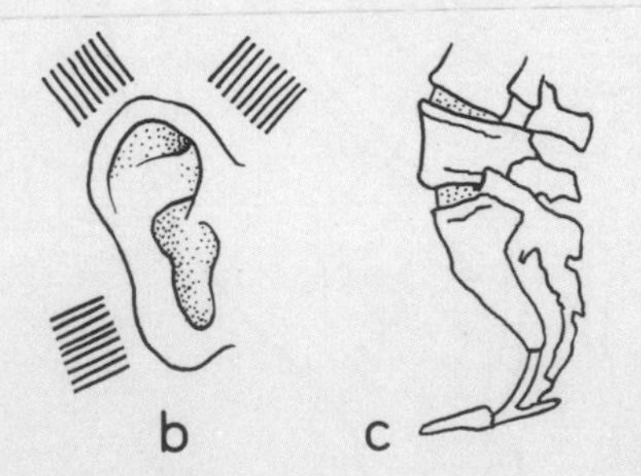

Parallels between human and other embryos are strong clues to our relationship with other forms of life.

In every animal from worms to humans, the body starts to grow from just two layers of dividing cells – the (outer) ectoderm and (inner) endoderm, corresponding to the double body layer seen in primitive invertebrates like jellyfish. Ectoderm produces skin and nervous tissue. Endoderm produces lining for the digestive tract, related organs like the pancreas and liver, membranes used in gills or lungs, and certain other structures. But in all vertebrates a third, middle, layer – the mesoderm – produces most parts of the body including bones, muscles, arteries, and veins, most of the genital and urinary systems, and a gristly rod, the notochord: precursor of the backbone.

Early on in embryonic life, mesodermal cells on each side of the notochord form blocks called somites – bilateral beginnings of a vertebrate's segmented body. At this stage a human embryo resembles an embryonic fish complete with fish-like brain, "gill pouches" in the neck, limb buds that look like fins, segmented muscles all along the trunk, a tail, a simple fish-like heart and kidneys, and several pairs of big blood vessels like those that link a fish's heart and gills.

Soon, though, our early fish-like structures alter, mammal fashion. Instead of turning into gills, gill bars contribute to a face, jaws, larynx, tongue, ears, and endocrine organs in the neck. Our potential fish's swim bladder develops into lungs. Most paired blood vessels linking heart to "gills" shrink and disappear. Limb arteries get rearranged to service legs and arms. The heart acquires partitions, and mammalian kidneys form. The trunk, by losing muscles, gains a waist. In time, only fused bones at the bottom of the spine remain as a reminder of our embryonic tail.

Between its fourth and sixth week, a human embryo changes from a fishlike organism to one indistinguishable from an embryonic monkey. By two months the fetus is unmistakably a tiny human being.

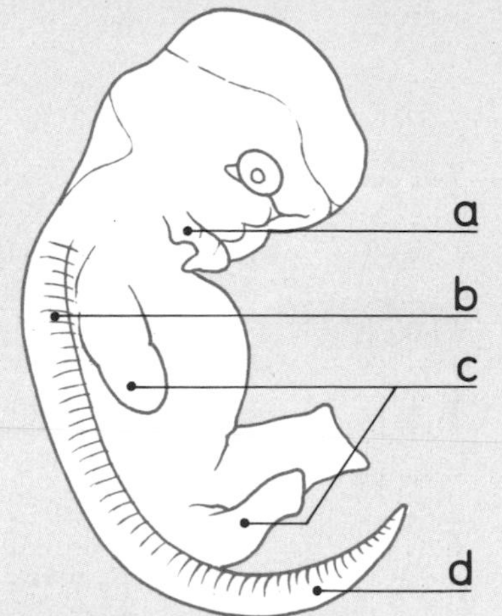

Human embryo (above)
Labeled features in this human embryo parallel those seen in other embryonic vertebrates.
a Gill pouches
b Somites
c Limb buds
d Tail

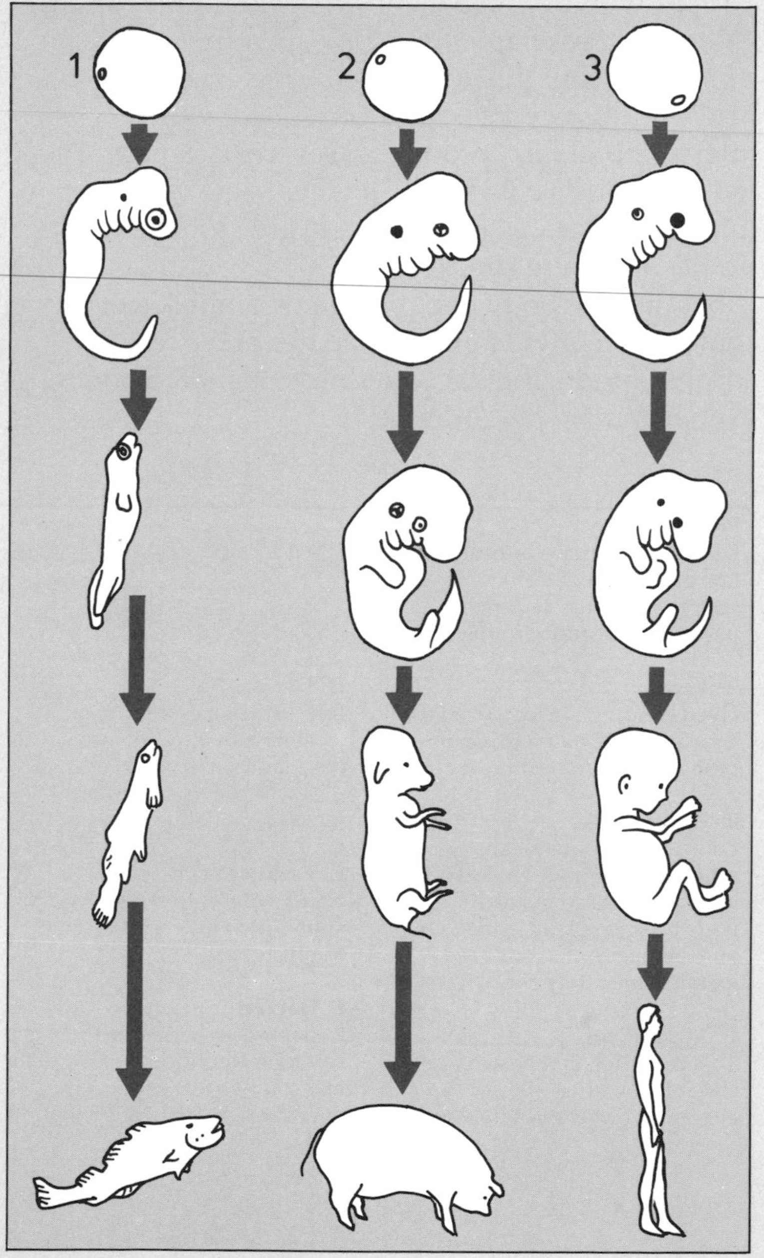

Evidence for evolution (left)
Three sets of illustrations trace the development of fish, pig, and human from egg through embryo to adult form. Each vertebrate passes through similar early stages, suggesting that all three evolved from the same ancient backboned stock.
1 Fish
2 Pig
3 Human

The makeshift mammal

Despite its many advantages, our body's form is flawed. One writer has even called man a "hodge-podge and makeshift creature."

Most major problems stem from upending a skeleton originally constructed as a cantilever bridge. To cope with new needs for twisting and bending we gained wedge-shaped vertebrae, pivoting on thick front edges. But this weakens the lower spine, so that heavy lifting can dislodge a vertebra or worn intervertebral disc, producing severe back pain. Then, too, the spine's basal curve pushes some vertebrae into the birth canal far enough to give some mothers problems during labor.

Because the body rests on just two limbs, feet sometimes suffer strain, and arches of the feet collapse producing flat footedness, distorted bones, even hammer toes and bunions.

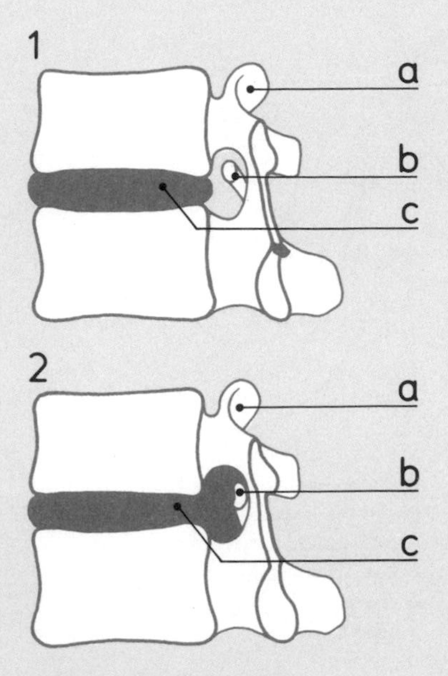

Slipped disk (above)
Prolapse of a cartilaginous disk buffering two spinal vertebrae can produce painful pressure on a spinal nerve and impair mobility.
1 Normal disk
2 Prolapsed disk
a Vertebra
b Nerve
c Disk

Problem areas
Below and right we indicate a few parts of the skeleton, blood supply, and other systems particularly prone to problems.

A Teeth
Overcrowding is a legacy of jaw shrinkage – a relatively recent product of evolution.

B Slipped disk
Lower back troubles usually reflect degenerative change with age compounded by load transmission via the spine to two limbs not four as in our early mammal ancestors.

C Appendicitis
This condition involves infection and inflammation of the appendix, a vestigial offshoot of the gut.

D Hernia
Intestine bulges through a weakness in the abdominal wall. Hernias tend to occur in slightly different sites in males and females.

E Varicose veins
Faulty valves allowing blood to "pool" in veins can affect both thigh and leg.

F Flat feet
Collapsed arches of the foot are a common condition, related to the fact that two feet support the entire weight of the human body.

Hernias are another penalty of redesigning for bipedal life. In four-legged animals, the gut is slung from the spine by a broad ligament; in man this is reduced and less well attached. Intestines may bulge out through our weakened abdominal wall.

Upright posture even hampers blood supply. Blood must overcome about 4ft (1.2m) of gravitational pressure to return from the feet to the heart. If faulty leg-vein valves allow blood to fall back, the result is heavy, tired legs, and other symptoms of varicose veins. "Milk leg" affecting the left leg in pregnancy is also due to poor venous drainage, as increased visceral pressure thrusts a vein against a sharp bone where two vertebrae join low in the spine.

A reduced birth canal (a result of bipedalism), and babies' enlarged skulls, causing childbirth problems are yet other penalties we pay for less than perfectly adjusted bodies.

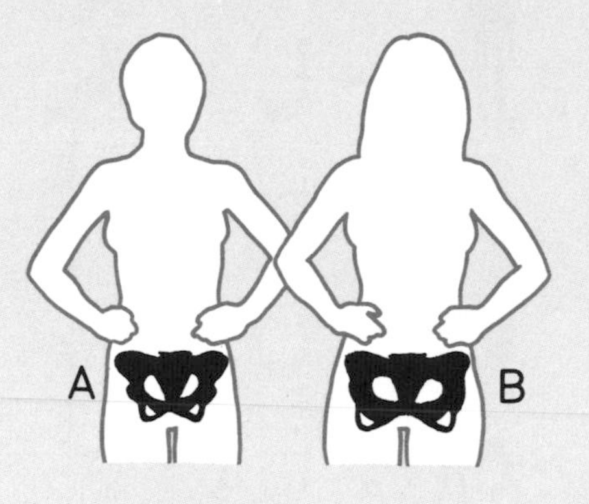

Narrow hips (above)
A Narrow hips imply a pelvis too narrow to allow normal birth – a common reason for operating to remove a baby.
B Broad hips imply a pelvic opening of normal width, unlikely to cause problems in giving birth.

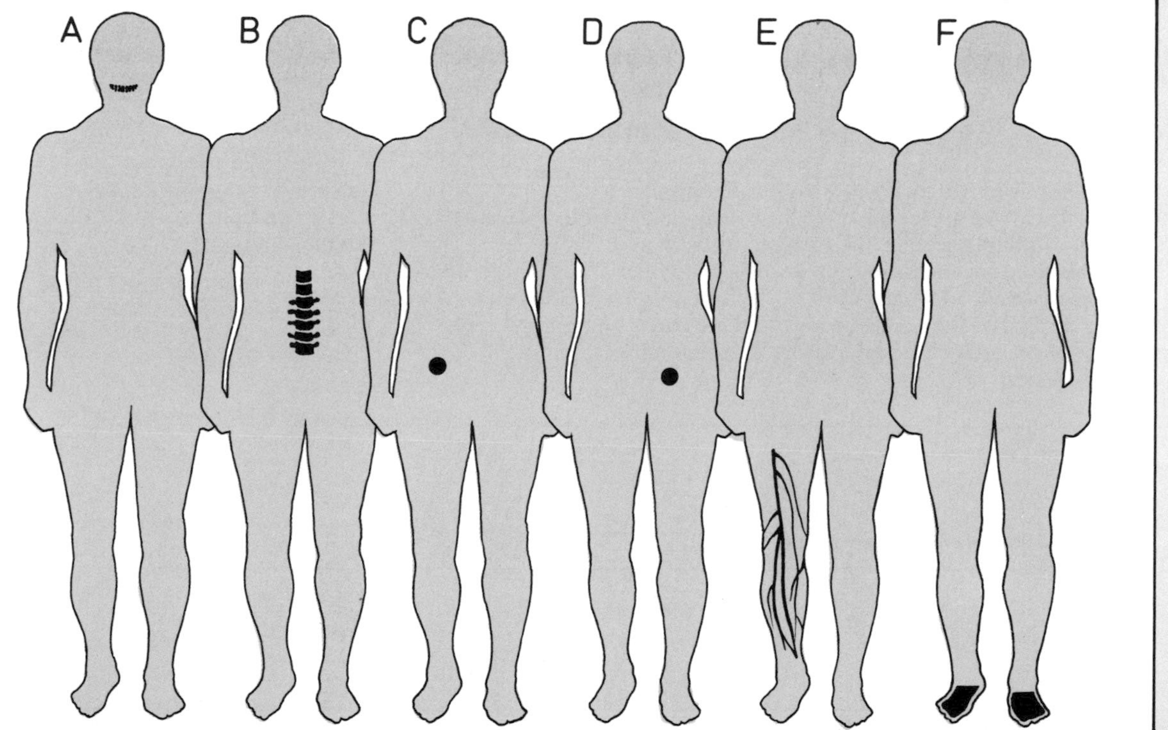

Clues in biomolecules

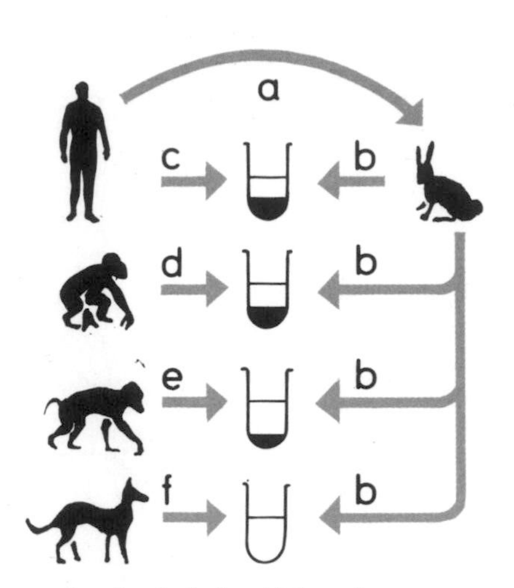

Serological clues (above)
The amount of antigen/antibody reaction produced by a foreign blood serum hints at biological links – the greater the reaction, the closer the supposed relationship.
a Human serum into rabbit
b Rabbit serum with antibodies to human serum
c Human serum (100% reaction)
d Chimpanzee serum (97%)
e Baboon serum (50%)
f Dog serum (0%)

Some of the strongest proofs of man's evolutionary origins come from comparing our own biochemicals – the body's building blocks – with those of other living organisms.

Like almost all of these our body's cells contain the energy-rich compound ATP (adenosine triphosophate,) used in the cells' energy-requiring processes. Like other animal and plant cells, our own contain DNA (deoxyribonucleic acid), the inherited material passed on in genes.

Differences between some of our own biomolecules and those of other organisms help to narrow down our relationship to certain groups. Thus phosphocreatine occurs in the muscles of man and other chordates (creatures with a skeletal rod, the notochord), but not in most invertebrates.

Experts have devised sophisticated batteries of tests to find which living chordates are biochemically our closest counterparts.

Immunological studies revealing how strongly different animals react to foreign blood sera indirectly show how greatly proteins from different species differ from each other.

Molecular tree of life (below)
Comparison of hemoglobin molecules in 48 placental mammals from seven groups suggests these evolutionary relationships. The time scale (in millions of years) is based on fossil evidence.

a Rodents
b Ungulates (hoofed placentals)
c Insectivores
d Carnivores
e Lagomorphs (rabbits etc)
f Tree shrews (arguably insectivores)
g Primates (lemurs, lorises, tarsiers, monkeys, apes, and – far right – man)

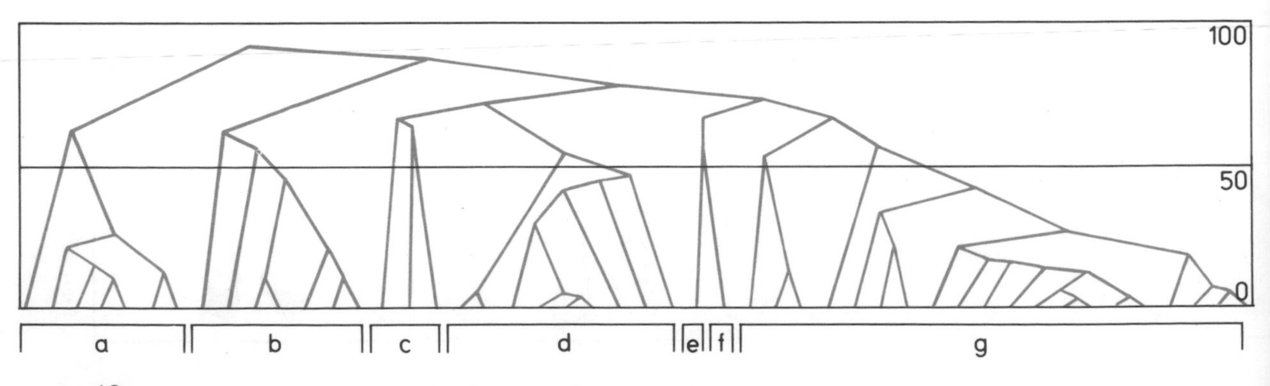

Other tests include comparing sequences of amino acids making up a given protein. The more similar the sequences in different animals, the closer their presumed relationship. More difficult to measure is variation in the sequence of nucleotides building long-chain molecules of DNA. As the so-called "ultimate ancestor," DNA enshrines what has been called the "ultimate truth of molecular evolution."

Between them, direct and indirect molecular comparisons confirm the findings of anatomy and physiology: that our closest living relatives are in the group containing apes and monkeys, so our origins must lie among its ancestors.

Later (pp. 84–85) we shall see just what molecular biology suggests – albeit controversially – about how recently the ape and human lines diverged.

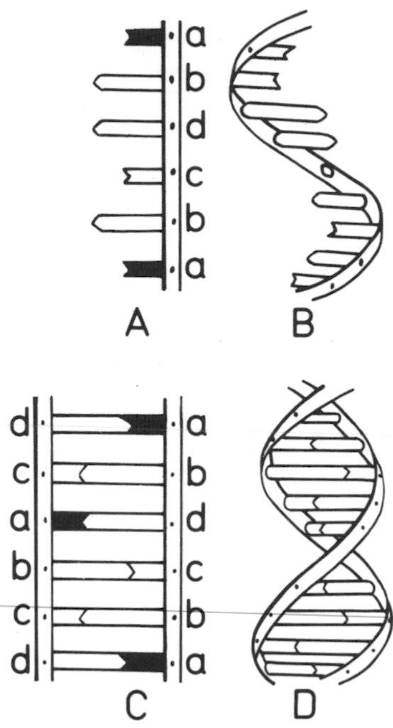

Genetic code (above)
The genetic code determining a species lies in the sequence of bases which link two strands of nucleotides to form the double helix that is DNA.
A Bases in part of one strand of nucleotides
B Part of one strand twisted
C Complementary bases joined
D Two strands joined and twisted
a Thymine
b Guanine
c Cytosine
d Adenine

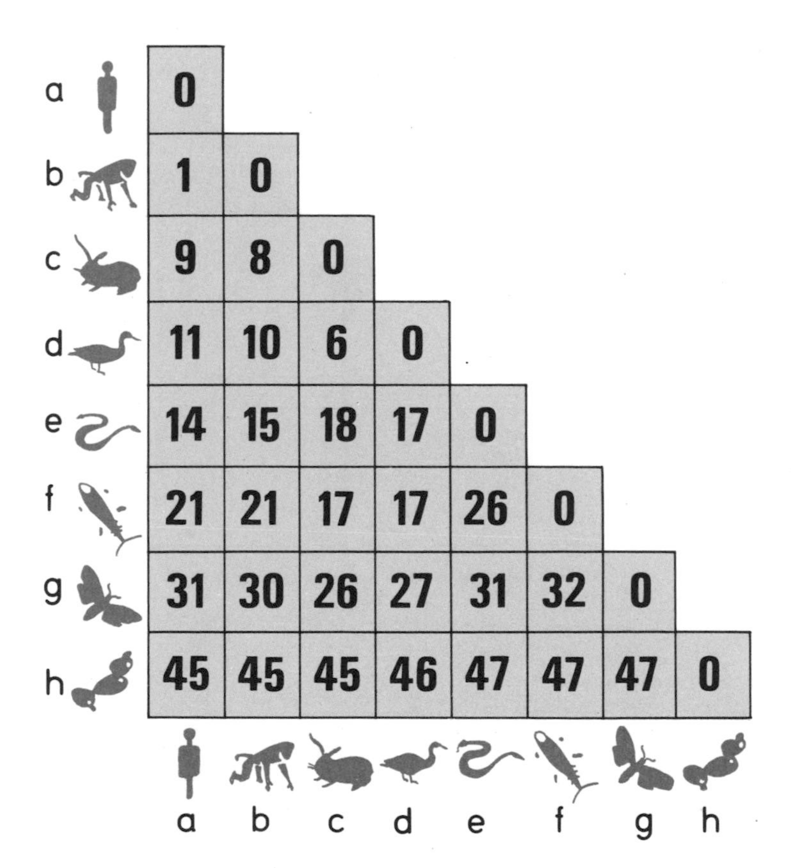

	a	b	c	d	e	f	g	h
a	0							
b	1	0						
c	9	8	0					
d	11	10	6	0				
e	14	15	18	17	0			
f	21	21	17	17	26	0		
g	31	30	26	27	31	32	0	
h	45	45	45	46	47	47	47	0

Amino-acid sequences (left)
Organisms differ in the sequence of amino acids in protein cytochrome c, an enzyme used in energy production. The higher the number shown the greater the difference between any two.
a Human
b Rhesus monkey
c Rabbit
d Duck
e Rattlesnake
f Tuna
g Moth
h Yeast

Man as a primate

While broad aspects of our bodies place us in the mammals, so detailed anatomy collectively identifies us with those agile mammals, primates – the order that includes the lemurs, monkeys, and apes. Most primate features seemingly evolved for life spent mostly up in trees.

First, each primate hand and food has five flexible digits as a rule, with nails not claws. Thumbs, and usually big toes, are opposable for grasping, and sensitive hands on long, mobile arms serve as tactile organs.

Cheek teeth with relatively low cusps (prominences), and an enlarged lower gut with a fermentation chamber – the cecum or appendix – are geared to break down and digest vegetable food, although many primates are omnivorous.

Treetop communication is easier by sight than scent, so primates tended to evolve shorter noses and a feebler sense of smell than many mammals. But their sense of sight enormously improved.

Primates' family tree
This shows likely links between living and extinct primate groups as classified by Szalay and Delson. Their scheme differs in some ways from traditional ones. Items **1–4** are suborders; **a–e** infraorders. Numbers show time in millions of years ago.
1 Ancestor of primates
2 Suborder Plesiadapiformes
3 Suborder Strepsirhini
a Adapiformes
b Lemuriformes (lemurs and lorises)
4 Suborder Haplorhini
c Tarsiiformes (tarsiers)
d Platyrrhini (New World monkeys)
e Catarrhini (Old World monkeys, apes, and men)

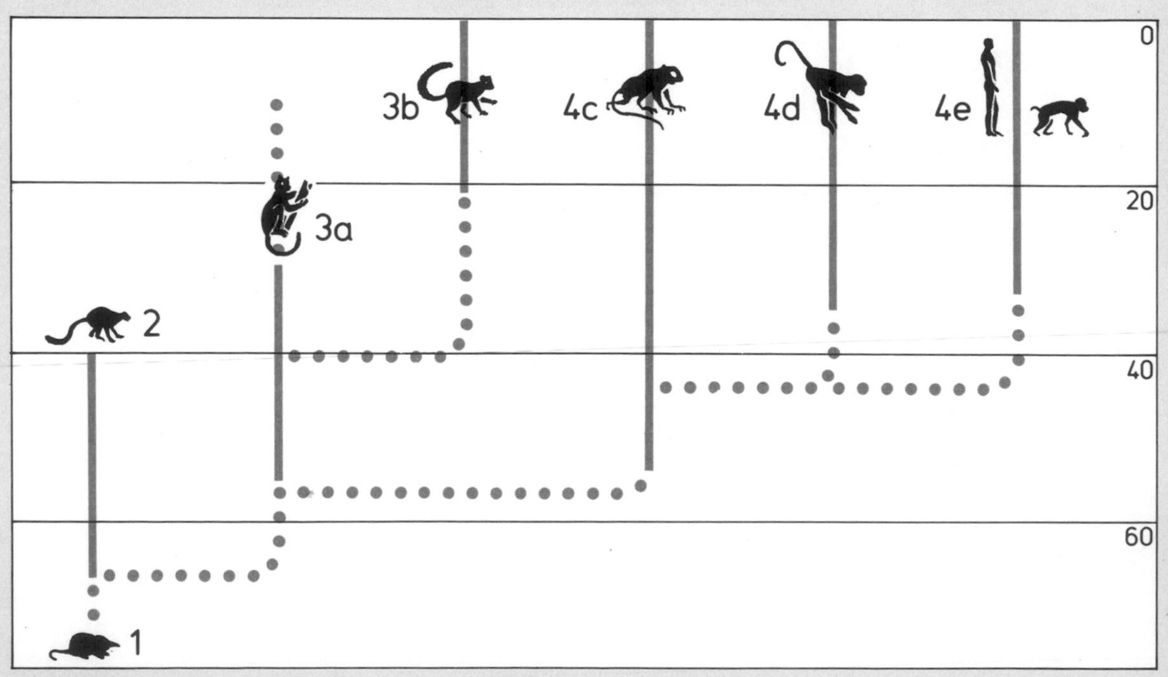

Big, keen, forward-facing eyes that see in three dimensions – in color in day-active primates – made it easy to judge leaps from branch to branch.

Enlarged brains show great development of sensory, motor, and coordination areas needed for split-second acrobatic timing.

Reduced muzzle, enlarged eyes, expanded brain, together with an upright sitting posture ideal for perching on a branch, redesigned the primate skull.

After a long gestation period most females produce a single baby which sucks milk from two glands on the mother's chest. Young stay long dependent on their mothers – a survival strategy evolved for life in trees by creatures whose behavior is learned rather than instinctive.

Lastly, most primates benefit from life in groups, where individuals communicate food finds, combine for mutual defense, and pass on learned behavior.

The next few chapters show changing trends in primate groups, whose evolution led to man.

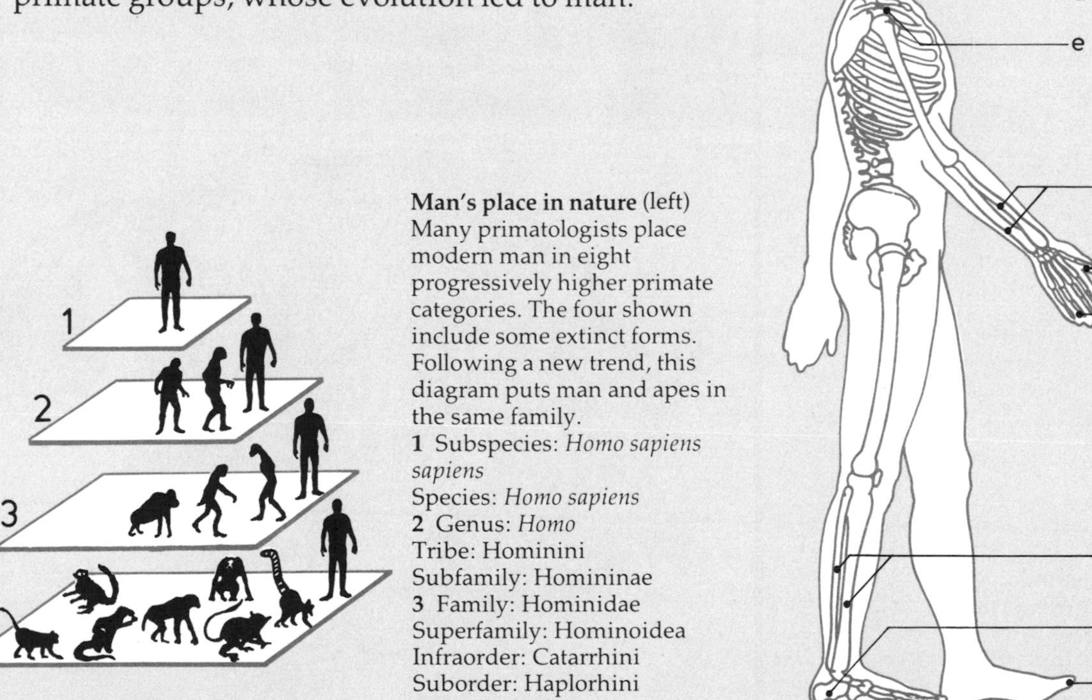

Primate features (below)
Man and other primates share these features (**e–h** also occur in various primitive mammals).
a Enlarged brain, with two unique creases
b Big, forward-facing eyes (with orbits ringed by bone)
c Distinctive middle ear structure
d "Primitive" teeth
e Collar bone
f Separate radius and ulna
g Separate tibia and fibula
h Five flexible digits per limb
i Thumbs usually opposable for grasping
j Distinctive heel bone
k Flat nail on big toe and usually nails (not claws) on other digits

Man's place in nature (left)
Many primatologists place modern man in eight progressively higher primate categories. The four shown include some extinct forms. Following a new trend, this diagram puts man and apes in the same family.
1 Subspecies: *Homo sapiens sapiens*
Species: *Homo sapiens*
2 Genus: *Homo*
Tribe: Hominini
Subfamily: Homininae
3 Family: Hominidae
Superfamily: Hominoidea
Infraorder: Catarrhini
Suborder: Haplorhini
4 Order: Primates

Chapter 2

PRIMITIVE PRIMATES

Beginning with the controversial origins of primates more than 60 million years ago, we give a brief account of lower primates, often called collectively prosimians. These pages include extinct early forms and their descendants the living lemurs, lorises, and tarsiers. Among their forebears were the ancestors of man.

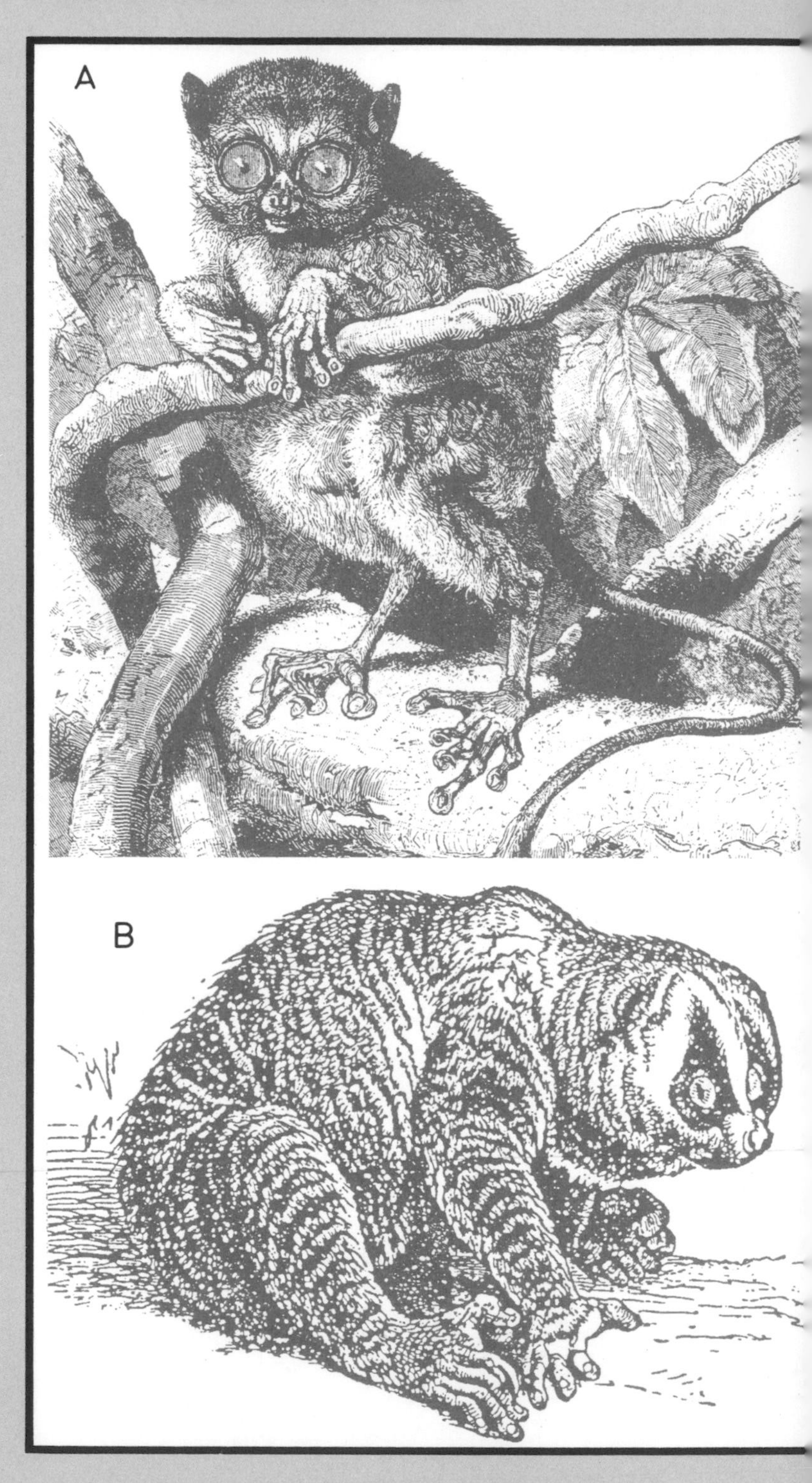

Old illustrations show three living lower primates:
A Tarsier (now often ranked with higher primates)
B Loris
C Lemur

C

How primates began

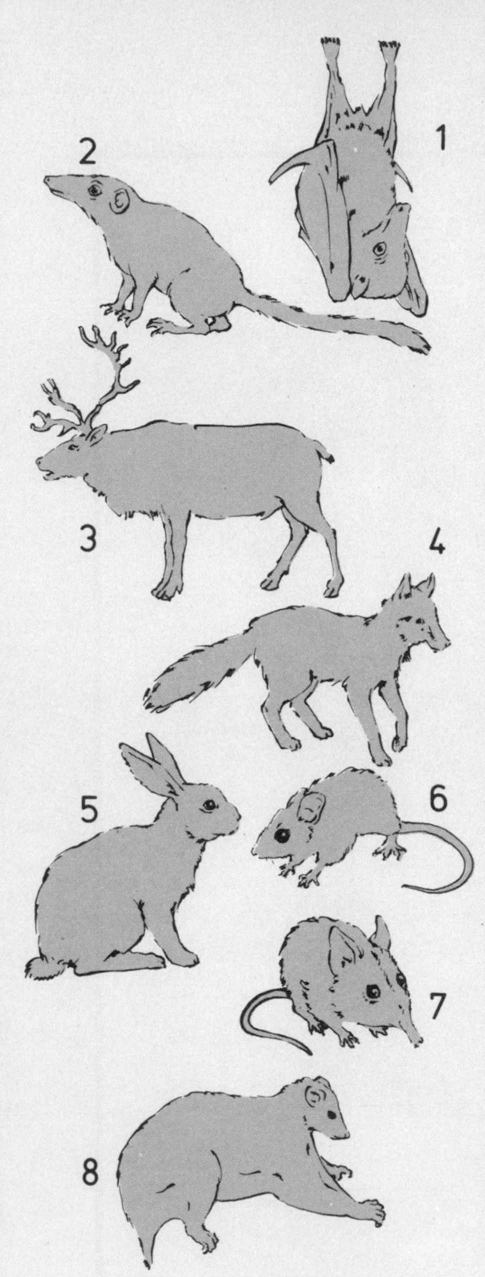

A gallery of candidates
Possible primate precursors include ancestors of:
1 Bats
2 Insectivores
3 Ungulates
4 Carnivores
5 Lagomorphs (rabbits etc)
6 Rodents
7 Elephant shrews
8 Colugos

Primates are one of the oldest, most primitive placental mammal groups, of doubtful ancestry. Some experts argue that they stemmed from some line of placental mammals with species still alive today. At one time or another anatomists have pointed out close similarities between primates and such diverse creatures as colugos, elephant shrews, lagomorphs (rabbits, hares, and pikas), rodents, carnivores, ungulates (hoofed placental mammals), bats, and insectivores.

Of all these, insectivores appear the primates' likeliest precursors. These primitive generalized mammals, including shrews and hedgehogs, have a low braincase, long snout and unspecialized limbs.

Of all insectivores, tree shrews – themselves once ranked as primates – might seem the most probable contenders. These small, agile inhabitants of Southeast Asian forests resemble squirrels with long, pointed faces more than apes or monkeys. But like primates, tree shrews have a large brain for the body size, large eyes, primitive cheek teeth, and thumbs that tend to diverge from the fingers. Close study of such features, though, suggests tree shrews and primates are less alike than people used to think, although their hemoglobin molecules show striking similarities.

Some experts would see the primates' forebears among long-extinct insectivores called microsyopids. Known species lived later than the first known primates, yet had a more old-fashioned inner ear design. Perhaps early microsyopids predated and produced the primate pioneers. But most experts are unconvinced by this suggestion, too.

Such inconclusive findings lead some scientists to believe that the primates' ancient order lacks any features that unquestionably link it with another, parent, group.

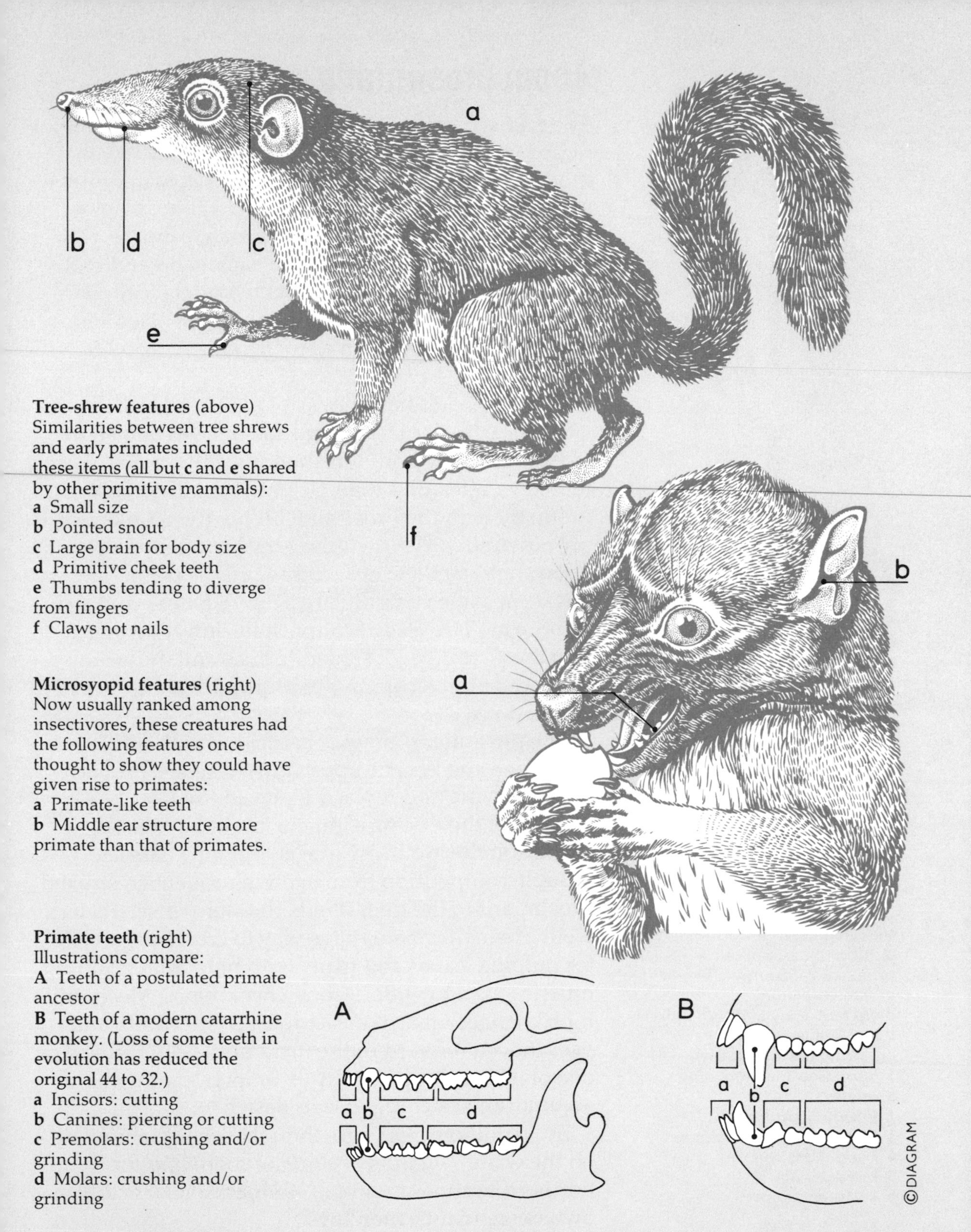

Tree-shrew features (above)
Similarities between tree shrews and early primates included these items (all but **c** and **e** shared by other primitive mammals):

a Small size
b Pointed snout
c Large brain for body size
d Primitive cheek teeth
e Thumbs tending to diverge from fingers
f Claws not nails

Microsyopid features (right)
Now usually ranked among insectivores, these creatures had the following features once thought to show they could have given rise to primates:

a Primate-like teeth
b Middle ear structure more primate than that of primates.

Primate teeth (right)
Illustrations compare:

A Teeth of a postulated primate ancestor
B Teeth of a modern catarrhine monkey. (Loss of some teeth in evolution has reduced the original 44 to 32.)

a Incisors: cutting
b Canines: piercing or cutting
c Premolars: crushing and/or grinding
d Molars: crushing and/or grinding

About prosimians

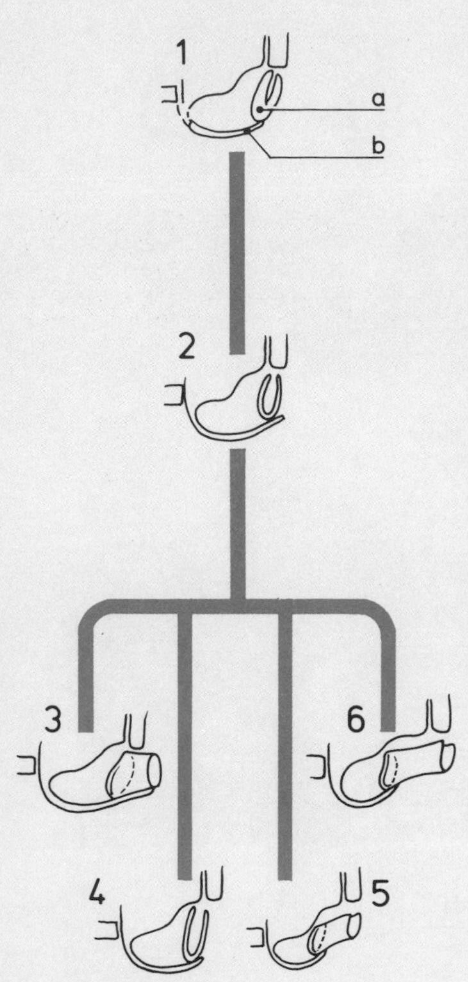

Evolving middle ears
For various prosimians we show possible evolutionary changes to the ectotympanum, a strong ring or tube which supports a primate's eardrum.
a Ectotympanum
b Section along auditory bulla (a bony, hollow priminence enclosing the middle ear)
1 Supposed ancestor of the primates
2 Primitive primate
3 Megaladapid lemuriform
4 Lorisid lemuriform
5 Plesiadapid
6 Early tarsiiform

The first primates probably evolved in North America or Europe about 70 million years ago. They were mouse-to-cat-sized and derived, perhaps, from tree-shrew-like insectivores (see pp. 46–47).

Like their remote descendant, man, primate pioneers retained five digits on each hand and foot: primitive features that stand us in better stead than the specialized limbs of mammals like whales and horses, locked forever in a single habitat or mode of life.

Early primates had claws not nails, and lacked most later primates' large, forward-facing eyes and short faces; but cheek teeth foreshadowed those of monkeys, apes, and man. Scientists traditionally group them in the Prosimii ("before the monkeys"), a primitive suborder including the living lemurs, lorises, and tarsiers, now found only in tropical forests in some parts of Africa, Madagascar, and Asia. A more modern classification splits lower primates into the primitive Plesiadapiformes and the more advanced Strepsirhini, with improved grasp-leaping adaptations.

The prosimians' heyday coincided with the Paleocene and Eocene epochs when scores of species were evolving and replacing one another.

Some of the first prosimians might have lived mainly on the ground. From 60 million years ago, though, competition from evolving rodents restricted prosimians to the trees. Here, the sharp insect-eaters' teeth of early primates gave way to crushers suitable for pulping leaves and fruits with help from grinding muscles fixed to wide, strong cheekbones. Meanwhile limb changes made it easier to leap and climb, and eyes moved forward in the head until both focused on one object, making branch-to-branch leaping more accurate and catching insects easier.

By 30 million years ago, though, prosimians were on the wane – victims of climatic cooling in the northern continents, and of competition from their own descendants, monkeys.

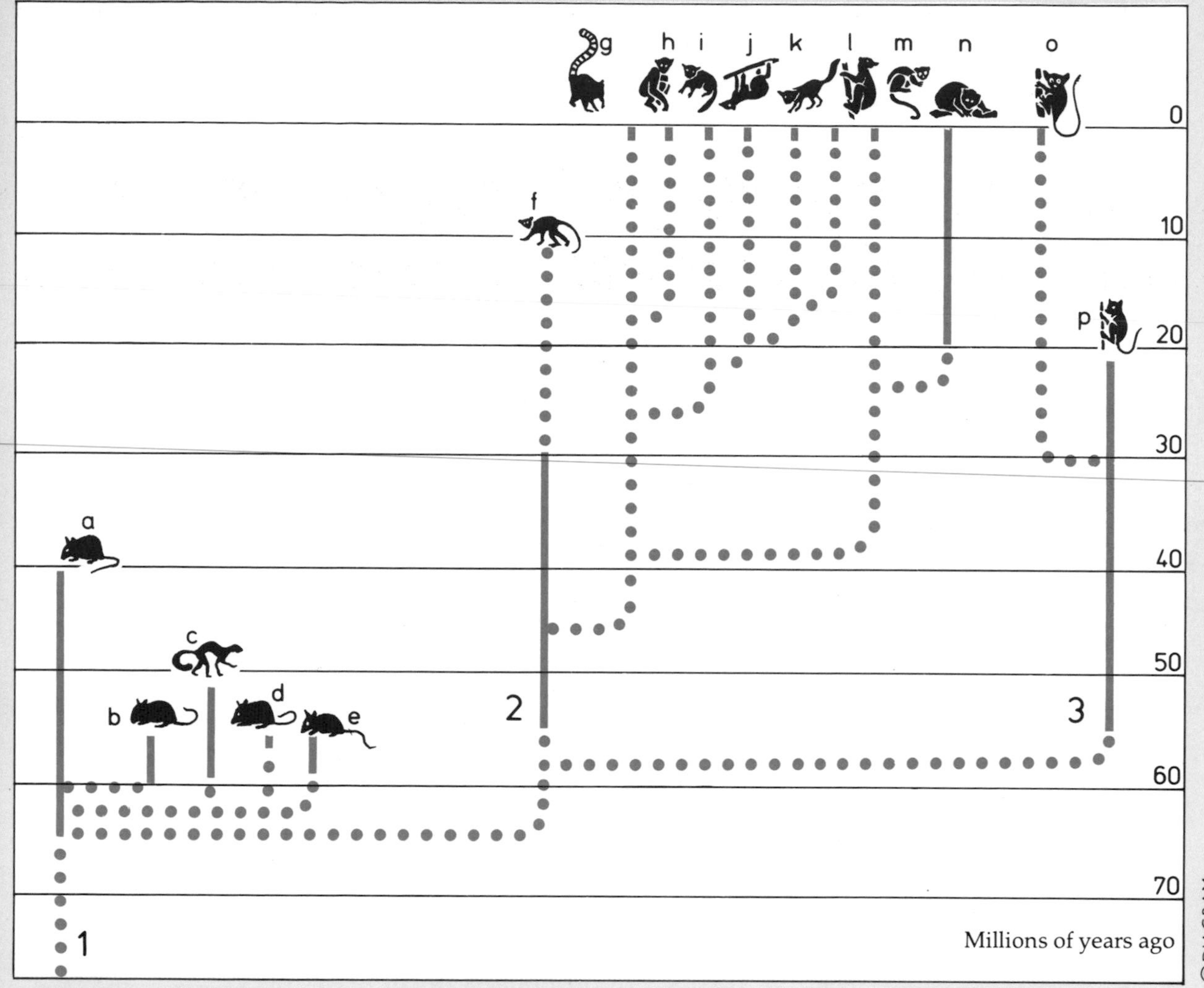

Prosimians' family tree
We show likely evolutionary links between 4 prosimian families. Numbers show suborders and letters show families and two in the Haplorhini (a revised grouping of higher primates). For some of these few fossil bones are known.

1 Plesiadapiformes
a Paromomyidae
b Picrodontidae
c Plesiadapidae
d Carpolestidae
e Saxonellidae
2 Strepsirhini
f Adapidae
g Lemuridae
h Megaladapidae
i Archeolemuridae
j Paleopropithecidae
k Daubentoniidae
l Indriidae
m Cheirogaleidae
n Lorisidae
3 Haplorhini
o Tarsiidae
p Omomyidae

World background

Early primates inherited a literally changing world. As the Cretaceous Period ended 65 million years ago, the two prehistoric supercontinents (northern) Laurasia and (southern) Gondwanaland were breaking up. In the Paleocene and Eocene epochs landmasses were taking on their present shapes and positions. North America and Europe separated, and South America became an island, with just a narrow strip of ocean between it and Africa at first. Meanwhile, shallow inland seas retreated and advanced in several continents, and the early Andes, Rockies, and other ranges rose.

Shifting continents affected climates. At first, North America and Europe basked in warmth. But by the Oligocene Epoch (36-26 million years ago), climates everywhere were cooling down.

This altered vegetation. Subtropical forest that had been flourishing in much of North America and Europe gave way to hardy trees and grasses.

All this affected the placental mammals, now fast evolving into different forms adapted for the modes of life and habitats left vacant when the dinosaurs died out as the Cretaceous Period closed. Early placentals could spread widely through Eurasia, and North America, where most early primates flourished. As oceans or mountains split one species into different populations some tended to evolve new forms, but others just became extinct.

By Paleocene times, rat-sized placentals had given rise to larger kinds, notably primitive hoofed herbivores called condylarths, and early flesh-eating mammals called creodonts. By Eocene times both faced increasing competition, from early modern ungulates and carnivores respectively. Meanwhile rodents ousted rodent-like primate pioneers from ground-based habitats.

1

a

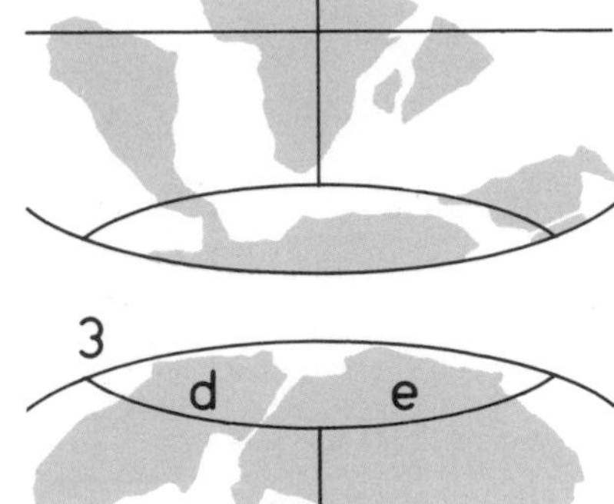

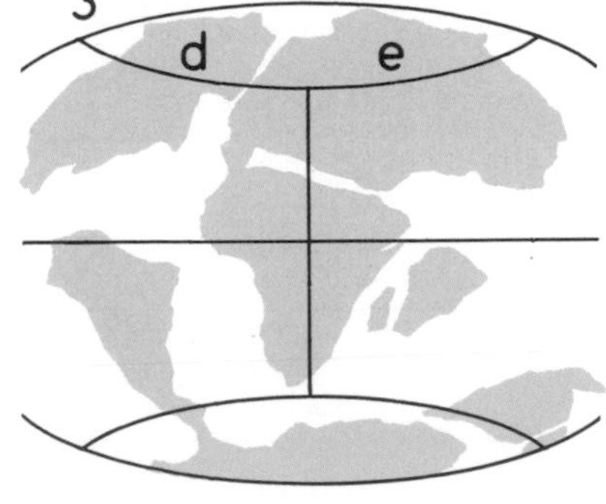

Drifting continents
Three maps show changes that separated continents, isolating some early primate populations.
1 200 million years ago
2 65–46 million years ago (Paleocene – Mid Eocene)
3 46–38 million years ago (later Eocene)
a Laurasia
b Euramerica
c Asiamerica
d North America
e Eurasia

Paleocene life
Here are nine prehistoric animals that lived in Paleocene time. They are not shown to scale. Items **c–i** represent the explosive evolution of the mammals.

INVERTEBRATES
a Tentaculate

AMPHIBIANS
b Cecilian

MAMMALS
c Amblypod
d Dermopteran
e Primate
f Condylarth
g Multituberculate
h Perissodactyl
i Rodent

Eocene life
These 18 prehistoric creatures were among those flourishing in Eocene time. They are not shown to scale.

INVERTEBRATES
a Coelenterate
b Nematode worm

MAMMALS
c Bat
d Tillodont
e Primate
f Creodont
g Carnivore
h Condylarth
i Amblypod
j Sea cow
k Proboscidean
l Perissodactyl
m Artiodactyls
n Edentate
o Whale

BIRDS
p Ratite
q Shore bird

Primate pioneers

The first primates arguably comprised the suborder Plesiadapiformes ("near Adapiformes"). Most are identified as primates by internal ear design. But skull, limbs, and long, and lower, middle incisor teeth that jutted forward show primitive features not found in later primates. The group probably evolved from early insectivores but many species ate tough vegetable foods. At least five families arose, with 20 very varied genera that ranged in size from mice to cats. Many might have looked like rodents, but most are known only from fossil teeth and bits of skull. Some early form might have given rise to all the other primates.

The group lived about 70–40 million years ago, and evidently spread from North America to Europe via Greenland, then a warm, wooded land bridge. Our examples represent four families.

Early primate sites (above)
Early Cenozoic primate sites of western North America:
1 Gidley Quarry, Montana
2 Mason Pocket, Colorado
3 Wind River Basin, Wyoming
4 Big Horn Basin, Wyoming
5 Uinta Basin, Utah
6 Bridger Basin, Wyoming

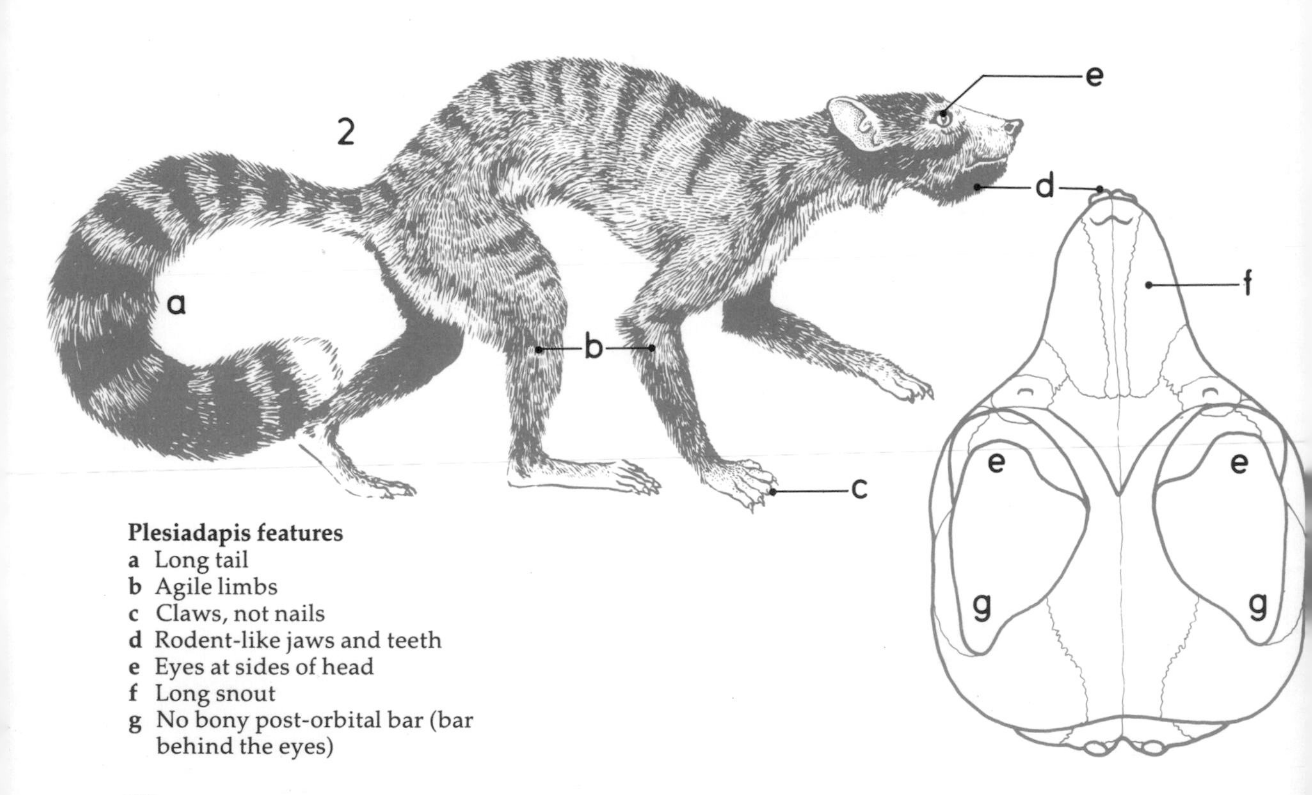

Plesiadapis features
a Long tail
b Agile limbs
c Claws, not nails
d Rodent-like jaws and teeth
e Eyes at sides of head
f Long snout
g No bony post-orbital bar (bar behind the eyes)

1 **Purgatorius** This oldest-known, rat-sized, primate had four premolar teeth (more than later primates) and sharp, three-pointed molars, like insectivores', but with an added "heel" to give a bigger chewing surface. Time: Late Cretaceous–Paleocene. Place: Montana (USA). Family: Paromomyidae.
2 **Plesiadapis** looked rather like a squirrel with side-facing eyes, long snout, projecting chisel-like incisor teeth, bushy tail, and claws on paws that could not grasp. It ate leaves, leapt well, and maybe lived in troops, often on the ground. Time: Mid–Paleocene-early Eocene. Place: Colorado (USA) and France. Family: Plesiadapidae.
3 **Carpodaptes** was mouse sized, with saw-edged cheek teeth and a huge premolar for shearing tough plant foods. Time: Paleocene–early Eocene. Place: North America. Family: Carpolestidae.
4 **Zanycteris** was a tiny mouse-like primate with cheek teeth like a bat's. It might have eaten nectar, fruit, or insects. Time: Late Paleocene. Place: North America. Family: Picrodontidae.

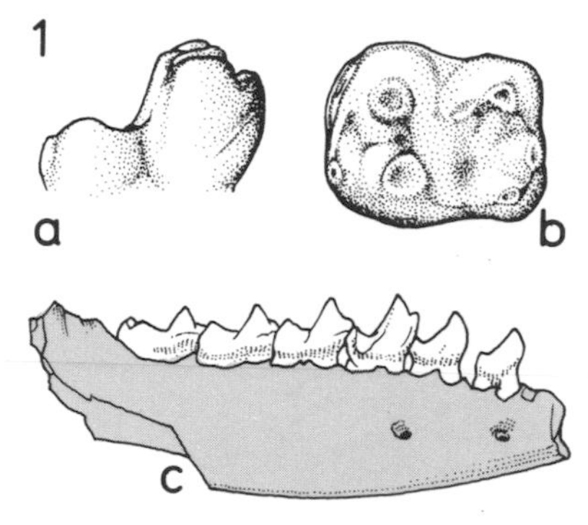

Purgatorius teeth
Above are (**a**) side view and (**b**) crown view of a *Purgatorius* molar tooth, also (**c**) an incomplete lower jaw, all much enlarged. Chewing-teeth with an added "heel" helped primates to crush plant foods.

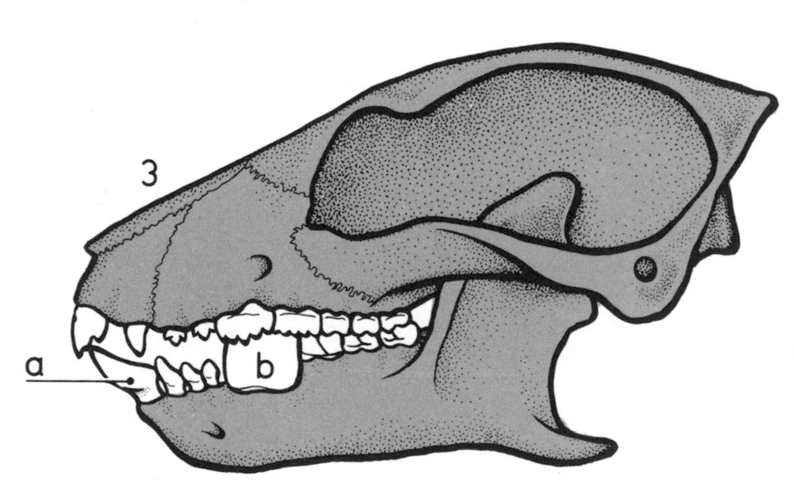

Carpodaptes skull (above)
Like other plesiadapids, *Carpodaptes* had long projecting lower incisors (**a**) but also huge lower premolars (**b**). Actual skull length: about 1in (2.54cm).

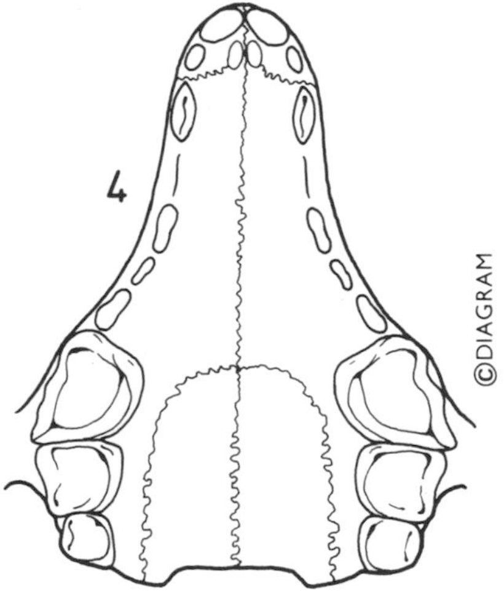

Zanycteris jaw (above)
This much-enlarged view of a lower jaw shows simple, low-crowned, bat-like molars. Some experts suggest that *Zanycteris* ate soft fruits, as living fruit-bats do.

Adapiformes

The lemur-like Adapiformes featured strongly in a second burst of primate evolution. Their infraorder comprised one family, Adapidae, with some 20 genera, all long extinct. Cat sized or smaller, these creatures had long hind limbs, short fore limbs, a long tail and a longish muzzle. Adapids were agile climbers but might have clung and leapt upright in search of fruits, shoots, insects and birds' eggs.

Unlike plesiadapids, adapids had premolars, and vertical front teeth. But they were more advanced than early primates in their relatively shorter snout and larger brain, more forward-facing eyes, protected by a bony bar behind, nails not claws, and grasping hands and feet.

The group flourished mainly in Eocene and early Oligocene North America and Europe (about 54–30 million years ago). They probably gave rise to lemurs, tarsiers, and, indirectly, to higher primates, too. Our examples come from both subfamilies.

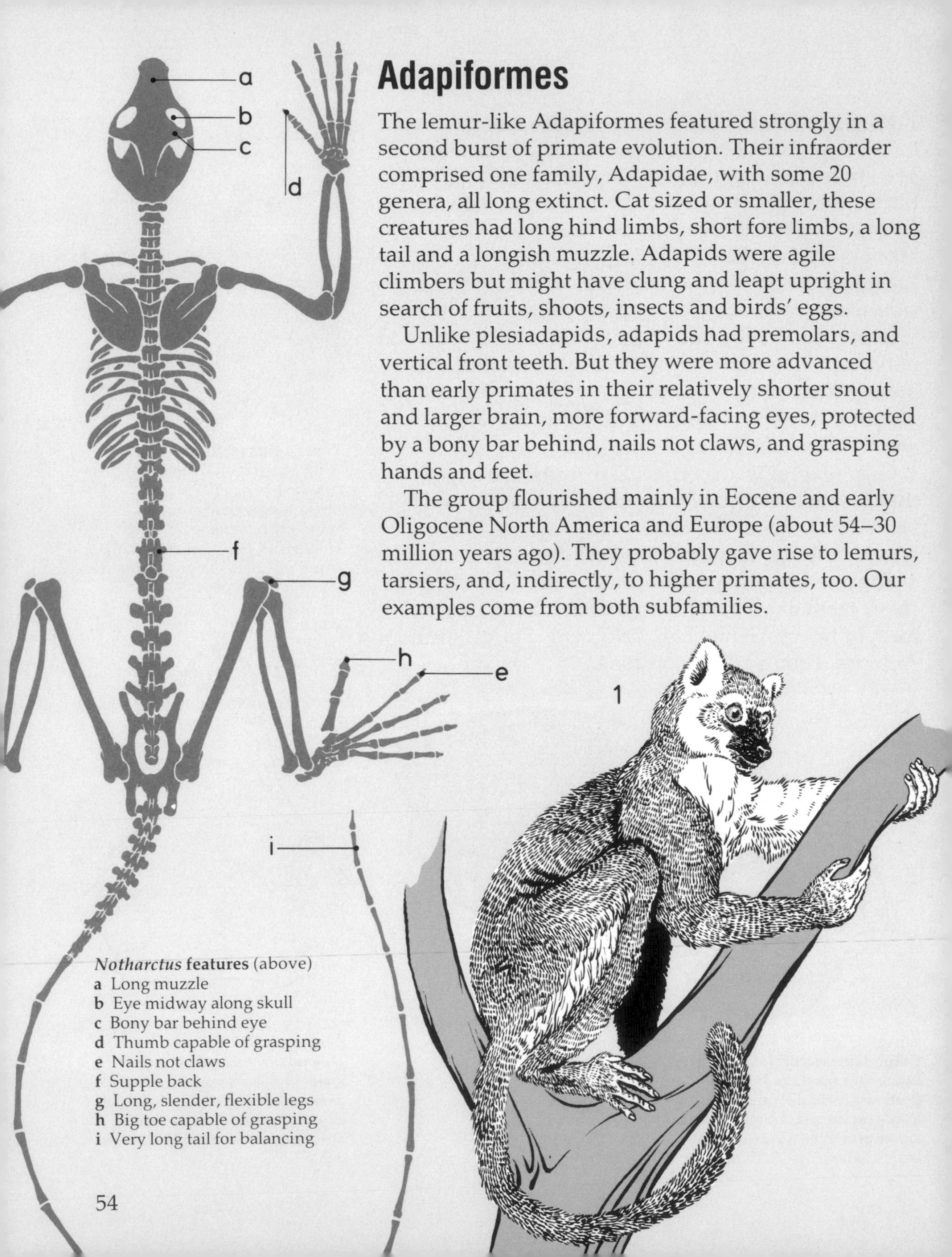

Notharctus **features** (above)
a Long muzzle
b Eye midway along skull
c Bony bar behind eye
d Thumb capable of grasping
e Nails not claws
f Supple back
g Long, slender, flexible legs
h Big toe capable of grasping
i Very long tail for balancing

1 **Notharctus** had a rather long muzzle, supple back, long, slender legs, thumbs and big toes separated from the other digits, and a long tail used for balancing in treetop climbing and leaping. Canine teeth were small and tusk-like; incisors small and vertical; and the (complete) set of cheek teeth included low-crowned molars capable of grinding leaves to pulp. Length: 33 in (84 cm). Time: Mid Eocene. Place: Wyoming (USA). Subfamily: Notharctinae (mostly from North America but also Europe, Africa, and Asia).

2 **Adapis** somewhat resembled *Notharctus* but its strong, short, jaws only bit up and down, with no sideways grinding action. Probably Adapis ate tough, fibrous plant food. The brain's cerebrum was somewhat primitive, but its temporal lobes were relatively larger than non-primate mammals'. Length: 16 in (40 cm). Time: Mid-Late Eocene. Place: Western Europe. Subfamily: Adapinae (from Europe and North America).

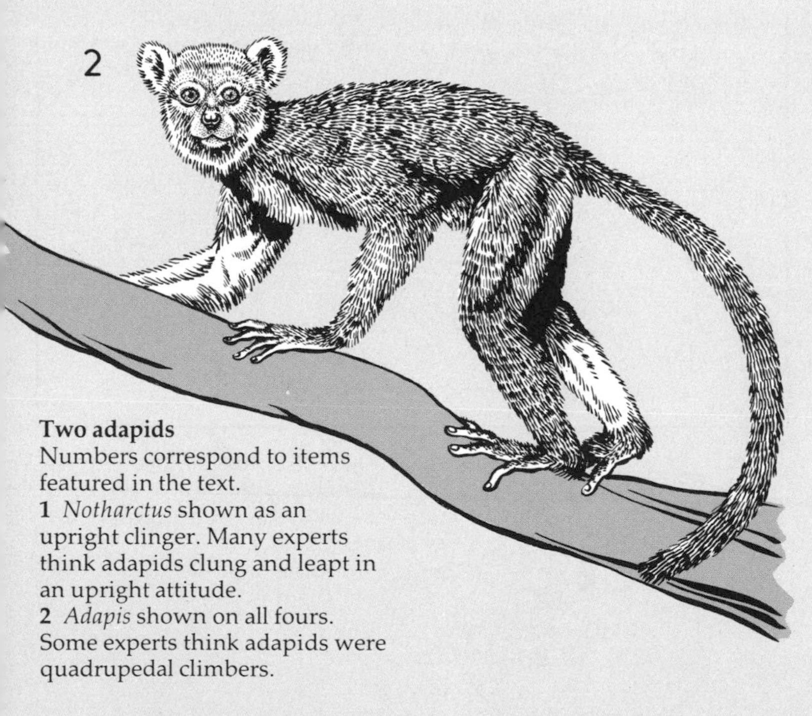

Two adapids
Numbers correspond to items featured in the text.
1 *Notharctus* shown as an upright clinger. Many experts think adapids clung and leapt in an upright attitude.
2 *Adapis* shown on all fours. Some experts think adapids were quadrupedal climbers.

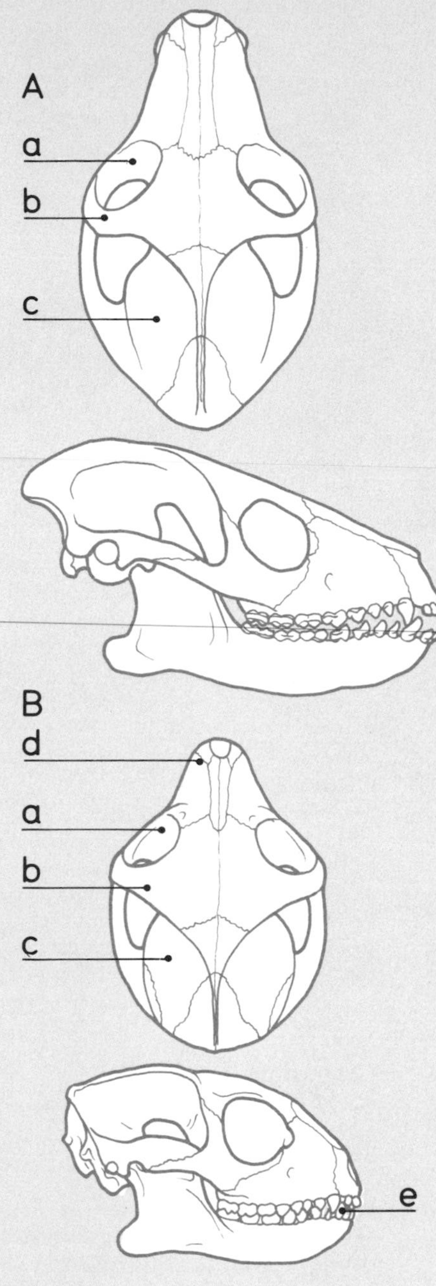

Adapid skulls (above)
Skulls of (**A**) *Notharctus* and (**B**) *Smilodectes*, a short-faced form, show adapid trends not seen in plesiadapids.
a Forward-facing eyes for stereoscopic vision
b Bony bar protecting eyes
c Relatively large brain
d Reduced snout
e More vertical incisors

Lemuriformes

Lemuriformes in many ways resemble their ancestors the Adapiformes (p. 54). The major difference between both groups is the Lemuriformes' long, lower, forward-jutting front teeth evolved into a comb for grooming fur.

Scientists know of nearly 30 genera, 9 of them extinct – most, recently. Lemuriformes probably arose about 25 million years ago in late Oligocene times, but early fossil finds are few.

Seven of the eight families appeared in Madagascar, evolving a variety of forms free from major predators until man arrived a few hundred years ago. Most are sharp-featured, long-limbed, long-tailed, agile climbers eating fruits, leaves or insects. The eighth family, the Lorisidae, includes the slow-moving Asian lorises and African pottos, and Africa's lively little galagos or bushbabies.

Our five examples show something of the Lemuriformes' diversity.

Giant lemuroid
Megaladapis was a recently extinct Madagascan lemuriform as heavy as a woman and as big as a large dog. Probably it browsed on leaves, and climbed trees slowly, making cautious jumps.

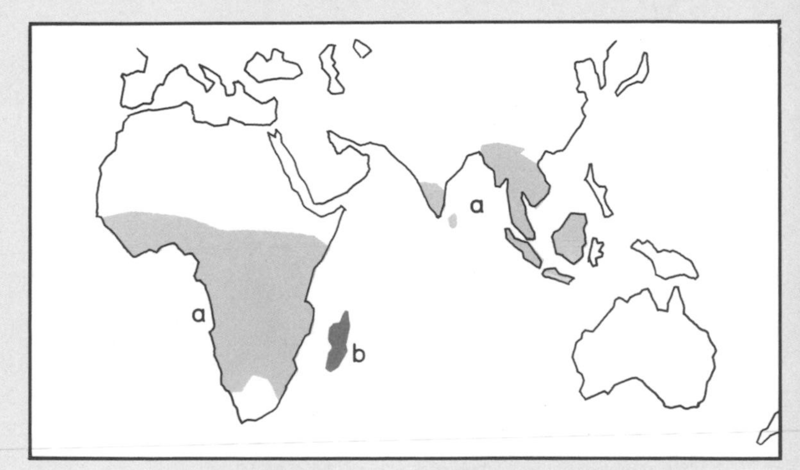

Where they live (above)
This map shows the distribution of living lemuriforms.
a Lorisids
b All other families

Contrasting skulls (below)
A Pig-like skull of *Megaladapis,* a browsing member of the extinct lemuriform family Megaladapidae.
B Deep, short hominine-like skull of *Hadropithecus*, a ground-dwelling grazer in the extinct family Archaeolemuridae.

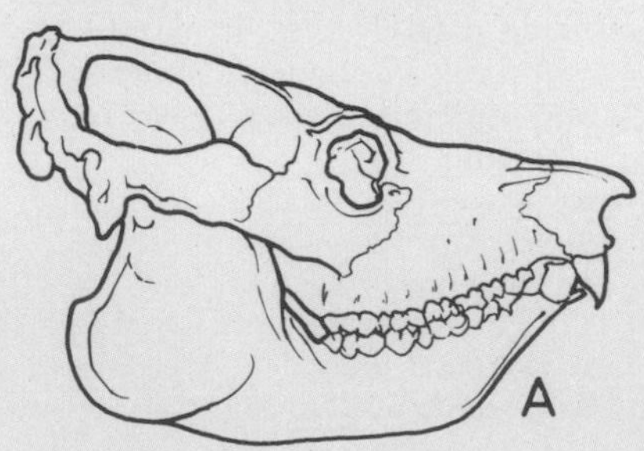

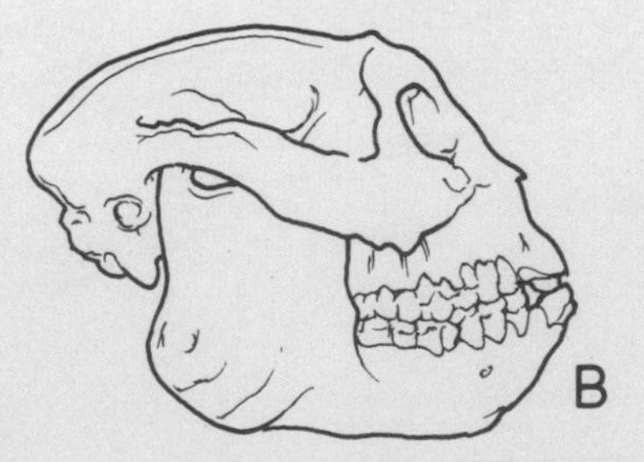

1 **Lemur catta,** the living ring-tailed lemur, has a typical lemuroid's dog-like face; 36 teeth; naked nose; big, furry ears; mobile limbs; grasping hands and feet with nails; and a long, bushy tail. Family: Lemuridae.
2 **Indri indri,** the indris, an indrioid lemuriform, has long limbs and stumpy tail, and clings and leaps upright. Family: Indriidae.
3 **Daubentonia madagascariensis,** the aye-aye, has a pointed muzzle, huge ears, bushy tail, clawed digits, and only 18 teeth. Its gnawing incisors open coconuts, and it squashes wood-boring insects with its wiry middle finger. Family: Daubentoniidae.
4 **Galago crassicaudatus,** the fat-tailed galago, has a dog-like muzzle; big eyes; mobile ears; long, bushy tail; and broad pads on the fingertips. Length: 25in (64cm). Place: Africa. Family: Lorisidae. (This largest living galago might resemble the early Miocene *Progalago*.)
5 **Nycticebus coucang,** the slow loris, has a short muzzle, big eyes, thin wrists and ankles, and no tail. Length: 13in (33cm). Place: South-east Asia. Family: Lorisidae.

Five lemuriforms
Numbers correspond to items featured in the text.
1 *Lemur catta*
2 *Indri indri*
3 *Daubentonia madagascariensis*
4 *Galago crassicaudatus*
5 *Nycticebus coucang*

Tarsiiformes

Tarsiiformes comprise more than 30 genera of small, short-faced primates, all but one extinct. Its three huge-eyed, nocturnal species look like goblins from a fairytale, yet might be our closest relatives outside the apes and monkeys. For tarsiers, monkeys, apes, and man all share similarities in nose and eye design, brain structure, fetal membranes, and biochemical ingredients. Indeed, many experts place the infraorder Tarsiiformes with the higher primates in the same suborder: Haplorhini ("single noses").

The Tarsiiformes evolved in Paleocene times from the Adapiformes (p. 54), or just maybe from the Plesiadapiformes (p. 52). They diversified and spread through northern continents and Africa. Their heyday was the Eocene, but most died out in the Oligocene, probably failing to compete with the evolving monkeys.

Last refuges (above)
The tarsiers' sole surviving genus, *Tarsius*, lives in low-lying forests, on only certain South-east Asian islands. These are the largest:
a Sumatra
b Borneo
c Sulawesi (Celebes)
d Mindanao

***Tarsius* features** (below)
Some features labeled in this skeleton of *Tarsius* occur in fossil tarsiiforms.
a Long, slim tail
b Long ankle (tarsal) region aids treetop leaping.
c Semi-fused tibia and fibula
d Long, slim toes and fingers
e Backbone articulated with skull from below, not behind (head usually held erect)
f Distinctive middle ear
g Huge eye sockets
h Small nose
i Short jaws

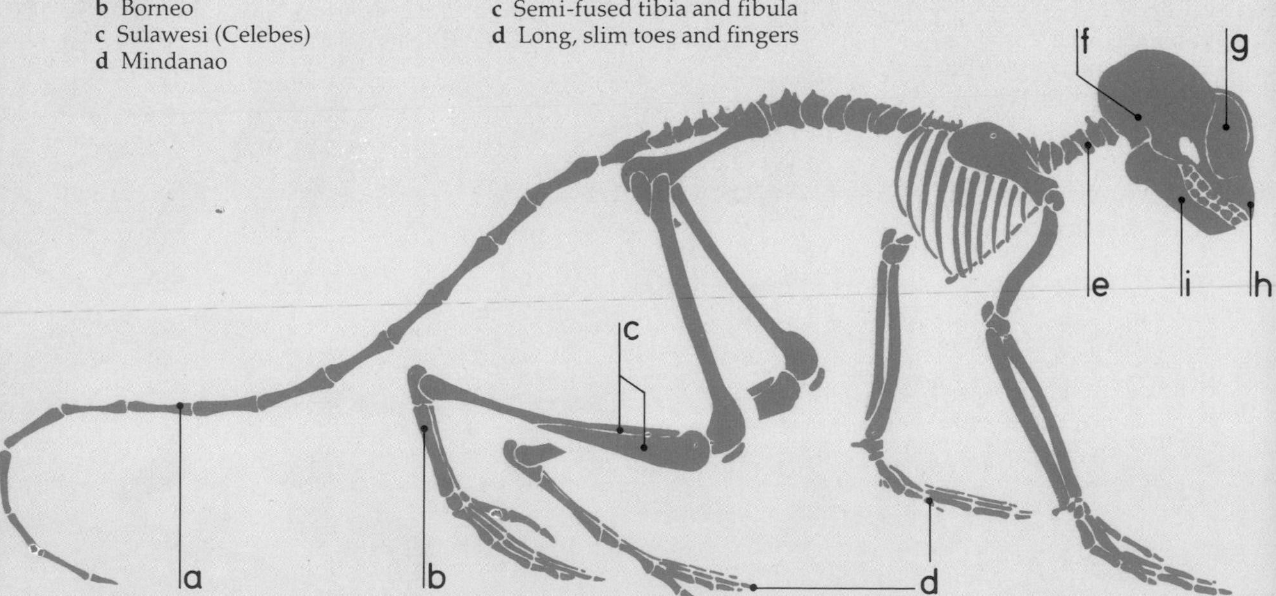

Our examples come from both families.

1 **Necrolemur** had a short face, big forward-facing eyes, heavy jutting lower front teeth, and long heel bones. It clung upright to tree trunks and might have leapt from branch to branch erect. Size: as 2. Time: Eocene. Place: Europe. Family: Omomyidae.

2 **Tarsius,** the living tarsier, has a dry, furry nose; 34 teeth; huge eyes; big ears; short arms; long, flexible legs with long ankle bones; disks on finger tips and short nails (but grooming claws on two toes per foot). Length: 4.7 in (12 cm) without the long, slim tail. It eats insects and lizards. Place: South-east Asia (but extinct *Afrotarsius* lived in Africa). Family: Tarsiidae.

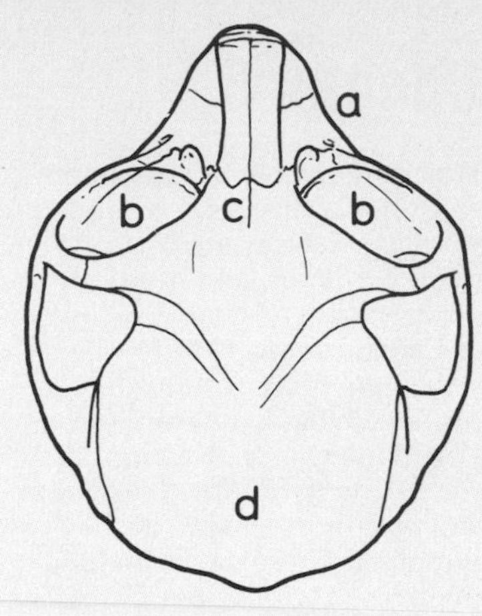

Tarsiiform skull (above)
Like other omomyids, extinct *Rooneyia* differed from modern tarsiids in tooth and middle ear design, but shared with them these features:
a Short face
b Big, forward-facing eyes
c Narrow gap between eyes
d Large braincase

Two tarsiiforms (not to scale)
Numbers correspond to creatures featured in the text.
1 *Necrolemur*
2 *Tarsius*

Chapter 3 EVOLVING ANTHROPOIDS

With this chapter we reach the so-called anthropoids or higher primates – the group including monkeys, apes, and humans.

Emerging in North America or Eurasia about 40 million years ago, early anthropoids produced the Platyrrhini (New World monkeys) and the Catarrhini (Old World monkeys and hominoids – apes and humans.) These pages include the living Hylobatidae or lesser apes, and early members of the Hominidae, a family taken here to comprise humans, great apes, and their closest prehistoric kin.

A 19th-century engraving depicts the Barbary ape, really a tailless Old World monkey, found in North-west Africa and Gibraltar.

About anthropoids

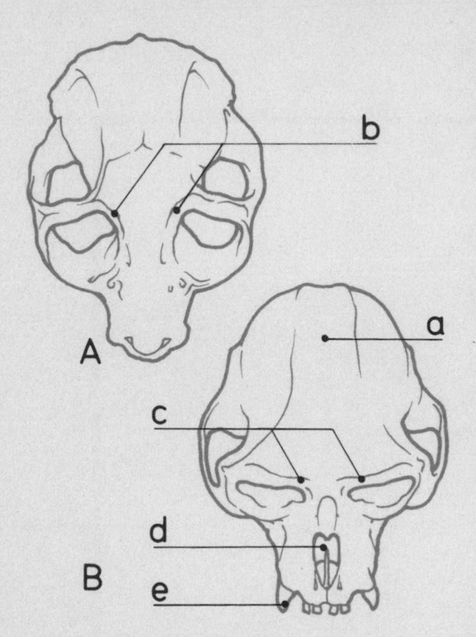

Contrasting skulls (above)
Illustrations contrast two types of primate skull.
A Lemur, a prosimian
B Langur monkey, an anthropoid
a Rounded cranium
b Bars behind eyes
c Bony caves behind eyes
d Narrow nasal cavity
e Well developed canines

Monkeys, apes, and men – the higher primates – traditionally form one great primate suborder: Anthropoidea. Anthropoids evolved from ancestors of the little living tarsiers (or maybe lemurs). They first appeared in North America or Eurasia some 40 million years ago. Two infraorders arose: the Platyrrhini (New World monkeys) and the Catarrhini (Old World monkeys, apes, and men).

Living anthropoids range from tiny squirrel-sized tamarins to gorillas three times as heavy as an average man. Some species walk on all fours, some swing from trees, by arms, legs, or tails. Only man regularly walks on hind limbs. Some monkeys have tails longer than their bodies; apes and men are tailless. One monkey moves around at night, other anthropoids by day.

Such variations apart, most anthropoids share certain basic features. Most can sit up, freeing hands for manipulating objects, helped by a thumb and big toe that can be turned in for grasping. Most species have flat nails on toes and fingers, and lack the lemurs' toilet claws.

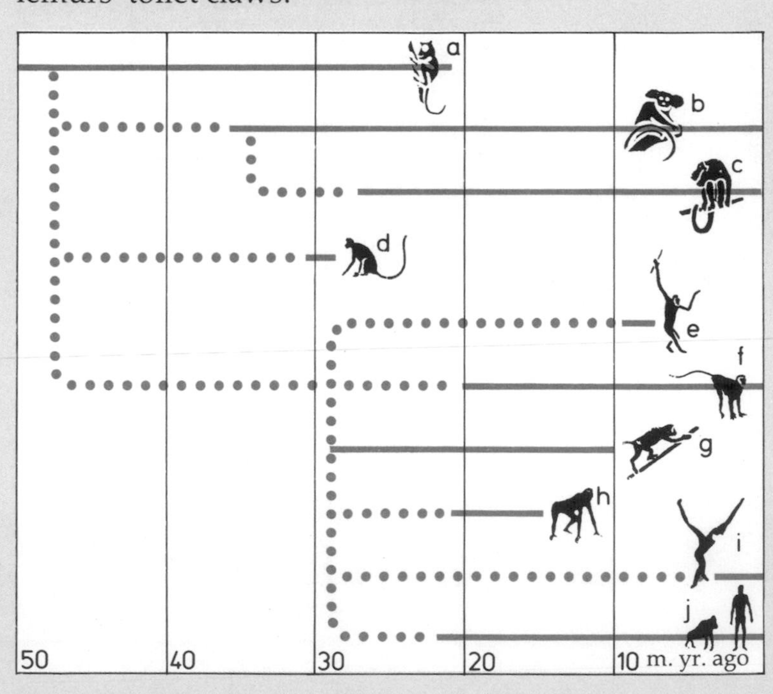

Anthropoids' family tree (right)
Lines show likely evolutionary links between nine families of anthropoids and their likely ancestors, the omomyid Tarsiiformes (see also page 49). All arguably form one suborder: Haplorhini.
a Omomyidae
b Cebidae
c Atelidae
d Parapithecidae
e Oreopithecidae
f Cercopithecidae
g Pliopithecidae
h Proconsulidae
i Hylobatidae
j Hominidae

The head is angled forward from the spine. The snout is generally short, but the eyes are enlarged and forward facing – each set in a protective bony cave, not just guarded by a bony bar as in the more primitive prosimians. The rounded skull shields a brain relatively bigger and more wrinkled than a prosimian's, and differently grooved, with a shrunken smell sector but enhanced visual and mental areas.

The 30–36 teeth are all vertical. They include enlarged, low-cusped cheek teeth designed for grinding tough food, not just shredding or "juicing," and low premolars that tend to mesh with upper canines, honing these into a point. Such teeth help make many anthropoids efficient omnivores.

Anthropoids have only two breasts, located on the chest. Anthropoid placentas differ from prosimians' and, compared with these primates, anthropoids have longer pregnancies and lives.

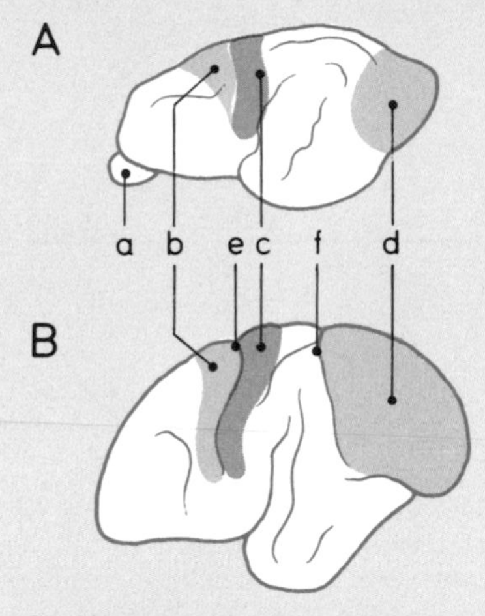

Brains compared (above)
Here are some functional areas and divisions in the brains of (**A**) lemur (a prosimian) and (**B**) macaque (an anthropoid).
a Olfactory bulb (smell)
b Voluntary movement
c Sensations
d Visual area
e Central groove
f Simian groove

Anthropoid features (right)
a Tail long, short, or absent
b Prehensile hands and feet
c Thumbs at least partly opposable to fingers in most
d Big toe turns in for use in grasping in most
e Flat nails on toes and fingers
f Only two nipples, on chest
g Large, rounded cranium
h Large, forward-facing eyes
i Face protrusion variable
j 30–36 vertical teeth, canines mostly large

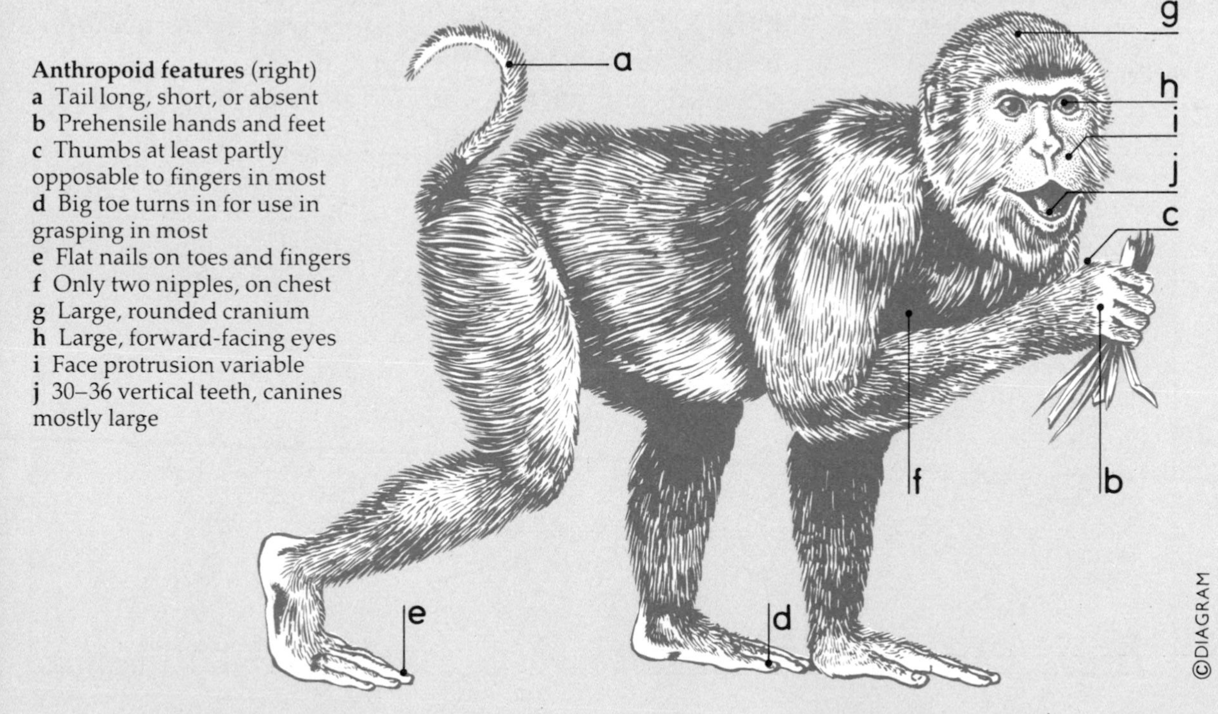

Miocene world background

Miocene world (above)
This world map shows that continents had taken up their present positions by Miocene times. Lines represent Equator, Tropics, and Polar Circles.

By Miocene times (25–5 million years ago) anthropoids were diversifying and colonizing Africa, Eurasia, and the tropical Americas.

The world changed greatly in this time. Ice blanketed Antarctica, India bumped into Asia, the world's great mountain ranges rose. Africa's Rift Valley yawned open to the belching of volcanoes. The vast prehistoric Tethys Sea shrank into big salty pools: the Mediterranean, Black, and Caspian seas. Animals easily migrated between Eurasia and Africa.

Deciduous and evergreen broad-leafed trees and conifers now covered northern continents. Farther south, some tropical forests shrank under the effects of drying climates. In East Africa, tropical forest still rimmed rivers and low mountain slopes, but great plains became savanna – open, park-like countryside of tall grasses and scattered trees.

The cooler northern climates dating from Oligocene times (38–25 million years ago) probably contributed to the evolution of the anthropoids. Where the year-round leaf and fruit supply dried up, prosimians mostly just became extinct or followed the retreating tropical forest south. Primates with flexible food demands and tolerance to cold seemingly gave rise to

Shrinking seas (below)
This shows part of Africa and Eurasia about 16 million years ago. Shrinkage of the Tethys Sea and a new land bridge allowed anthropoids to migrate between Africa, Europe, and Asia.
a Tethys Sea
b Para-Tethys Sea
c Arabian land bridge

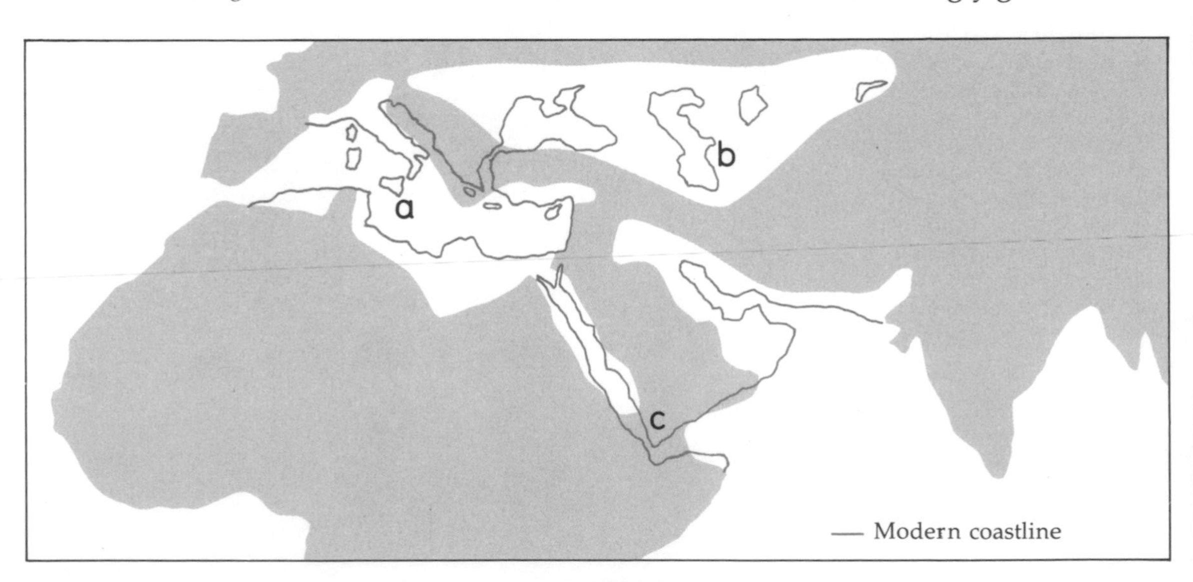

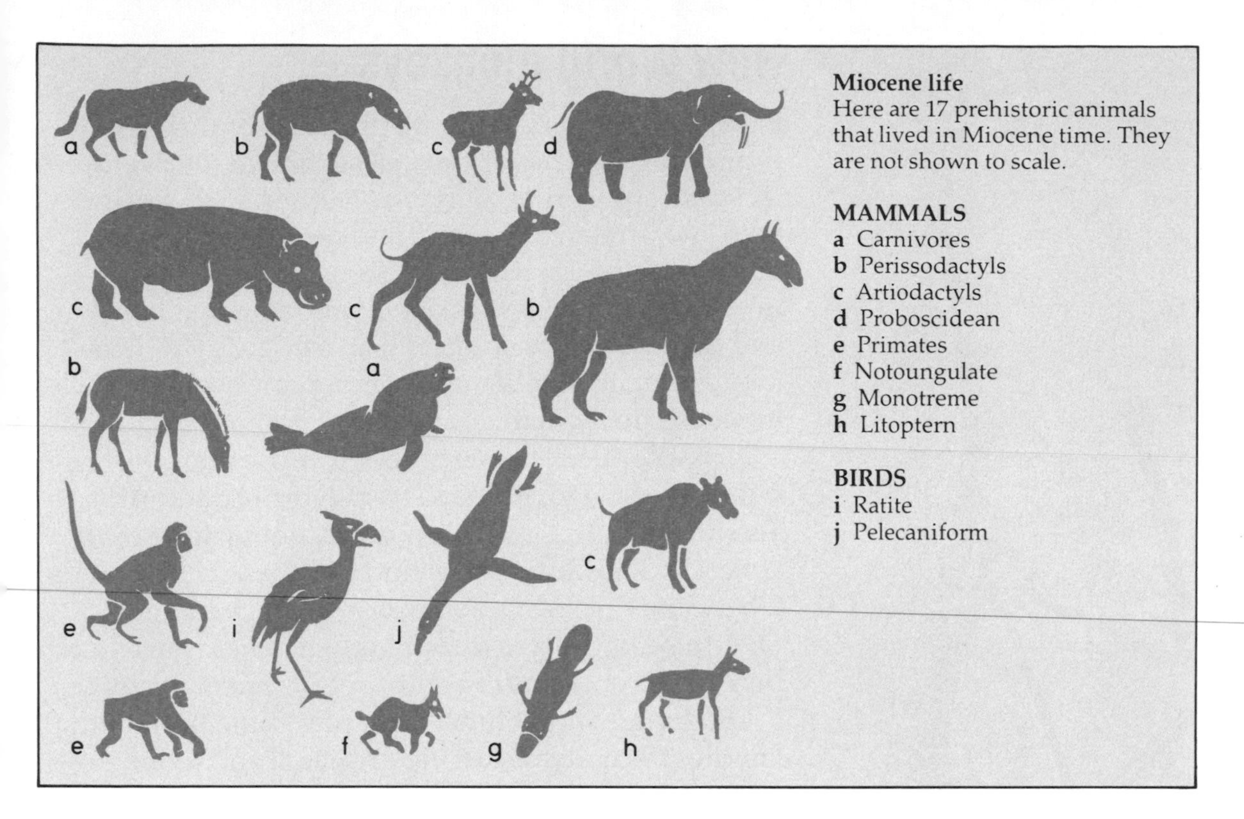

Miocene life
Here are 17 prehistoric animals that lived in Miocene time. They are not shown to scale.

MAMMALS
a Carnivores
b Perissodactyls
c Artiodactyls
d Proboscidean
e Primates
f Notoungulate
g Monotreme
h Litoptern

BIRDS
i Ratite
j Pelecaniform

anthropoids. Influential trends would have included eating less fruit where this grew seasonally scarce, but more bark and leaves of evergreens. Another tendency was increased body size, conserving body heat. This probably went hand in hand with longer pregnancies, smaller litters, and a shift from night to day activity, with sight increasingly outranking smell.

Animal migrations (below)
This diagram gives the pattern of two-way mammal migrations in the Miocene, when land linked northern continents. Yet Old World apes and monkeys never entered North America; perhaps climatic barriers defeated them. New World monkeys evolved in isolation on South America, an island. No primates reached Australia.

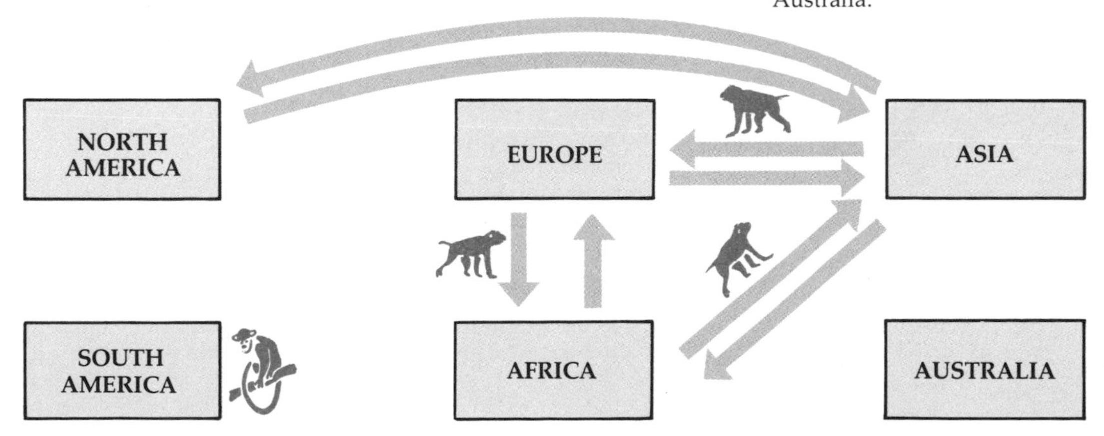

New World monkeys

Platyrrhini ("flat noses") are monkeys of the New World tropics. They have well separated, outward-facing nostrils, three premolar teeth in each side of each jaw, a relatively larger and more convoluted brain than the Tarsiiformes, some distinctively interlocking skull bones, long tails (some prehensile), and poorly developed thumb and finger grip. They range from minute tamarins and marmosets to howler monkeys that weigh about 22lb (10kg).

New World monkeys probably arose about 40 million years ago, perhaps from North American Tarsiiformes, but colonized only Central and South America. The oldest fossil find is *Branisella* from Early Oligocene Colombia, some 35 million years ago. One-third of the two dozen known genera are extinct, but 35 per cent of living anthropoids are platyrrhines.

Our living and defunct examples come from both families, recently reclassified by skull and tooth design.

1 **Neosaimiri** was a likely ancestor of the squirrel monkeys, with small, expressive faces, and a tail three times longer than the body. Length: 20in (50cm). Time: Mid Miocene. Place: Colombia. Family: Atelidae.

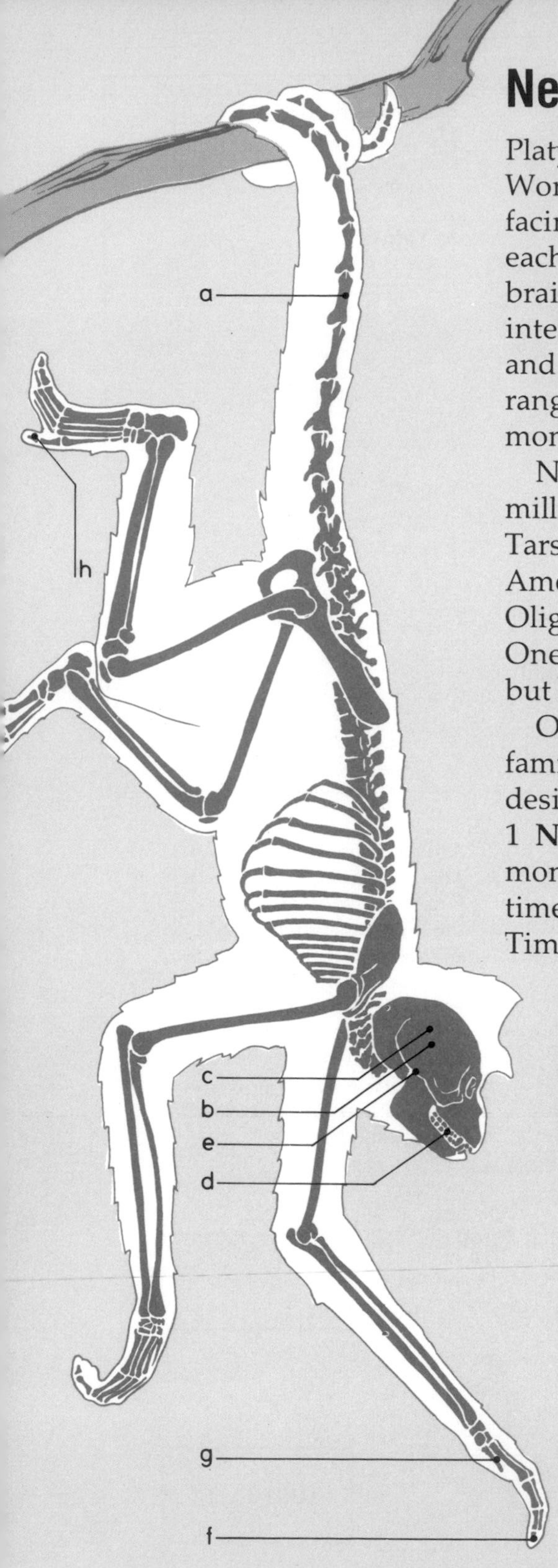

Ateles features (left)
Ateles, the living spider monkey, shows these typical platyrrhine features:
a Long, prehensile tail
b Distinct pattern of sutures between skull bones
c Relatively larger, more complex cerebrum than that of the tarsiiforms
d Thirty-six teeth, with three premolars per side per jaw.
Features found in its family, the Atelidae, include:
e Naked, not tufted, ears
f Short, curved, but nail-like claws
g Non-opposable thumb
h Opposable big toe

Where they live (above)
This map of the Americas gives the present distribution of New World monkeys.

2 **Callithrix jacchus**, the common marmoset, has ear tufts, 32 teeth with long lower incisors, and claws but flat nails on the (small) big toes. Family groups eat forest plants and insects. Length: 20in (50cm). Time: Present. Place: Brazil. Family: Cebidae.
3 **Stirtonia**, a likely ancestor of the howler monkeys, was probably a large, hunched, bearded leaf-eater with a prehensile tail, and a howling call to threaten rival groups. Length: 3ft 7in (1.1m). Time: Mid Miocene. Place: Colombia. Family: Atelidae.
4 **Cebupithecia** had long, narrow, jutting incisor teeth, big canines, and small, crowded premolars. This fruit eater might have resembled today's bearded sakis. Length: 31in (80cm). Time: Mid Miocene. Place: Colombia. Family: Atelidae.
5 **Tremacebus** had huge eyes and maybe resembled the douroucouli, today's sole nocturnal monkey, a fruit and insect eater. Length: 30in (75cm). Time: Late Oligocene. Place: Argentina. Family: Atelidae.

Five platyrrhines
Numbers correspond to items featured in the text.
1 *Neosaimiri* (here shown like a squirrel monkey)
2 *Callithrix jacchus*
3 *Stirtonia* (here shown like a howler monkey)
4 *Cebupithecia* (here shown like a bearded saki)
5 *Tremacebus* (here shown like a douroucouli)

Old World anthropoids

The Catarrhini ("down-facing noses") are the Old World's monkeys, apes, and men. Some 40 genera are known, more than half of them extinct.

Catarrhines have nostrils close together and facing down or forward. They tend to have a better thumb-and-finger grip than New World monkeys but none has a prehensile tail. Many later monkeys evolved shortened tails; modern apes and man have none. Catarrhines share a relatively large brain and similar bony ear passages, and all but primitive species have only two premolar teeth in each side of each jaw. Catarrhines range in size from squirrel-sized *Apidium* to the immense ape *Gigantopithecus*, both long extinct.

The group probably evolved from the Tarsiiformes of 40 million years ago. Southern Asia yields the earliest known catarrhines, but these diversified and multiplied in Africa and through Eurasia.

Apidium skull (above)
The actual length of this reconstructed face and lower jaw is little more than lin (2.5cm).

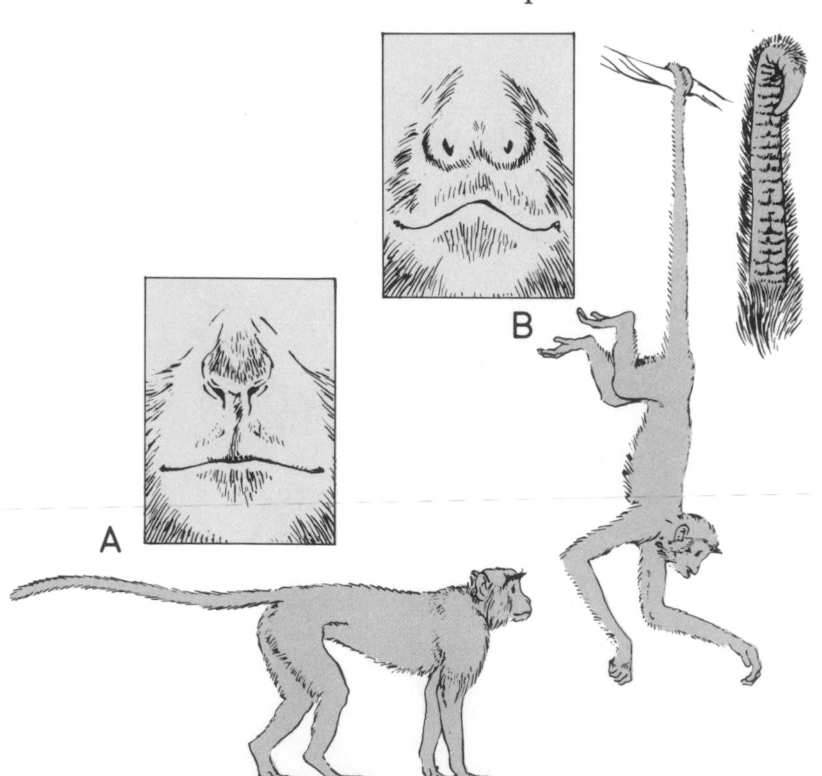

Tails and nostrils
Illustrations (left) compare key features of New and Old World anthropoids.
A Old World monkey, with a non-prehensile tail and down-facing nostrils.
B New World monkey, with a prehensile tail and well separated nostrils

Where they live (below)
This map shows the present distribution of Old World anthropoids.

Examples given here include some possible precursors of apes and Old World monkeys.

1 **Pondaungia** is known only from weathered bits of fossil jaw, with broad, thickly enameled molars. It might have resembled *Propliopithecus* (see pp. 74–75). Probably it climbed trees and ate fruit. Time: Late Eocene. Place: Pondaung, Burma. Family: uncertain.

2 **Oligopithecus**, called alternatively "the first unquestioned ape" and "part prosimian," was perhaps ancestral to the long-limbed gibbons. Crested cheek teeth hint at leaf and insect eating. Size: that of a small monkey. Time: Early Oligocene. Place: Egypt. Family: uncertain.

3 **Apidium** was no bigger than a squirrel monkey, which it possibly resembled. Probably it ran up branches on all fours, and munched fruit and insects with its many-cusped cheek teeth. Time: Mid Oligocene. Place: Egypt. Family: Parapithecidae (monkey-like anthropoids, extinct by the end of the Oligocene).

Three early catarrhines (above)
Numbers correspond to items featured in the text.
1 *Pondaungia* (poorly known)
2 *Oligopithecus*
3 *Apidium*

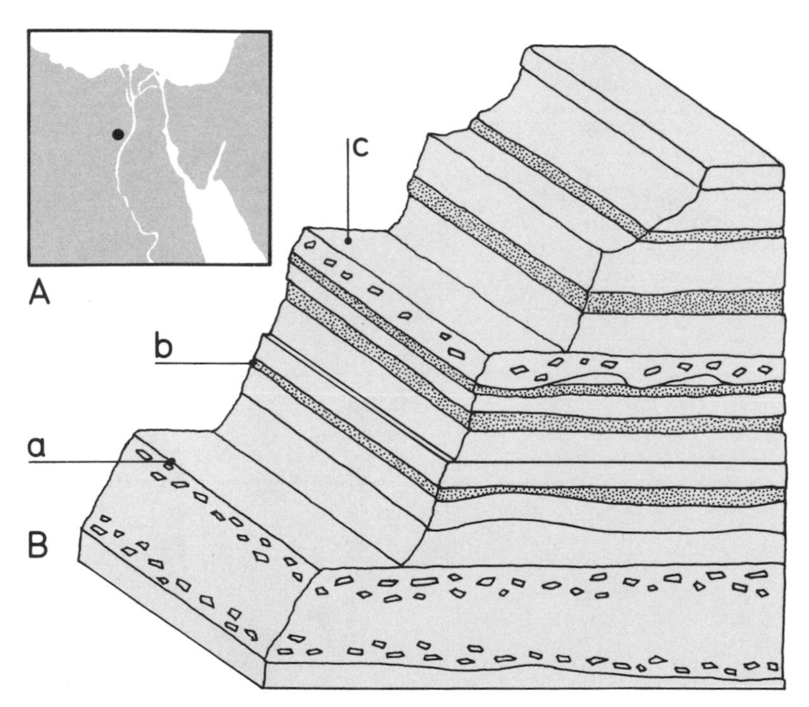

Early fossil finds (left)
A Northern Egypt, showing the Fayum Depression, site of major finds of early Old World anthropoids.
B Section through Fayum deposits with discoveries dating from Oligocene times.
a Lower fossil wood zone quarries: *Oligopithecus*
b Quarry G: *Apidium*
c Upper fossil wood zone: *Apidium, Parapithecus, Propliopithecus (Aegyptopithecus)*

Old World monkeys 1

Old World monkeys – the family Cercopithecidae – with strange, extinct *Oreopithecus* comprise the anthropoid superfamily Cercopithecoidea, possibly evolved from pliopithecids (see p.74).

Miocene and later fossils give only glimpses of the Old World monkeys' history. Most are small to medium-sized mammals, with a large, well-developed brain, large, rounded braincase, only two premolars on each side of each jaw, small external ears, and a long or short tail that is not prehensile.

There are two subfamilies: Cercopithecinae and Colobinae. Cercopithecines have cheek pouches, tend to be more omnivorous and less tree-bound than colobines, and live in Africa and Asia. They include guenons, baboons, macaques, and geladas. Here we show *Oreopithecus* and three cercopithecines.

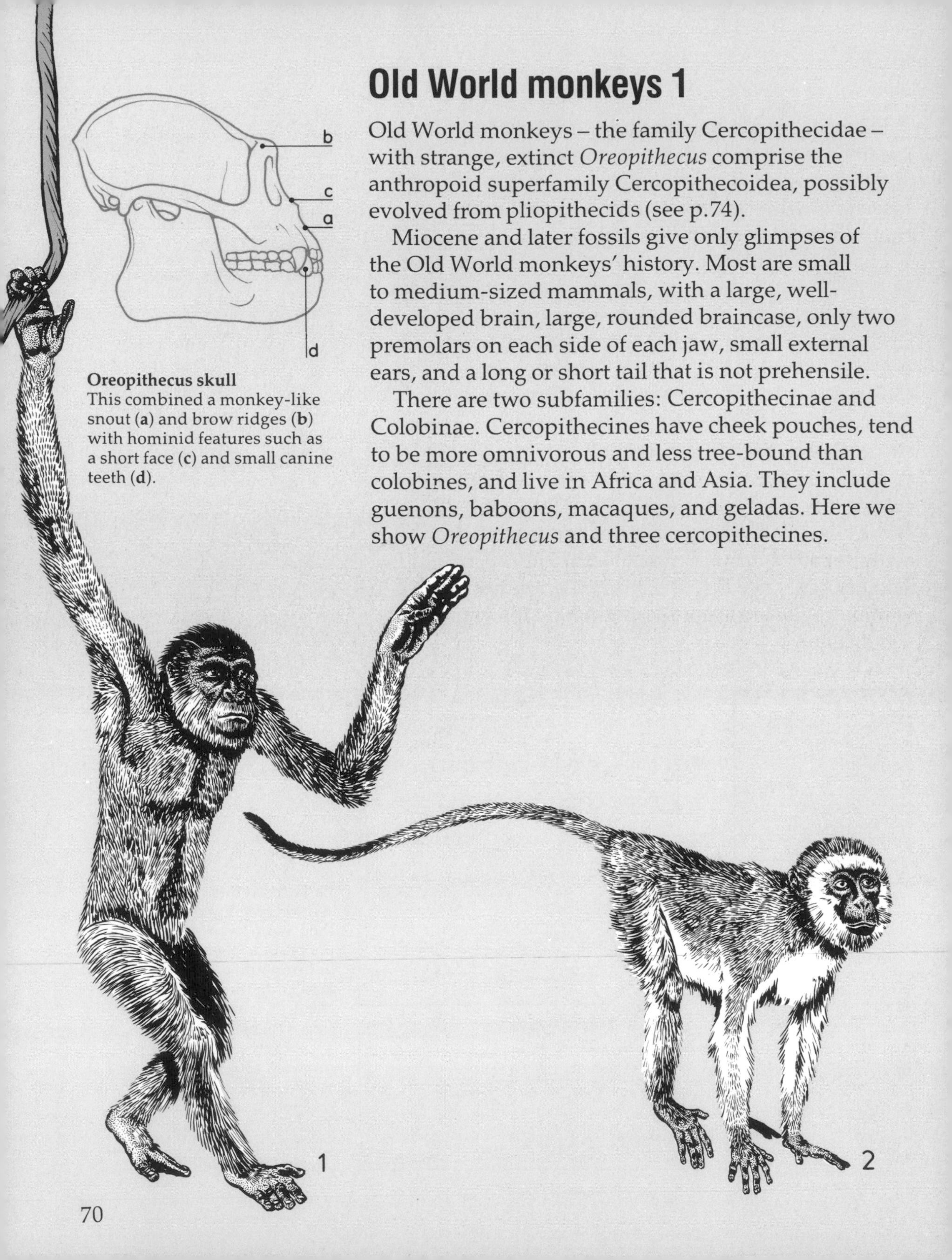

Oreopithecus skull
This combined a monkey-like snout (**a**) and brow ridges (**b**) with hominid features such as a short face (**c**) and small canine teeth (**d**).

1 **Oreopithecus**, a large early anthropoid, had monkey, ape, and hominid features. It hung from its long arms, but sometimes might have walked on two legs. Height: 3ft 11in (1.2m). Time: Late Miocene. Place: north Italy and maybe South-west Russia. Family: Oreopithecidae, an evolutionary dead-end derived from ancestors of Old World monkeys.

2 **Cercopithecus**, the guenons, are small, quadrupedal monkeys with tails longer than the body. Most live in trees and eat fruit. Time: Late Pliocene to modern. Place: Africa.

3 **Papio**, the baboon genus, features large, quadrupedal ground dwellers (males far bigger than females) with powerful arms and legs of equal length, arched tail, long, dog-like face, and massive canines. Baboons live in troops, eating plants and small animals. Time: Late Pliocene to modern. Place: Africa and South-west Arabia.

4 **Macaca**,the macaques, are robust, small to medium monkeys with long, rounded muzzles, and arms and legs of similar length; some lack a tail. Time: Late Miocene to modern. Place: North-west Africa to Japan.

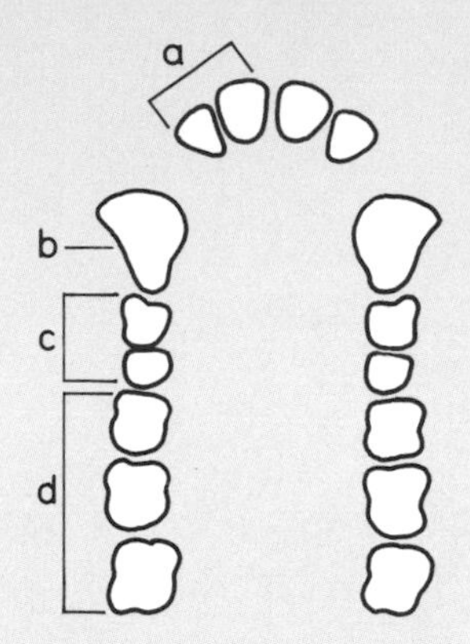

Tooth formula
Each side of a male baboon's upper jaw shows the typical tooth formula of Old World anthropoids.
a Two incisors
b One canine
c Two premolars
d Three molars

Four cercopithecoids (below)
Numbers correspond to items featured in the text.
1 *Oreopithecus*
2 *Cercopithecus*
3 *Papio*
4 *Macaca*

Old World monkeys 2

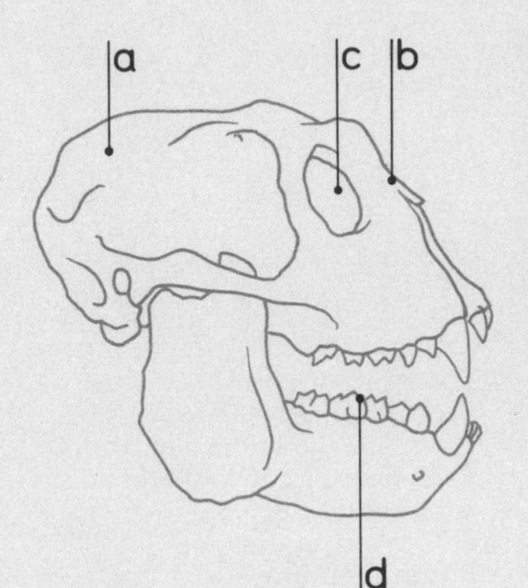

Old World monkeys with reduced or absent thumbs form the cercopithecid subfamily Colobinae, named from *kolobus*, the Greek for "maimed." Colobines tend to have long bodies, limbs, and tails, and most are expert climbers making tremendous leaps high among the trees. Their teeth and stomachs are adapted to a bulky diet mostly made of leaves.

Colobines probably evolved in Africa about Mid Miocene times, and spread across Eurasia. Half their 10 or so genera died out, but the survivors include many species, most in southern Asia.

Colobine features
Collectively these features help identify the fossil skeleton of *Mesopithecus* as a colobine.

a Globular neurocranium (with, in modern colobines, a brain geared more for limb control than visual ability)
b Short face, wide not high (but some colobines have long muzzles)
c Eyes wide apart
d Distinctive teeth, with high molar crowns
e Long tail
f Slim long bones
g Long toes and fingers for grasping branches
h Reduced thumb
i Long body

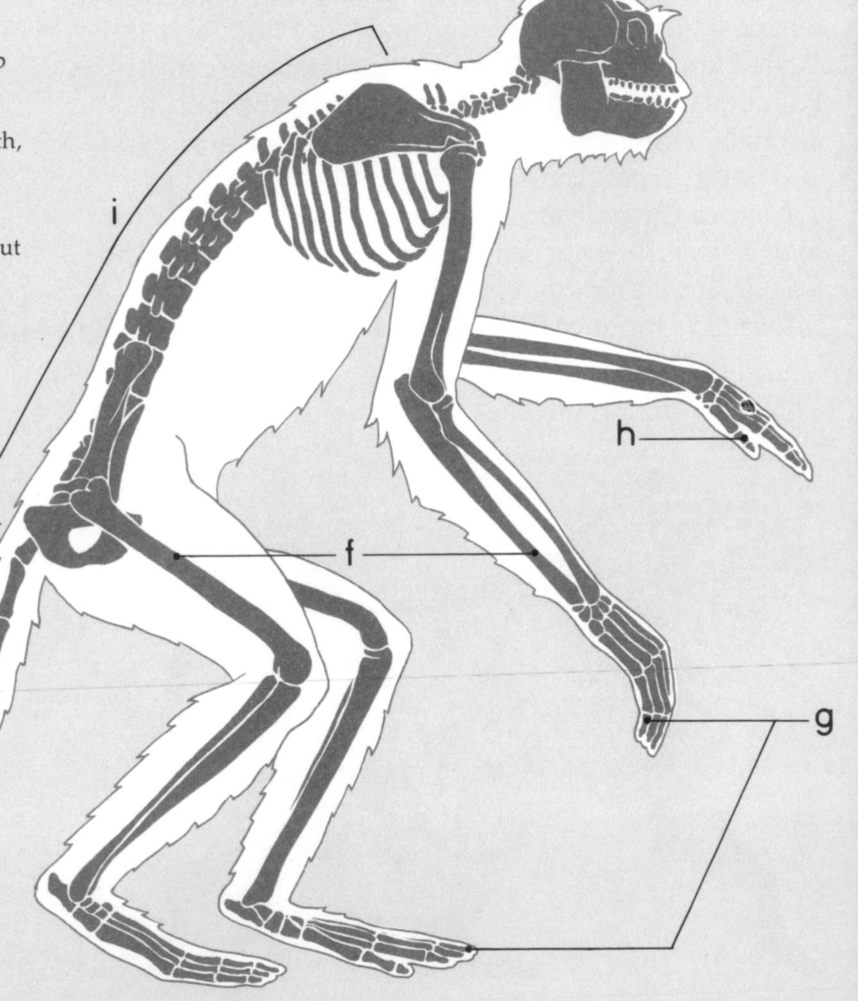

Our examples represent different subgroups.
1 **Victoriapithecus**, one of the earliest known cercopithecids, was possibly ancestral to the colobines, but some fossil finds that bear its name derive from several genera. Size: small. Time: Early to Mid Miocene. Place: East Africa.
2 **Mesopithecus** had long, strong limbs, and a longer thumb than any living colobine. It might have given rise to the slim, long-legged, long-tailed langurs of modern Asia. *Mesopithecus* lived in park-like countryside, and often walked on all fours on the ground. Size: small to medium. Time: Late Miocene to Late Pliocene. Place: Europe and South-west Asia.
3 **Colobus** is the generic name for long-limbed, agile colobines from Africa, also called guerezas. They lack thumbs and have a short muzzle and slightly overhanging upper lip. Most live high among the trees. Size: 18–24in (46–60cm) plus an even longer tail. Time: Late Miocene to modern. Place: North, East, and Central Africa.
4 **Presbytis**, the langurs and other Asian leaf-eaters, resemble *Colobus* but have short thumbs, and many have a hairy cape or crest. Langurs live in day-active troops. Length: 16–31in (40–80cm) plus even longer tail. Time: Late Miocene to modern. Place: Pakistan to Indonesia.

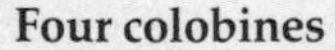

Four colobines
Numbers correspond to items featured in the text.
1 *Victoriapithecus* (restoration based on only scanty fossil evidence)
2 *Mesopithecus*
3 *Colobus*
4 *Presbytis*

About hominoids

Apes and men belong to the catarrhine superfamily Hominoidea. Front teeth differ from the Old World monkeys', and arms and shoulder girdle are designed for hanging and have great mobility. Living hominoids lack tails and have larger brains and bodies than monkeys. Also their rates of reproduction and development are slower.

Hominoids and Old World monkeys probably evolved in Africa by the late Oligocene, from small, early anthropoids in the family Pliopithecidae. Experts differ about the identity, rank, and kinship of many fossil forms. Here we recognize three hominoid families: Hylobatidae (the gibbons and siamang), Proconsulidae ("stem" hominoids), and Hominidae (great apes, men and their closest ancestors).

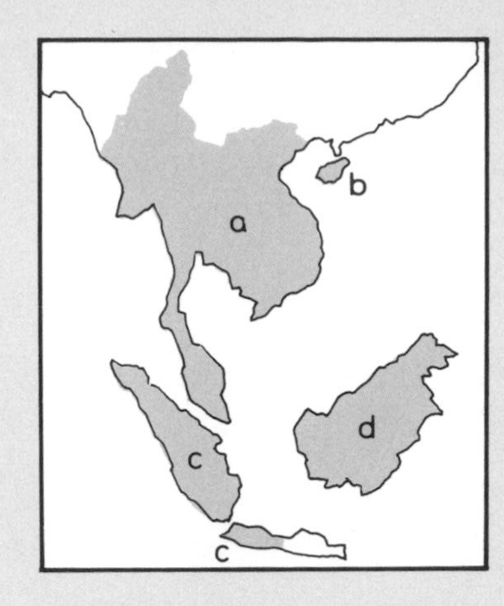

Where hylobatids live (above)
a Mainland South-east Asia
b Hainan Island
c Sumatra and Java
d Borneo

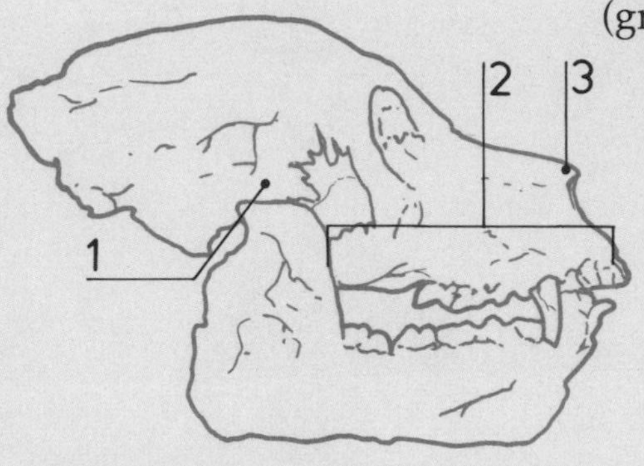

Propliopithecus **skull** (above)
A confusing meld of features suggests perhaps a primitive ancestral anthropoid.
1 Ancestral catarrhine ear
2 Teeth with features found both in apes and monkeys
3 Primitively long snout

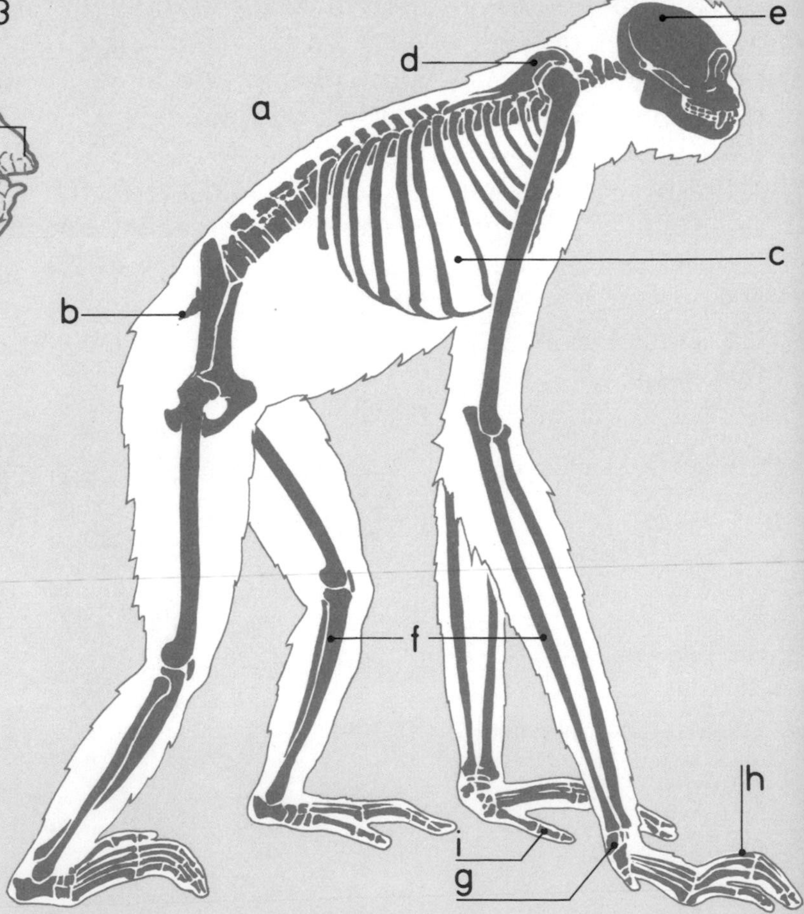

Gibbon features (right)
Most of these occur also in other apes and man.
a Large size
b No tail
c Broad chest
d Very mobile shoulder joint
e Large braincase
f Arms longer than legs
g Mobile wrist
h Long, prehensile hands
i Opposable thumb

Here are three pliopithecids and a hylobatid.

1 **Pliopithecus**, once thought ancestral to the gibbons, was gibbon sized, with a gibbon-like short face, big eyes, and tall, sharp canines, but lacked the long arms of a gibbon and might have had a tail. It ran, climbed, leapt, and hung from trees. Time: Miocene. Place: Europe. Family: Pliopithecidae.

2 **Propliopithecus** (also called *Aegyptopithecus*) was small to medium, built like a howler monkey, and had a long, low face. It ran on all fours and climbed with grasping hands and feet. Time: Mid Oligocene. Place: Egypt. Family: Pliopithecidae.

3 **Dendropithecus** was no bigger than a gibbon and built slimly like a spider monkey. It climbed, leapt, and hung from trees. Time: Early Miocene. Place: Kenya. Family: Pliopithecidae.

4 **Hylobates**, the gibbons, are slim, broad-chested, "lesser apes." Family groups eat fruit, call loudly, and swing from tree to tree by arms far longer than the legs. Gibbons briefly walk bipedally. Head-body length:16–26in (40–65cm). Time: Pleistocene to modern. Place: South-east Asia. Family: Hylobatidae.

Members of two families (above)
Numbers correspond to items featured in the text.
1 *Pliopithecus*
2 *Propliopithecus*
3 *Dendropithecus*
4 *Hylobates*

Arm swinging (below)
A gibbon swings from one hold to another by its arms.

Dryopithecines and others

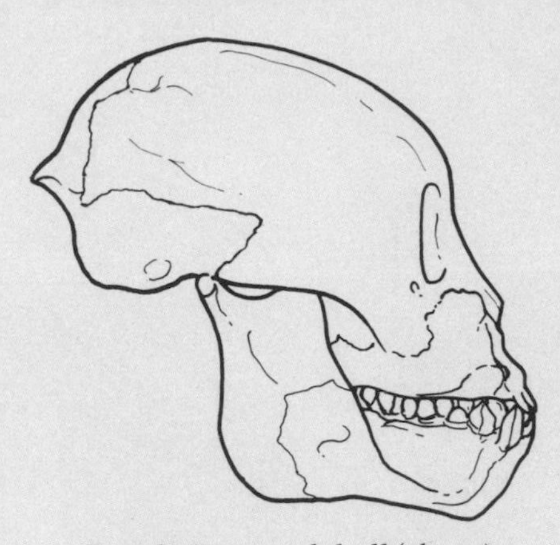

Female *Proconsul* skull (above)
This skull's small size, short face, and small teeth suggest that female early hominoids were smaller and less powerfully built than males. Such sexual dimorphism has featured in most or all later hominids.

This book ranks the extinct Dryopithecinae as one of three subfamilies forming the family Hominidae. (The other two include great apes, men, and extinct "near men.") Hominids tend to have arms, shoulder girdles, and wrists evolved to some extent for swinging through trees, plus a broad chest and shortened small of the back, but long spinal base (the fused bones of the sacrum), vestigial tailbone (coccyx), and broadened upper hipbones (ilia). Teeth and skull, with its broad and often longish face, are much like those of other Old World anthropoids, but body size and relative brain size have tended to increase.

Dryopithecines ("oak apes") were early apes that probably evolved in Miocene Africa and reached Europe during shrinkage of the prehistoric Tethys

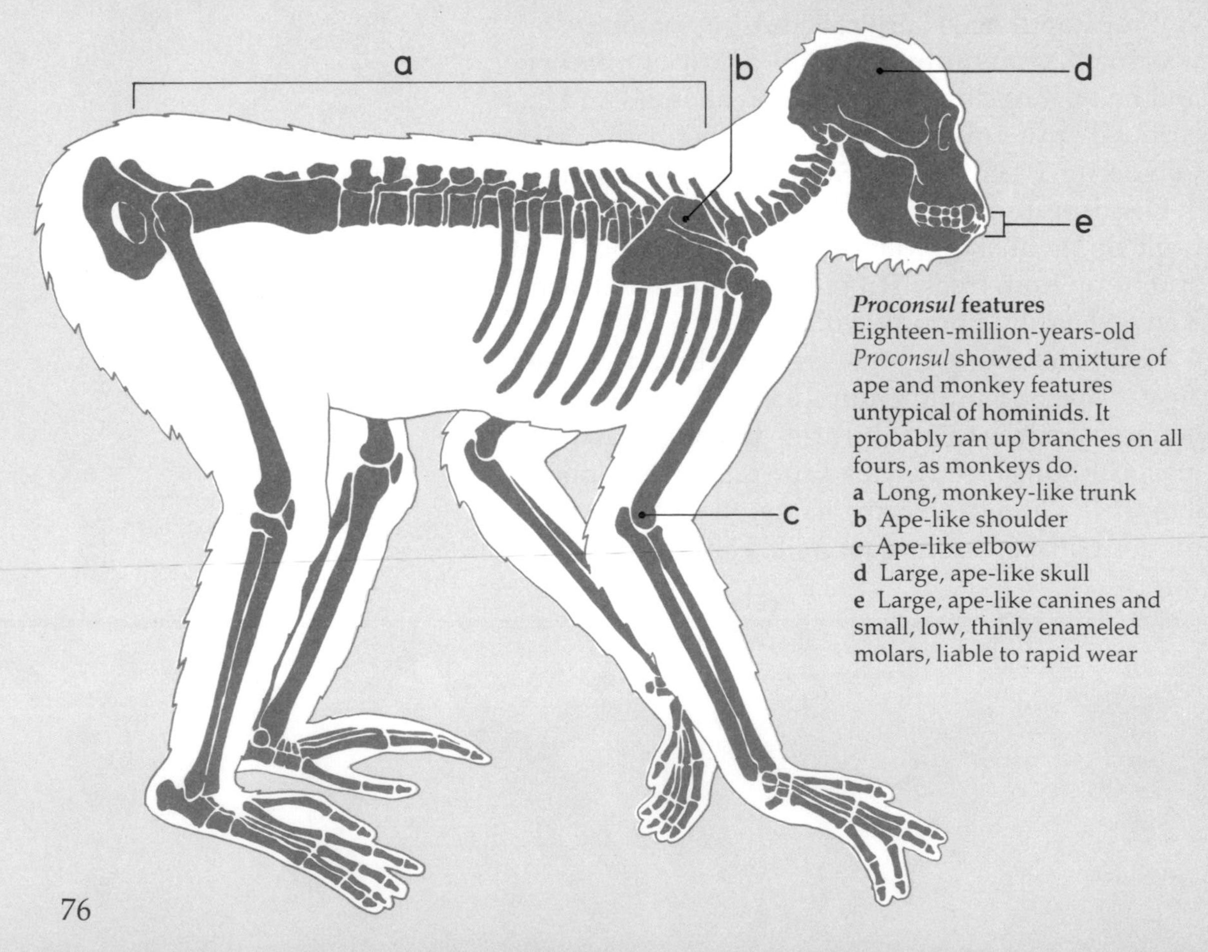

***Proconsul* features**
Eighteen-million-years-old *Proconsul* showed a mixture of ape and monkey features untypical of hominids. It probably ran up branches on all fours, as monkeys do.
a Long, monkey-like trunk
b Ape-like shoulder
c Ape-like elbow
d Large, ape-like skull
e Large, ape-like canines and small, low, thinly enameled molars, liable to rapid wear

Sea. Groups climbed and swung from oaks and subtropical trees. These apes seemingly ate fruit – their thinly enameled cheek teeth were not designed to chew tough food. The dryopithecine *Dryopithecus* of 11.5–9 million years ago may rank with Africa's *Kenyapithecus* (not shown) as an early member of the family Hominidae. The other creatures pictured here have been identified with *Dryopithecus* but resembled large quadrupedal monkeys, differing from living apes in limb and jaw design. They belong in the Proconsulidae, an early hominoid family of 22–15 million years ago perhaps ancestral to apes and man.

1 **Dryopithecus** had broad, low incisors, tall lower canines, long upper cheek teeth, and short lower molars. There were two species, one larger than the other. Time: Mid-Late Miocene. Place: Europe.

2 **Proconsul** included some large early hominoids, bigger than a monkey, and with a somewhat jutting, chimpanzee-like face. Time: Early-Mid Miocene. Place: East Africa.

3 **Limnopithecus** was small and more gibbon-like than other proconsulids. Time: Early-Mid Miocene. Place: East Africa.

4 **Rangwapithecus** had long upper cheek teeth with much wrinkled crowns. Size: variable. Time: Early-Mid Miocene. Place: East Africa.

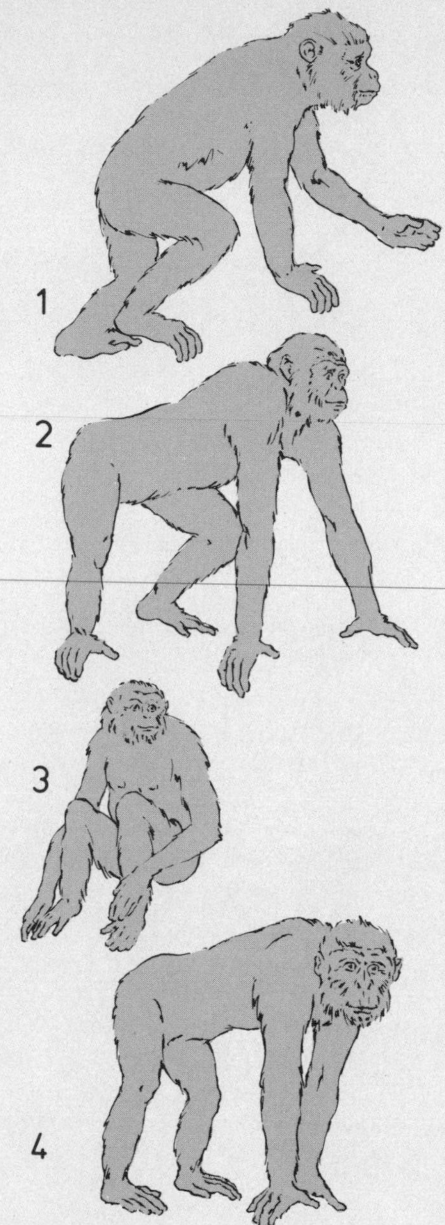

Four hominoids (above)
Numbers correspond to items featured in the text.
1 *Dryopithecus*
2 *Proconsul*
3 *Limnopithecus*
4 *Rangwapithecus*

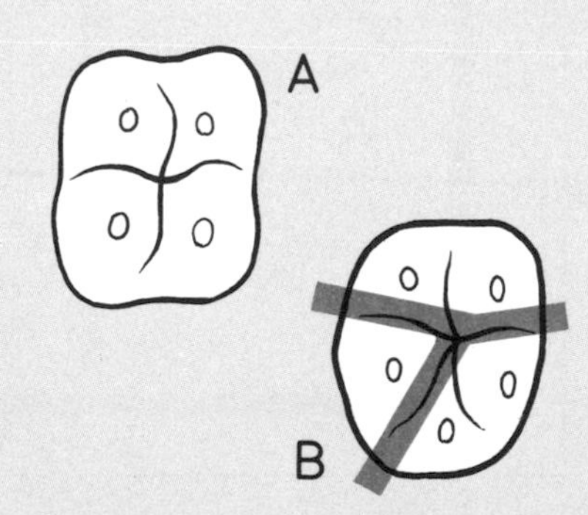

The Y–5 molar pattern (right)
Molar crowns help experts distinguish fossil monkeys from hominids including man.
A Monkey molar, with two pairs of cusps (prominences)
B Dryopithecine (hominid) molar, with five cusps separated by a Y-shaped fissure pattern. One cusp may be suppressed in modern man.

Early pongines

The hominid subfamily called Ponginae includes the ancestors of Asia's sole living great ape, the orangutan (see pp. 82–83). The Ponginae possibly evolved in Africa but spread to Europe and Asia and persisted from about 17–1 million years ago. Most were smaller than a man, but some prehistoric pongines grew larger than a male gorilla.

Fossil finds including reduced front teeth and big cheek teeth reinforced with thick enamel suggest these animals lived in more open country than the dryopithecines, eating drier, tougher foods than those could tackle – a likely adaptation to open woodland that replaced rain forest where rainfall dwindled in the later Miocene.

This "ground ape" hypothesis, coupled with supposed similarities to human tooth, jaw, and skull design, led many experts to believe that *Ramapithecus* walked upright, used tools, and led directly to the human line. Skeptics disagree, although one expert has identified *Ramapithecus* with *Sivapithecus* as an ape en route to man.

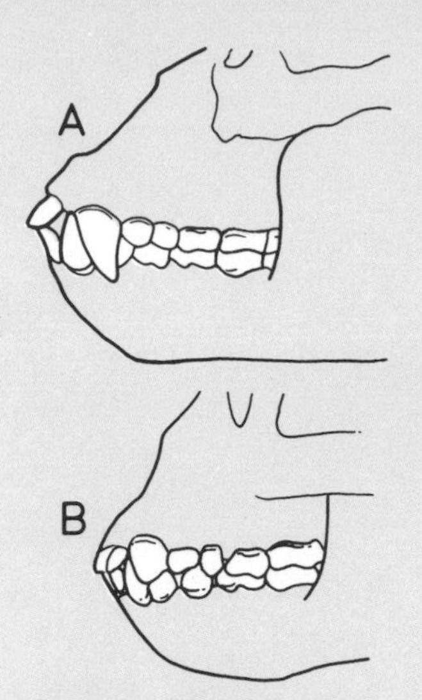

Snouts and teeth (above)
A A gorilla's relatively long snout and large canine teeth
B *Ramapithecus*'s shorter snout and smaller canines

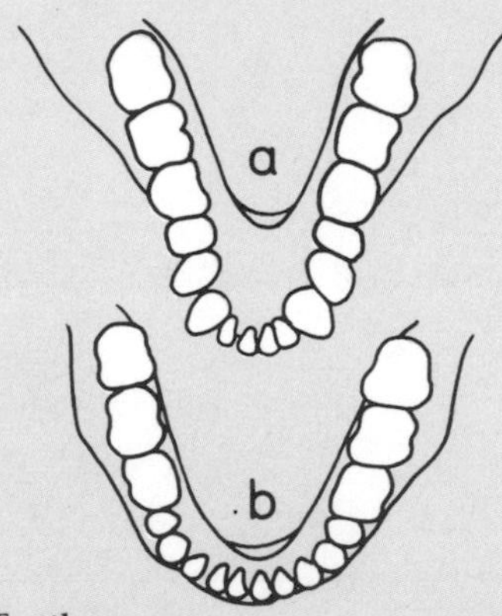

Tooth rows
a *Ramapithecus* had a V-shaped tooth row, unlike the U-shaped row of modern apes or:
b Human tooth row, a parabola

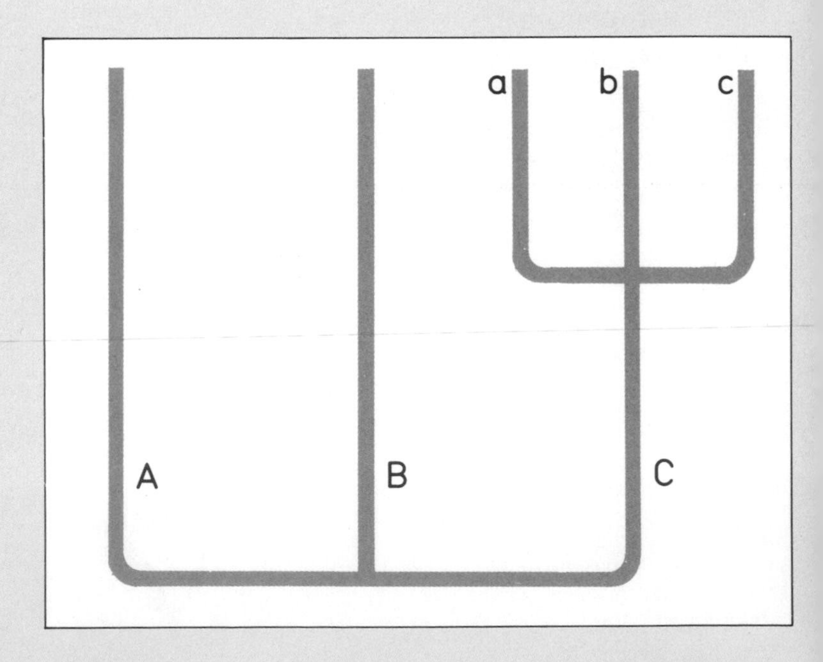

Pongines in perspective (right)
In this simple hominoid family tree letters **A-C** show families, **a-c**, subfamilies.
A Proconsulidae
B Hylobatidae (gibbons)
C Hominidae (apes and men)
a Dryopithecinae
b Ponginae
c Homininae

1 Sivapithecus had a face similar to an orangutan's, chimpanzee-like foot bones, and wrists it could rotate. Probably it climbed, hung from trees, and walked on all fours. Size: about that of an orangutan. Time: about 15–9 million years ago. Place: Europe and southern Asia.

2 **Ramapithecus** supposedly resembled a smaller version of *Sivapithecus* but with a short, deep face, and smaller, flattish teeth with relatively larger biting surfaces than those of man or great apes, and set in a deep jaw, shorter and more pointed than those of man, great apes, or Dryopithecinae. Weight: about 31lb (14kg). Time: 15–7 million years ago. Place: Europe and Asia.

3 **Gigantopithecus**, the last and largest of its group, was a huge ape, with a deep, short face, powerful jaws, rather small canines used like grinding cheek teeth, and immense molars. Height: up to 9ft 10in (3m). Time: 10–1 million years ago. Place: Asia, from Pakistan to southern China.

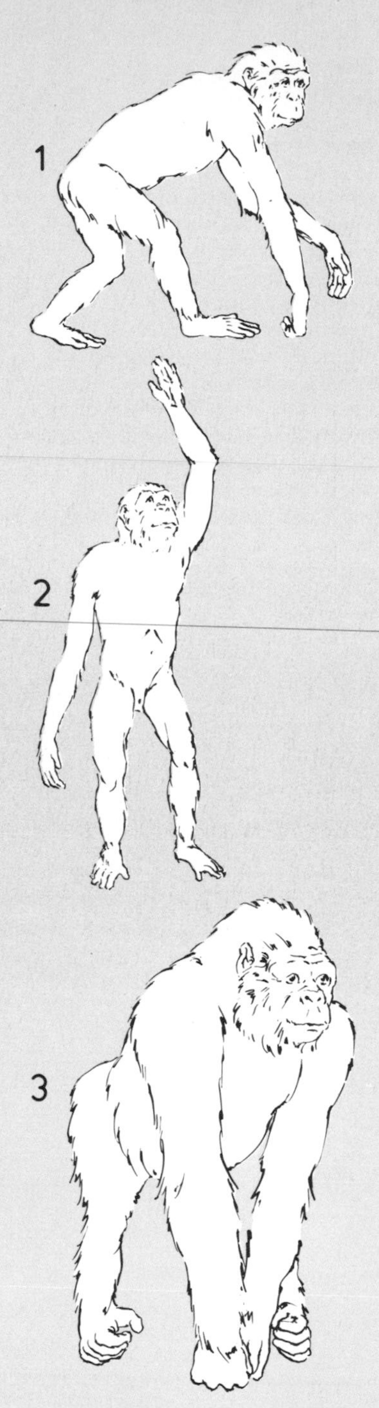

Three fossil pongines (above)
Numbers correspond to items featured in the text.
1 *Sivapithecus*
2 *Ramapithecus*
3 *Gigantopithecus*

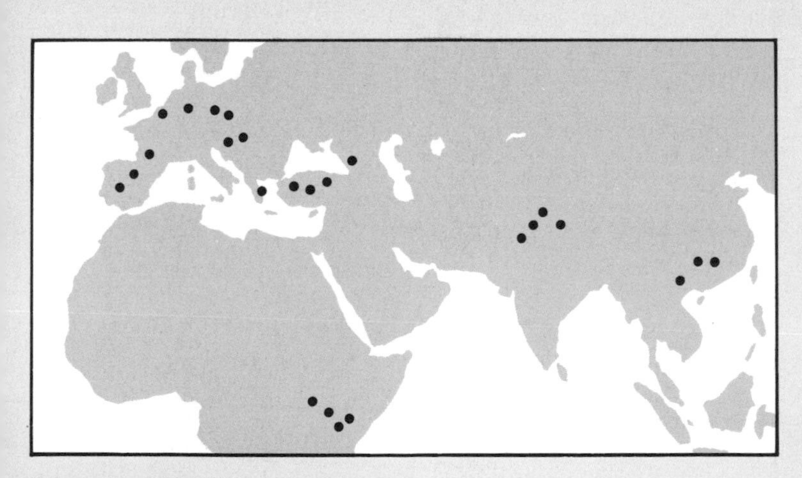

Where they lived (above)
This map of Europe, Africa, and Asia show fossil finds of pongines and other hominoids dating from the Miocene epoch (25–5 million years ago).

Chapter 4

APES AND MAN

By the mid 1980s scientists lacked firm fossil evidence for a common ancestor of modern apes and man. But shared ancestry is obvious in similarities of body build, behavior, and, above all, in biochemical ingredients. Biomolecular studies point to the great apes of Africa as our closest living relatives. And molecular evidence suggests that the man-ape evolutionary split occurred a mere five to eight million years ago – far more recently than people once supposed.

Sixteen engravings stress the almost human attitudes and movements seen among the living apes. (Illustration from *Brehms Tierleben*)

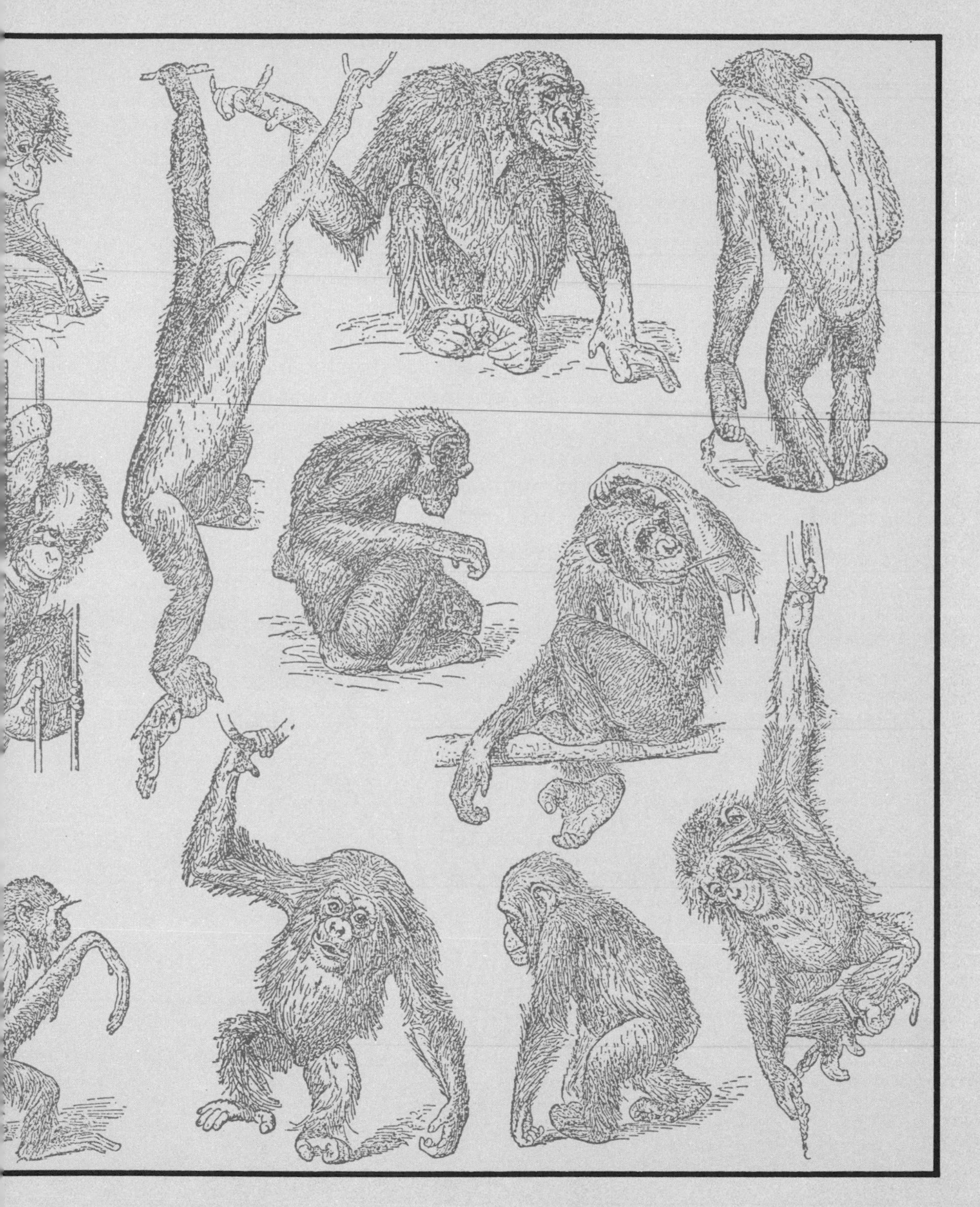

The great apes

The extinct dryopithecines and pongines of the previous chapter undoubtedly included ancestors of man and the living great apes – those large, hairy, intelligent inhabitants of tropical forests in Africa and South-east Asia. The great apes' fossil record is poor apart from finds connecting the orangutan with the fossil group including *Ramapithecus*. But biological studies have proved that great apes and man must share a recent ancestor.

Zoologists traditionally set aside one family, Pongidae, to hold all three great apes – orangutan, chimpanzee, gorilla – placing modern man with fossil men and "ape men" in another family, Hominidae. But close study of body build, behavior, and molecular biology now persuades most experts that the Hominidae includes great apes as well as men. This chapter reveals similarities so close that zoologists now even tend to rank gorilla, chimpanzee, and man in one subfamily: Homininae.

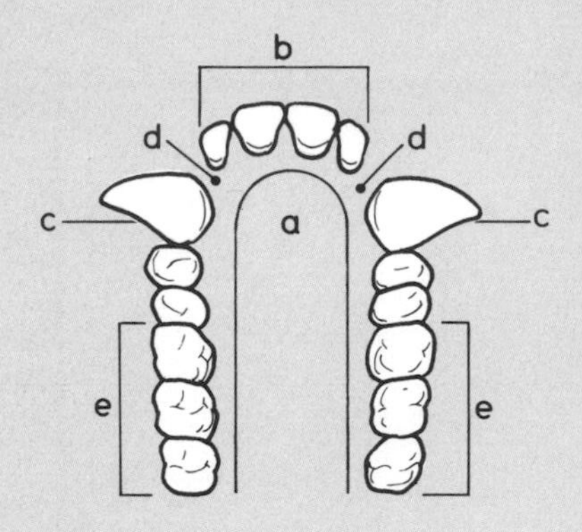

Gorilla upper tooth row
a U-shaped tooth row
b Large, spoon-shaped incisors
c Large conical canines
d Gap between incisors and canines
e Very large molars

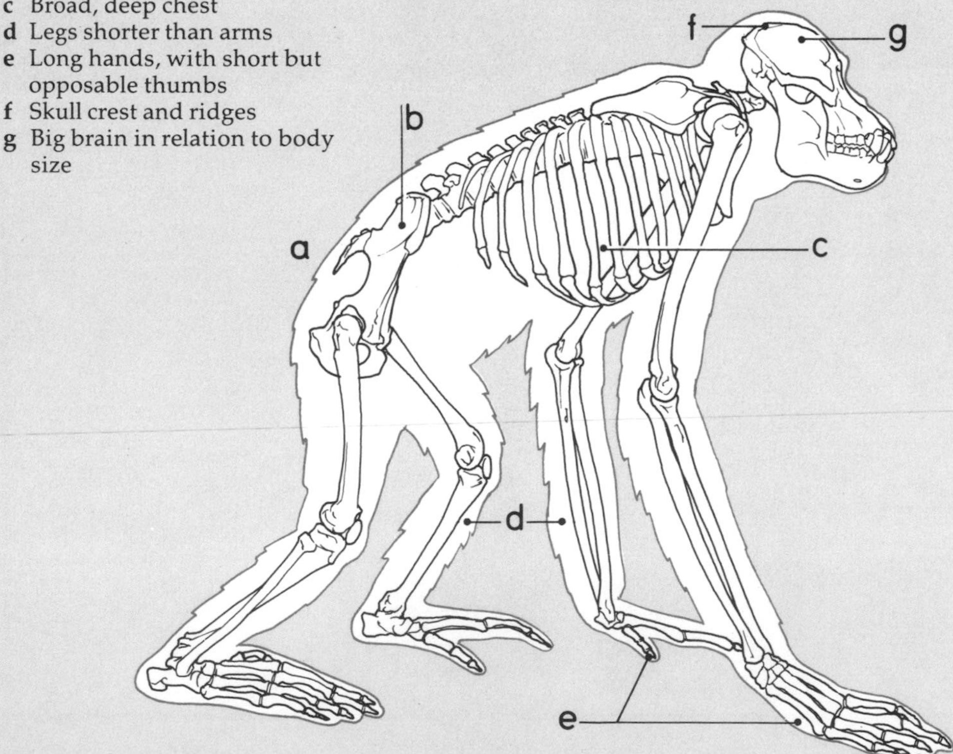

Orangutan features
a No tail
b Elongated pelvis
c Broad, deep chest
d Legs shorter than arms
e Long hands, with short but opposable thumbs
f Skull crest and ridges
g Big brain in relation to body size

1 **Pongo,** the orangutan, has a shaggy, reddish coat, long arms, relatively short legs, short thumbs and toes, and big, low-crowned cheek teeth. It eats fruit, swings from trees, and hangs by hands or feet. Size: males up to 55in (1.4m) tall, and 165lb (75kg) – twice as heavy as females. Time: modern. Place: Borneo, Sumatra, and once South-east China. Subfamily: Ponginae.
2 **Pan**, the chimpanzee, has a long, shaggy, black coat, longer arms than legs, bare face with big brow ridges, large jutting ears, flat nose, and mobile lips. It eats fruit, leaves, seeds, and small animals; climbs trees with hands and feet; swings from branches; and knuckle-walks on the ground. Size: up to 5ft (1.5m) erect, and 110lb (50kg). Time: modern. Place: Equatorial Africa. Subfamily: Homininae. Tribe: Panini.
3 **Gorilla**, the gorilla, is the largest living ape. Males, twice the size of females, reach 6ft (1.8m) and 397lb (180kg). This strong, stocky ape has thick black hair, bare face and chest, large, flaring nostrils, and relatively small ears. Adult males have a bony skull crest and massive jaws. Gorillas knuckle-walk, and eat leaves, shoots, stalks, and roots. Time: modern. Place: Equatorial Africa. Subfamily: Homininae. Tribe: Panini.

Living great apes
1 *Pongo*
2 *Pan*
3 *Gorilla*

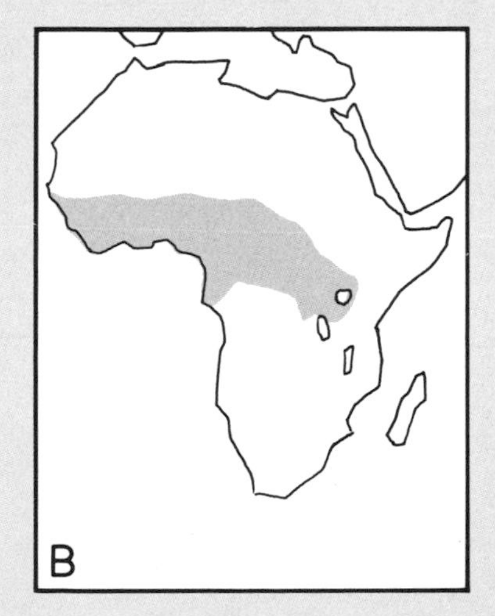

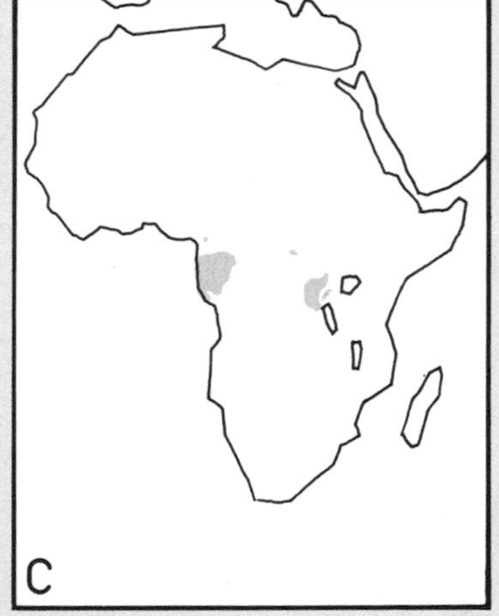

Where they live
A Orangutan
B Chimpanzee
C Gorilla

Man, apes, and molecules

Molecular and other studies show astonishingly close relationships between man and the great apes – especially Africa's gorilla and chimpanzee.

Take DNA annealing for example. Scientists mix "unzipped" strands from double helices comprising DNA from animals of different species. Mixed strands join to make hybrid double helices, but only matching units in two strands unite. Next, heat tests to separate the strands show each hybrid's thermal stability – the more heat needed, the more matching units there must be, and so the closer the relationship between the animals they came from. This test reveals that chimpanzee and human DNA are 99 per cent identical.

Comparing the percentage of differences in the amino-acids making up a protein gives a similar result. Testing the immunological reactions of primates to antibodies produced by rabbits in

What biochemicals reveal
Below, three sets of diagrams depict biochemical techniques used to prove Africa's great apes to be our closest kin.

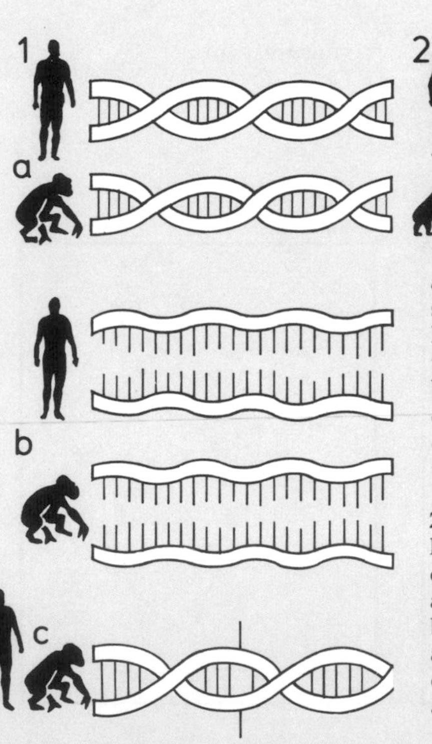

1 DNA annealing
a In man and chimpanzee, two strands of simple compounds have matching units that zip up to form a DNA double helix.
b Double helices unzipped
c Mixed strands form a hybrid double helix of DNA. Only one unit per 100 is a mismatch.

2 Protein sequencing
Man and gorilla show only two differences in the sequences of amino acids forming the red-blood protein, hemoglobin. (Man and chimpanzee show no differences. Man and all other animals show more than two).

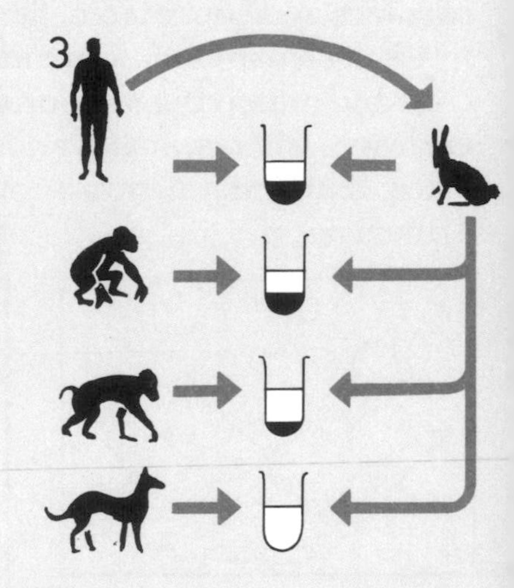

3 Immunological testing
For details, see page 40.

response to albumen from human blood just confirms these findings.

All tests show man, chimpanzee, and gorilla related more closely to one another than to the orangutan, their next nearest relative. Some tests show man closer to the chimpanzee than to the gorilla.

Traditional wisdom based on scanty fossil finds put the evolutionary man–ape split as much as 20 million years ago. Biochemists reckoned it came much later. They based this notion on the so-called molecular clock whose ticks are random mutations thought to build up a constant rate as changes to ingredients in proteins. By measuring degrees of difference in molecular structure or immunological response between creatures with an already securely dated common ancestor, the scientists believed they could work out when major primate lines diverged.

According to this theory, great apes and gibbons split 10 million years ago, while man, chimpanzee, and gorilla shared a common ancestor a mere 6. million years back – 8 million at the very most.

Objectors argued that the theory was untestable, but supporters claimed that molecular clock dates matched prehistoric dates that could be verified by other means. Fossil finds have since confirmed our recent ancestry among the apes.

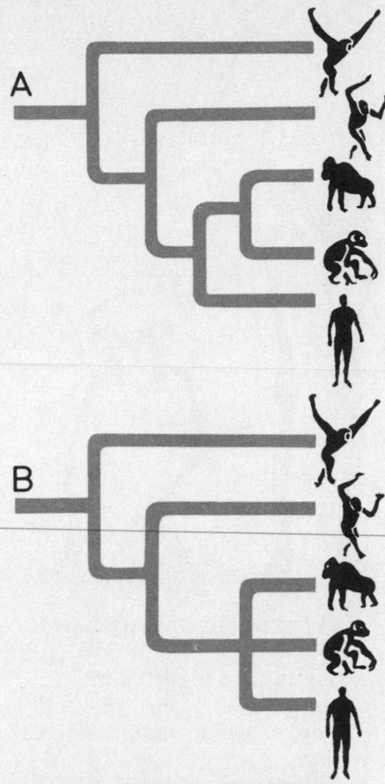

Rival evolutionary tree (above)
A Fossil finds once suggested long separation of the human line from the ancestor of the chimpanzee and gorilla.
B Molecular studies suggested all three were close enough to share a common ancestor.

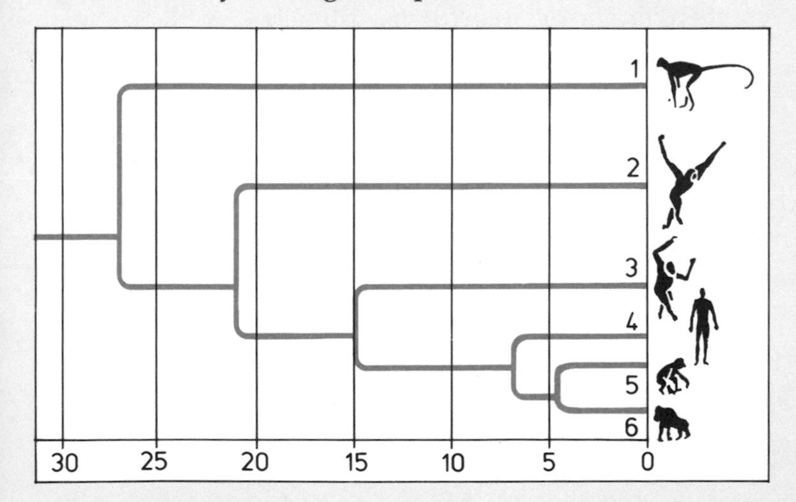

A DNA clock (left)
This family tree uses genetic differences based on DNA studies to estimate divergence dates for different anthropoids. Figures represent millions of years ago.
1 Old World monkeys
2 Gibbons
3 Orangutan
4 Humans
5 Chimpanzee
6 Gorilla

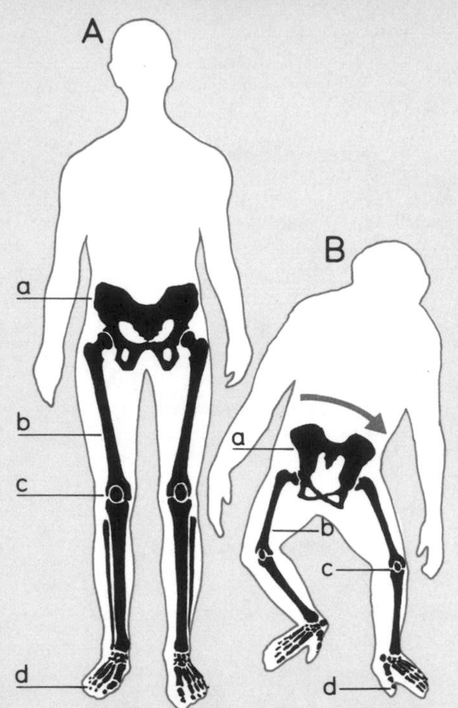

Walking and waddling (above)
A Human lower limbs. Flaring pelvis, inward-angled femur, strong knee joint, and "platform" foot are made for smooth bipedal walking.
B Chimpanzee's lower limbs. Long pelvis, outward-angled femur, knee joint, and grasping toes aid quadrupedal walking but produce a bow-legged, body-rocking bipedal waddle.
a Pelvis **c** Knee joint
b Femur **d** Foot

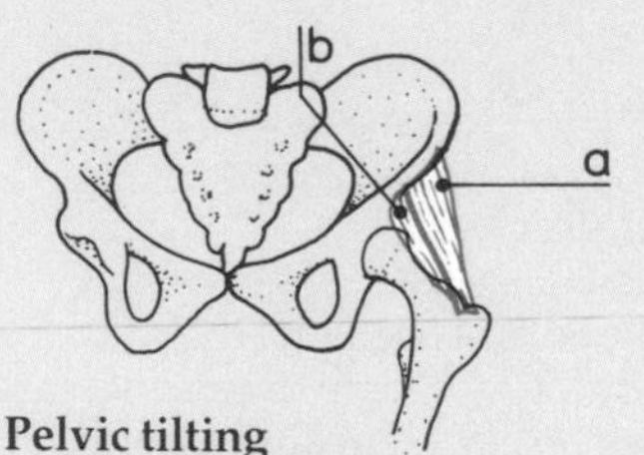

Pelvic tilting
In humans, contracting gluteus medius and gluteus minimus muscles on one side of the pelvis tilt that side down, and lift the other leg, yet help hold the body upright.
a Gluteus medius
b Gluteus minimus

Man–ape anatomy compared

Anatomical comparison suggests strongly that our bodies are apes' bodies redesigned as bipeds. Our arms and shoulders differ little from a chimpanzee's. Unlike apes, however, our legs are longer than our arms, and our hips, spine, thighs, legs, feet, and toes have all been modified for standing and walking in an upright position. (Great apes can only stand on two legs with knees bowed, and walk as bipeds with a sideways lurching motion.)

Redesigning of the feet means we can no longer use big toes as extra thumbs. Our thumbs are relatively longer than an ape's and swing across the palms to meet fingers tip to tip for the precision grip we use in making and exploiting tools.

Bipedal walking, increased intelligence, and an omnivorous diet all contribute to differences between our own and apes' skulls, brains, jaws, and teeth. In relation to body size, man's brain and braincase are far larger than an ape's and the brain is more highly organized, with relatively bigger frontal, parietal, and temporal lobes, collectively concerned with thinking, social behavior, and producing and understanding speech.

Omnivorous modern man has much shorter, weaker jaws than the chiefly vegetarian apes, with their shock-absorbing brow ridges and bony skull crests to anchor powerful jaw muscles. We lack the thick neck muscles needed to support an adult ape's forward-jutting face. Our parabolic tooth row differs from the U-shaped tooth row of the apes. Their canines are much larger and their molars' cusps much higher than our own. But human molars wear a thicker coating of enamel to resist wear caused by chewing harder foods. Then, too, differences between our tongue and pharynx and a chimpanzee's enable us to make more kinds of sounds, though chimps and people share expressive faces.

Our next chapters show that some ape-like features persisted in early members of the human tribe.

Anatomies compared (below)
Lettered items pinpoint some anatomical differences between man and a great ape.

A Human anatomical features:
a Short-faced skull with large rounded braincase, balanced upright on the spine.
b Small jaws, small, thickly enameled teeth, low-crowned molars and parabolic tooth row.
c Long thumb, meeting fingers for precision grip.
d Short lower back
e Broad, short pelvis
f Legs longer than arms
g Big toe aligned with others transmits weight in walking.

B Gorilla's anatomical features:
a Long-faced skull jutting forward from the spine and ridged to take strong jaw and neck muscles.
b Massive jaws with large canine teeth, high-crowned thinly enameled molars and a U-shaped tooth row.
c Long fingers, short thumb
d Lower back relatively shorter than man's
e Long pelvis
f Legs shorter than arms
g Grasping, divergent big toe

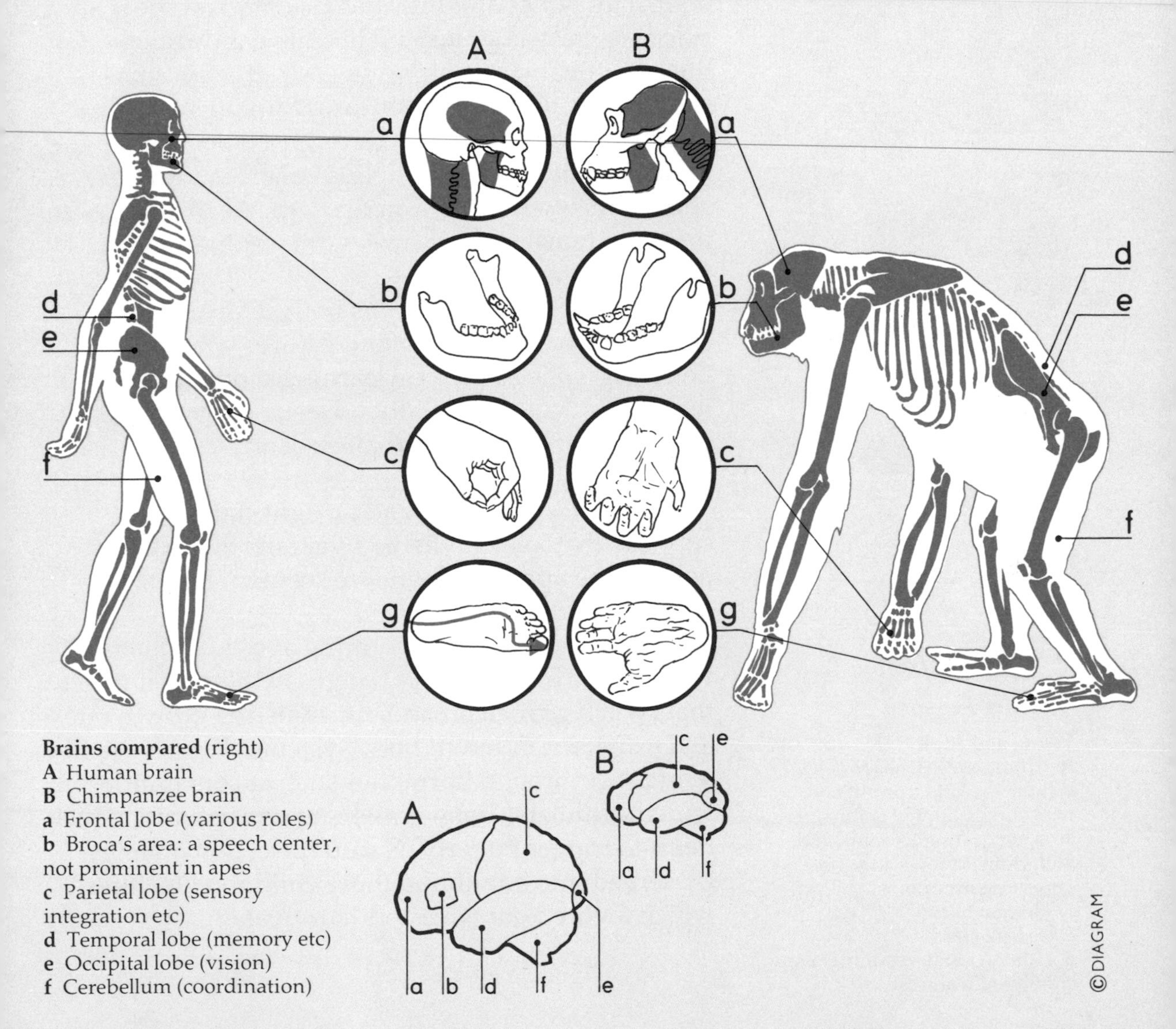

Brains compared (right)
A Human brain
B Chimpanzee brain
a Frontal lobe (various roles)
b Broca's area: a speech center, not prominent in apes
c Parietal lobe (sensory integration etc)
d Temporal lobe (memory etc)
e Occipital lobe (vision)
f Cerebellum (coordination)

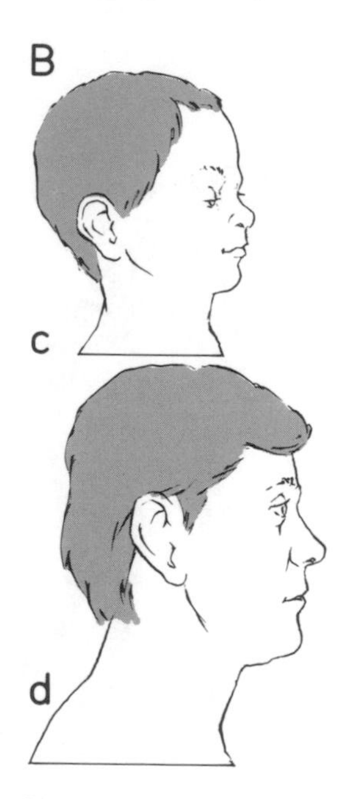

Young and adult
A Chimpanzees' heads
a Juvenile, held upright and with almost human features
b Adult, jutting forward, and with ridged brows and projecting muzzle
B Human heads
c Human child
d Human adult, retaining many childhood features

Man, the neotenous ape

Impressive circumstantial evidence for man's shared recent ancestry with great apes comes from neoteny: the survival into adulthood of features found in many species only in their young. Adult humans share numerous features found in baby chimpanzees, but lost as these grow up.

Like humans, baby chimpanzees have a sparse covering of body hair. Like us, they have a relatively big brain, shielded by a bulbous cranium. Like ours, their skull bones are thin, and lack marked brow ridges or crests. Because no muzzle has developed, the face in baby chimpanzees is short, with small jaws and teeth, but a protruding chin. In both, the brain stem joins the brain through a hole beneath the middle of the skull, which thus balances above the spine if its owner walks on two legs. As in women, so in young female chimpanzees, the vagina faces forward instead of back.

Humans and chimpanzees both share extended childhoods – advantages for creatures whose behavior will be based on learning more than instinct. Indeed our childhood lasts so long that it prolongs our lives beyond those of any other mammal.

Big brain, small jaws, and upright stance – supposedly the most distinctive traits that separate us from the apes – all therefore seem at least part products of neoteny.

The mechanisms transforming apes to humans would have been simply changes to the sets of genes that switch growth on and off, adjusting growth rate and extent for different body systems. These switches tend to act through hormones such as somatotrophin, which stimulates bone growth, and melatonin, which helps to trigger puberty. Natural selection merely preserved into adulthood those childhood features which favored our ancestors' survival.

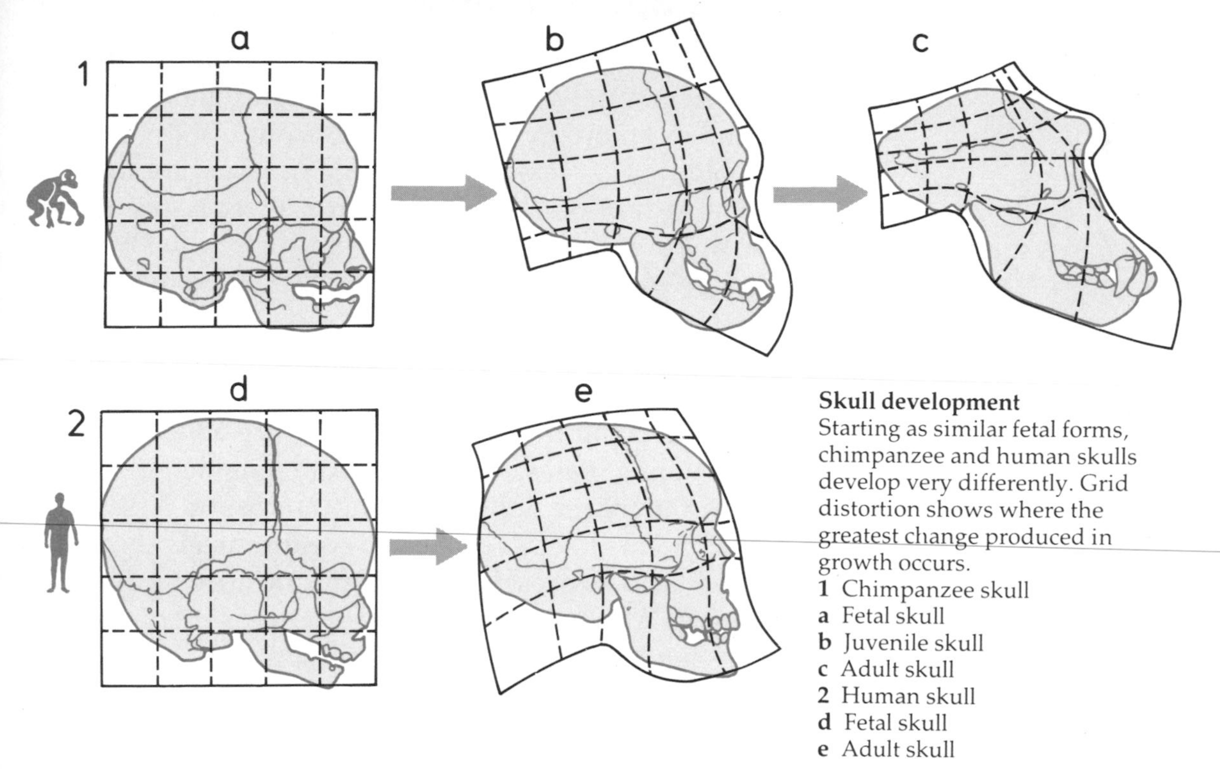

Skull development
Starting as similar fetal forms, chimpanzee and human skulls develop very differently. Grid distortion shows where the greatest change produced in growth occurs.
1 Chimpanzee skull
a Fetal skull
b Juvenile skull
c Adult skull
2 Human skull
d Fetal skull
e Adult skull

Brain and body growth
This diagram shows how brain and body weights increase (in kilograms) with age, among nine animals. Brain growth tapers off at adulthood, so the human brain may owe its relatively large size to an unusually extended human childhood.
a Mouse
b Rat
c Guinea pig
d Rabbit
e Cat
f Dog
g Old World monkey
h Chimpanzee
i Human

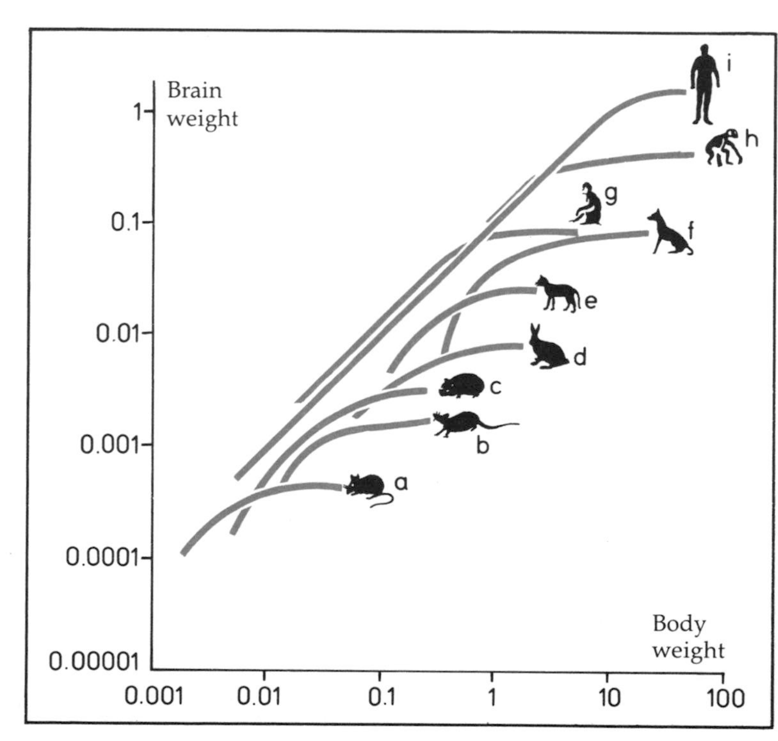

Play
Like human children, young apes exercise their limbs through play. A common ancestry equips both for swinging by the arms.

How apes behave

How nonhuman primates live shows further similarities between the great apes and ourselves. Apes and monkeys live in social groups obeying rules involving age, sex, and a pecking order, but unlike social groups of animals like birds, these social groups are largely permanent, embrace both sexes, and involve behavior that is largely learned, not just instinctive.

The closest man–ape parallels are seen in chimpanzees – intelligent, strongly social animals living in communities whose cores are bands of males with young females recruited from outside.

Chimpanzees develop strong, enduring attachments to each other and communicate through many facial expressions, gestures, postures, and sounds. Although with no exact equivalent of speech they signal greeting, reassurance, deference, or aggression in almost human ways.

Some chimpanzees make tools, an ability once thought unique to man. Females, especially, eat termites caught on twigs trimmed until they can be poked down holes in termite mounds. Individuals also suck water from sponges made of leaves, use sticks as levers, and brandish natural clubs at enemies.

In some groups males *cooperate* in hunting monkeys, then *share* the meat. And although human warfare has no true parallel among these apes, a group of raiding males has killed members of

Facial expressions
Although lacking speech, chimpanzees can communicate emotions by almost human facial expressions. Four are pictured here.
1 Relaxed
2 Greeting
3 Smiling: showing only the bottom teeth
4 Anger: showing top and bottom teeth

Tool using
A Poking a twig into a termite nest to capture termites
B Sucking water from a bunch of leaves that have been moistened in a pool.

another group – behavior influenced perhaps by overcrowding due to loss of habitat, where farmers had felled forest.

In short, then, chimpanzees form relatively fluid social groups living in defended ranges, where members gather food, hunt, breed, and sometimes manufacture simple tools. To these ingredients add upright walking and tool dependence and you approach the likely lifestyle of prehistoric creatures that had crossed the threshold from ancestral ape to early man.

Family life
Like humans chimpanzees develop strong attachments to each other. The mother-child relationship helps each infant through its long learning period. Also, the need to protect their young helps to stabilize each troop of chimpanzees.

Chapter 5

"MEN-APES" AND EARLY MAN

By about four million years ago apes in Africa had given rise to hominids that walked on two legs only. Bipedalism fostered hand-eye coordination and brain development. So emerged the human tribe: Hominini. Its oldest members seemingly comprised the genus *Australopithecus*. Many scientists believe one such "man-ape" gave rise to little *Homo habilis*, the first-known species of our genus *Homo*-"man." This chapter explores the forms and lifestyles of these creatures of a million years and more ago.

Australopiths and early *Homo* clash in this reconstruction painted by the artist Maurice Wilson. The scene is Olduvai, East Africa, about two million years ago.

MWilson

Mankind in the making

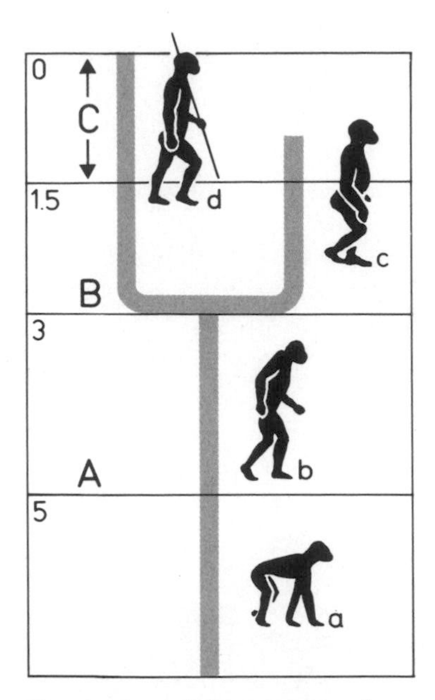

Hominids and climate (above)
Landmarks in human evolution perhaps reflect climatic change caused by Ice Age events, shown in millions of years ago.
a Ancestral ape
b Early australopiths
c Later australopiths
d *Homo* species
A Antarctic Ice Cap formed
B Arctic Ice Cap formed
C Glacial-interglacial climatic fluctuations

By 3.5 million years ago, intensifying Ice Age cold had locked up so much water in the form of ice that worldwide rainfall dwindled, and tropical grasslands had spread at the expense of shrinking forests. In Africa big grazing herbivores and other animals adapted to savanna life now multiplied, while forest creatures suffered loss of habitat.

Incompletely designed for arboreal life, certain creatures derived from dryopithecines seemingly became adjusted to life in open countryside. These "man apes" were the australopiths featured later in this chapter, with the first "true man," their likely offshoot. Hominization – the making of mankind – seemingly involved a complex of mutually reinforcing changes, some seen in fossil finds and archeological discoveries, others just inferred.

Fossil footprints 3.8 million years old reveal that by then bipedal walking had freed hands for the making and habitual use of tools, of which the earliest discovered date from 2.5 million years ago. Canines shrank, arguably as tools increasingly performed their

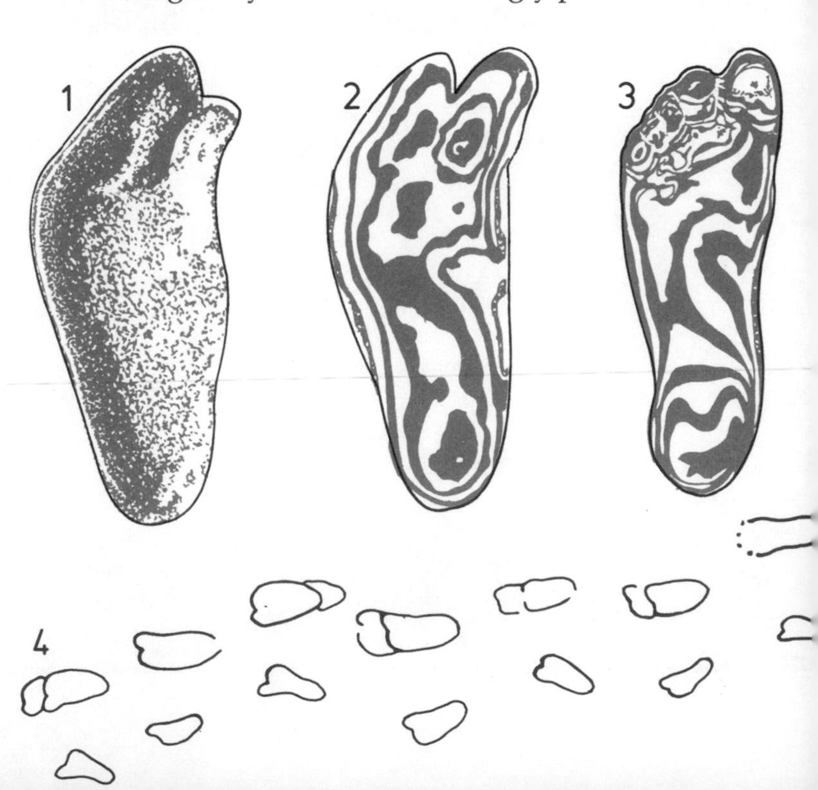

First footprints (right)
1 Fossil footprint from Laetoli, Tanzania. It formed in newly fallen, soft, volcanic ash that soon hardened into solid rock.
2 Contours of fossil footprint.
3 Contours of modern human footprint. (**2**) and (**3**) reveal similarities of shape and weight distribution.
4 Overlapping fossil footprints show where three hominids walked upright more than three and a half million years ago.

tasks. Toolmaking and upright walking stimulated brain development, and offered new survival strategies involving foraging for plants and game. In time some hominines lost body hair and gained sweat glands – body changes that helped stop overheating on hot, shadeless plains.

Hunting big meaty prey or scavenging their corpses, killed by carnivores, encouraged food-sharing by whole groups at "home bases" where communal activities fostered communication that preceded speech. Here, young learned from elders how to use and manufacture tools. For the first time in the history of evolution a premium was placed upon intelligence: brain not brawn or speed decided which hominines survived and which became extinct.

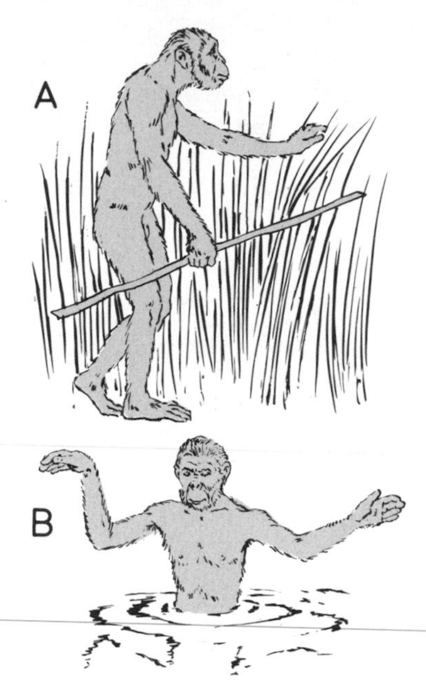

Bipedal walking (above)
Pictures illustrate rival ideas about why our ancestors took to walking on their hindlimbs.
A Standing upright to see over tall savanna grasses.
B Standing upright to wade through shallow water.

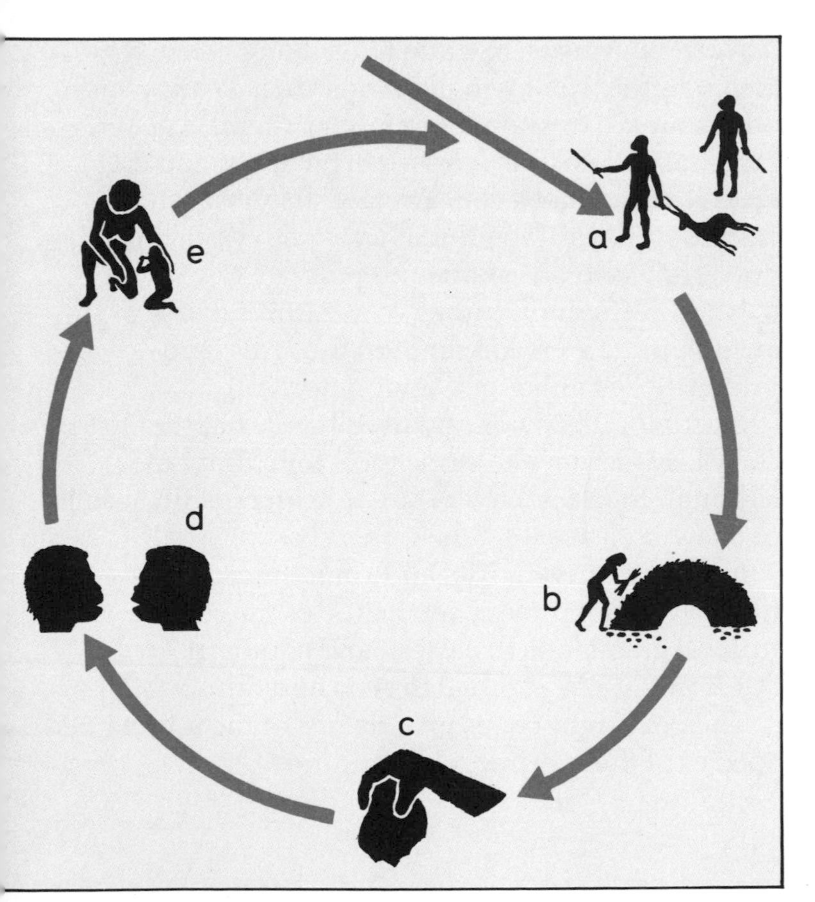

Feedback processes (left)
Environmental change probably promoted hominid evolution by triggering these self-reinforcing changes.
a Group cooperation helped survival on open grasslands.
b A home base where food was shared cemented social ties.
c Toolmaking and tool use promoted hand-eye coordination, food-gathering, hunting, and bipedalism.
d Increasing brain complexity improved communication and the ability to learn to make and use new tools.
e Prolonged infant care and childhood learning aided survival.

The human tribe

Increasing brain size, remodeled teeth, and hips and lower limbs redesigned for walking are major hallmarks of the human tribe, Hominini. This holds at least two genera: the extinct *Australopithecus* ("southern ape"), and *Homo* ("man"). Some scientists add *Ramapithecus* (described on pp. 78–79).

Australopiths – the so-called "ape men" – evolved in Africa 4 million years ago, or even earlier. They evidently came from a dryopithecine ape ancestor, yet to be identified. There arose three or four species (experts disagree). By 2 million years ago one probably gave rise to the first species of our own genus (see p. 106). Both genera endured side by side for another million years before the last australopith died out, possibly exterminated by its brainier successor.

Australopiths were quite likely hairy creatures – some slightly built and no bigger than a chimpanzee, others muscular and nearer modern-man size. Brain–body ratio was little better than an ape's, and the ape-like head had a concave face, flat nose, and chinless muzzle. Jaws and cheek teeth were big and powerful, with a V-shaped or pointed-U-shaped tooth row. But unlike apes' lower limbs, those of australopiths were longer than the arms, and primarily designed for walking upright.

Australopiths evidently used hands habitually to carry loads and make stone tools for cutting meat, although their teeth were suitable for crushing seeds or chewing leaves.

Foraging, scavenging, and sometimes maybe even hunting for their food, small groups ranged the tropical grasslands of eastern and southern Africa. Some might have spread to Asia and Europe.

The next eight pages give details of the four named species of these primeval hominines.

d
e

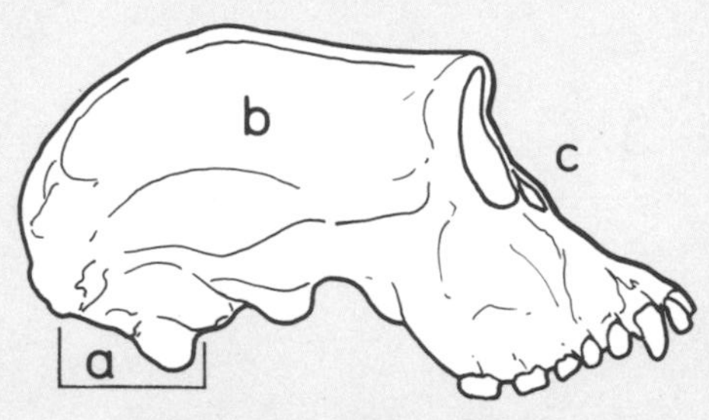

***Australopithecus* skull** (above)
Two views show some features not found in our own genus.
a Little of skull behind hole for spinal cord
b Small brain capacity
c Large, dished face
d Wide midface
e Distinctive tooth row

Ancient finds (below)
Illustrations depict two finds perhaps from very early australopiths.
A Jaw 5.5 million years old from Lothagam in Kenya
B Elbow joint 4 million years old from Kanapoi in Kenya

A

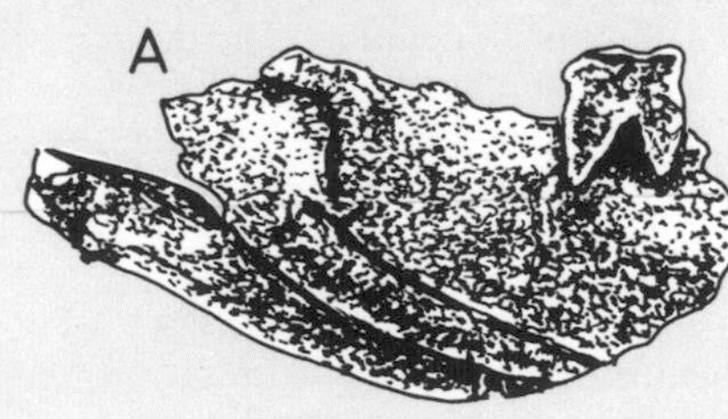

B

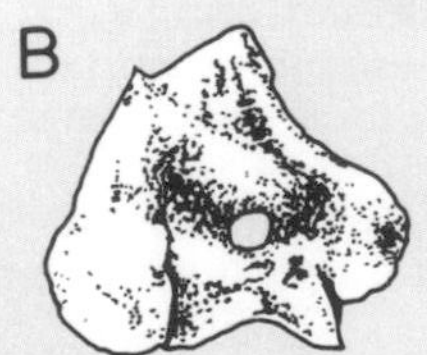

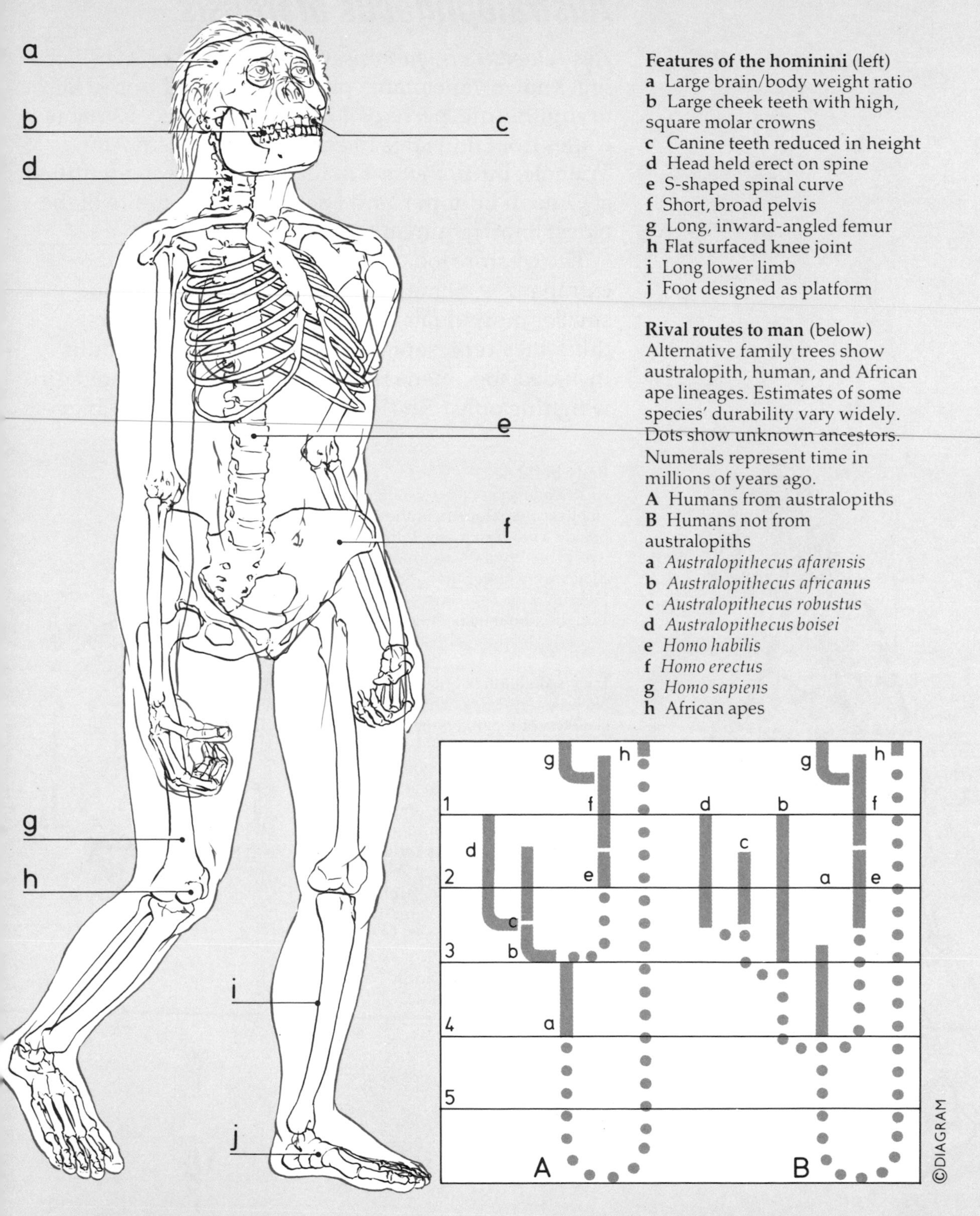

Features of the hominini (left)
a Large brain/body weight ratio
b Large cheek teeth with high, square molar crowns
c Canine teeth reduced in height
d Head held erect on spine
e S-shaped spinal curve
f Short, broad pelvis
g Long, inward-angled femur
h Flat surfaced knee joint
i Long lower limb
j Foot designed as platform

Rival routes to man (below)
Alternative family trees show australopith, human, and African ape lineages. Estimates of some species' durability vary widely. Dots show unknown ancestors. Numerals represent time in millions of years ago.
A Humans from australopiths
B Humans not from australopiths
a *Australopithecus afarensis*
b *Australopithecus africanus*
c *Australopithecus robustus*
d *Australopithecus boisei*
e *Homo habilis*
f *Homo erectus*
g *Homo sapiens*
h African apes

Australopithecus afarensis

Australopithecus afarensis ("southern ape of Afar"), the first known "ape man," probably evolved from a late dryopithecine, perhaps 4 million years ago. Its name comes from find sites in Ethiopia's northern Afar Triangle, but *afarensis* fossils have also been identified at Omo in Ethiopia, and Laetoli, Tanzania, site of the oldest known human footprints.

The creature looked like a small yet upright chimpanzee. Some experts interpret larger and smaller individuals as males and females; others think they represent quite separate species. Adults included specimens no bigger than a six-year-old girl, weighing only 65lb (30kg). The brain was little bigger

Body build (left)
Australopithecus afarensis, the smallest australopith, is shown beside a modern man. *Afarensis* was probably dark and hairy. Males were larger than females.
Height: 3–4ft (1–1.3m).
Weight: about 65lb (30kg)

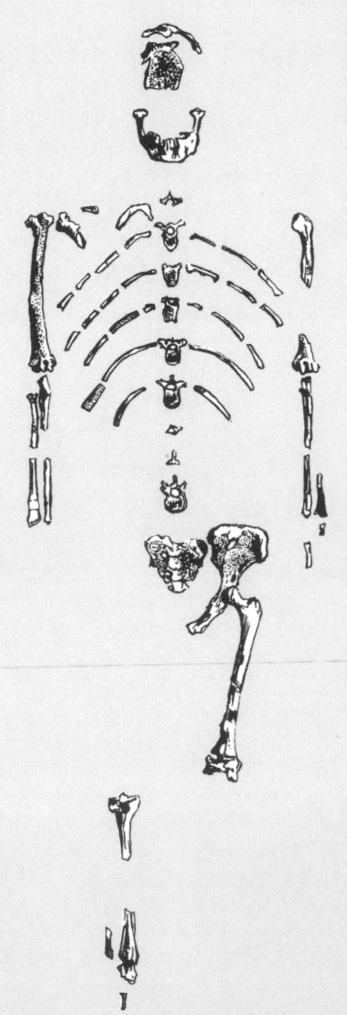

Lucy's skeleton (right)
Nicknamed "Lucy," this partial skeleton of *Australopithecus afarensis* reveals a bipedal creature with smaller body, and relatively smaller brain and longer arms than ours, and slimmer hips than a modern woman's. (Other finds show that *afarensis* had slightly curved toe and finger bones). American anthropologists discovered it in North-central Ethiopia in 1974. Lucy's three-million-year-old skeleton became the oldest known for any hominine.

than a chimpanzee's and probably could not organize speech. The face was ape-like, with a low forehead, brow ridge, flat nose, no chin, but jutting jaws with massive back teeth. Front teeth were chipped, perhaps through use as gripping tools.

Afarensis walked slightly bow legged, and the somewhat chimpanzee-like hips and curved toe and finger bones suggest it spent much time in trees, perhaps sleeping high among the branches out of reach of predators. Females had much slimmer hips, and therefore narrower birth canals, than modern women, and must have given birth to young with relatively far smaller heads and brains than those of modern newborn human babies.

Family groups would have foraged for plant foods including tough, hard, or fibrous fruits and seeds. Individuals might have made crude tools of wood and stone to scavenge meat from carnivores' kills.

Before dying out by 2.5 million years ago, *afarensis* almost certainly gave rise directly or indirectly to the other australopithecines and to our genus, *Homo*.

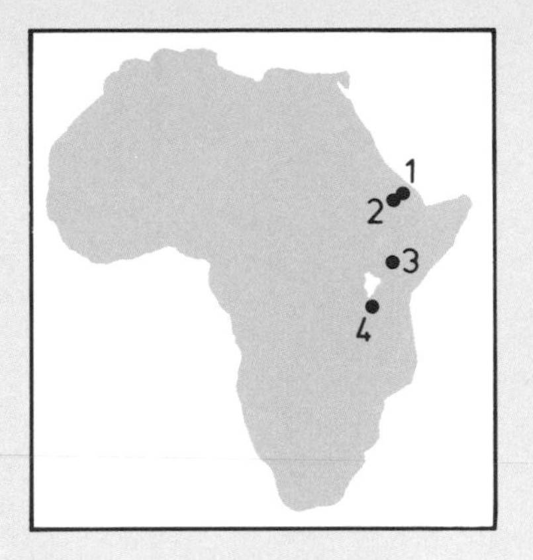

Where *afarensis* lived
1 Hadar
2 Middle Awash
3 Baringo
4 Laetoli

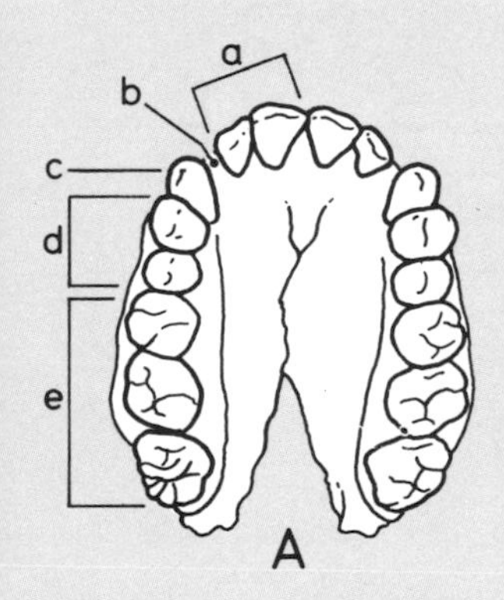

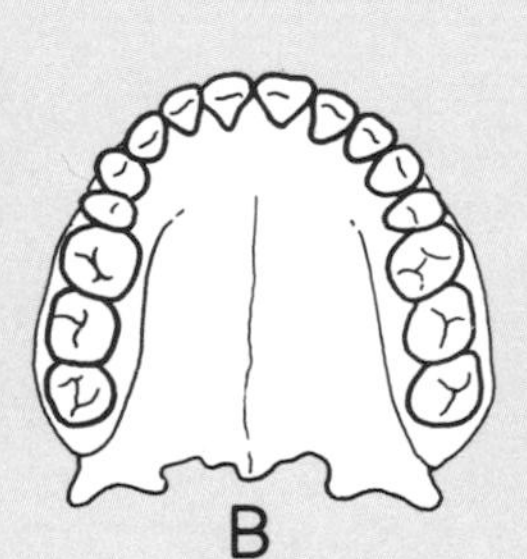

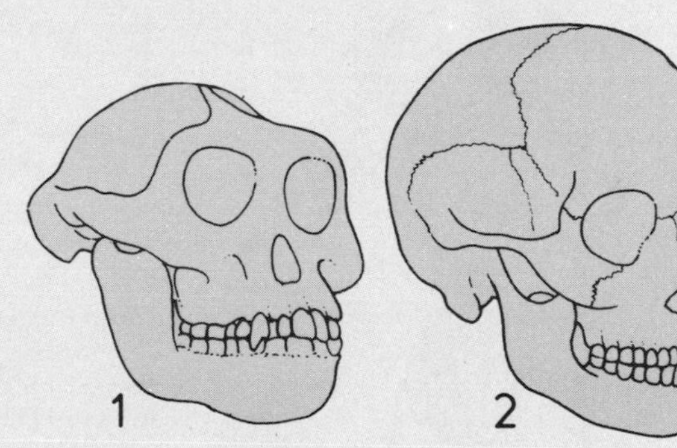

Two tooth rows (above)
Two pictures contrast (**A**) *afarensis*'s somewhat ape-like tooth row with (**B**) the more smoothly curved tooth row seen in members of the genus *Homo*.
a Big, ape-like incisors
b Diastema (gap) between incisors and canines in many specimens of *afarensis*
c Canines larger than in later hominids
d Premolars more primitive than later hominids'
e Large, thickly enameled molars, worn rather flat

Skulls compared (above)
1 Three-quarter view of *A. afarensis* skull showing relatively small size, small braincase, low forehead, brow ridge, flat nose, jutting jaws, and no chin. Brain capacity: about 410cc.
2 Three-quarter view of skull of *Homo sapiens sapiens*. Brain capacity: 1400cc.

Australopithecus africanus

Australopithecus africanus ("southern ape of Africa") lived perhaps from three million to one million years ago. It probably evolved from *Australopithecus afarensis*, and some writers half-jokingly suggest it gave rise to the chimpanzee.

A small, slight, ape-like creature, *africanus* stood as high as a small African bushman and weighed no more than a 12-year-old European girl. It walked upright, although the leg muscles differed from ours. The arms were relatively long, and the thumb and fingers may have handled objects less skillfully that we can.

The lower face jutted forward, but face and jaws were deeper and shorter than an ape's. Some skulls show traces of a crest that anchored strong neck muscles. The brain was no bigger than a gorilla's, but casts show that brain structure differed somewhat from an ape's. For relative brain-body size, *africanus* ranks midway between modern apes and modern man.

Body build (left)
Australopithecus africanus is shown beside a modern man, for scale.
Height: 3–4ft (1–1.3m).
Weight: 45–90lb (20–40kg).

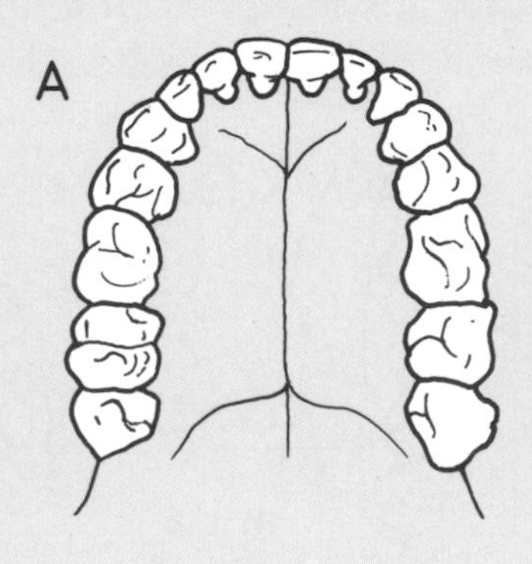

Tooth rows compared (right)
Illustrations contrast (**A**) *africanus's* large tooth row, and very large back teeth with (**B**) the smaller tooth row and smaller teeth of modern man.

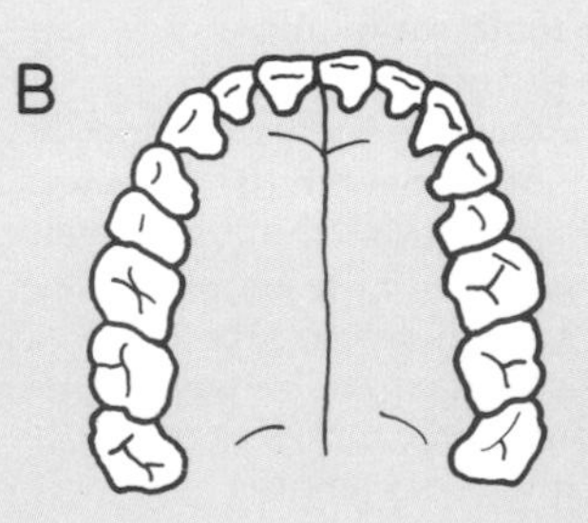

Experts disagree about its likely mode of life. Tooth and jaw design suggest this ape-man chewed plant foods but also might have scavenged meat from carnivores' kills. Experts dispute its ability to manufacture tools: most bone "tools" found near *africanus* fossils have proved to be remains of meals consumed by hyenas or other carnivores.

Some writers have argued that *africanus* fossils are just female robust australopithecines. But robusts seemingly postdated most *africanus* specimens. The earliest alleged *africanus* fossil is a 5.5-million-year-old jaw fragment from Lothagam in Kenya. A suggested end date of 700,000 years ago seems equally unlikely.

Most fossils come from Sterkfontein Cave in South Africa. Others show that *africanus* also lived in Ethiopia, Kenya, and Tanzania.

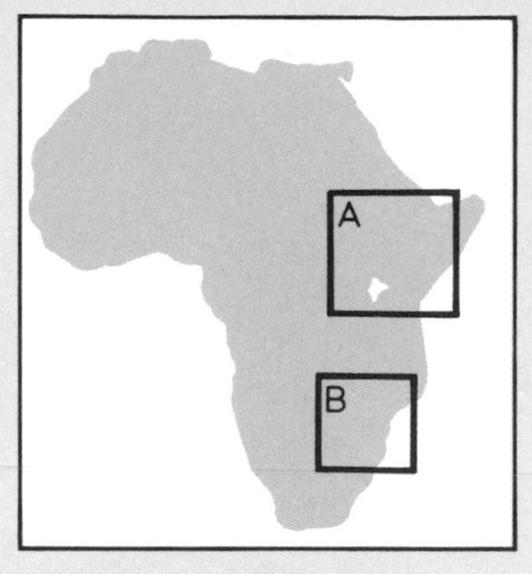

Where *africanus* lived
East African finds might be those of other hominines.
A East Africa
1 Omo River
2 Koobi Fora
3 Lothagam
4 Olduvai Gorge
B South Africa
5 Makapansgat
6 Sterkfontein
7 Taung

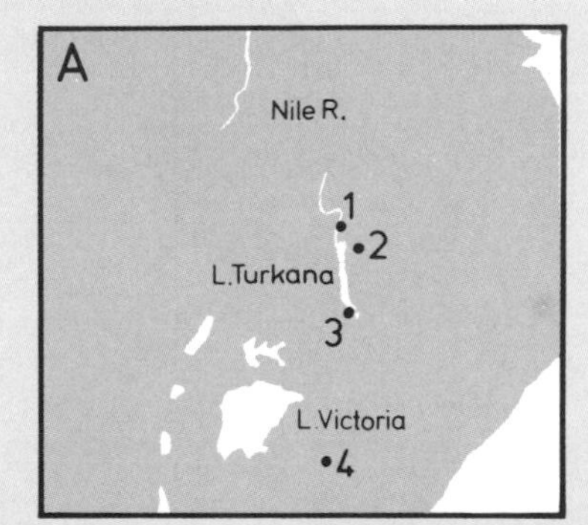

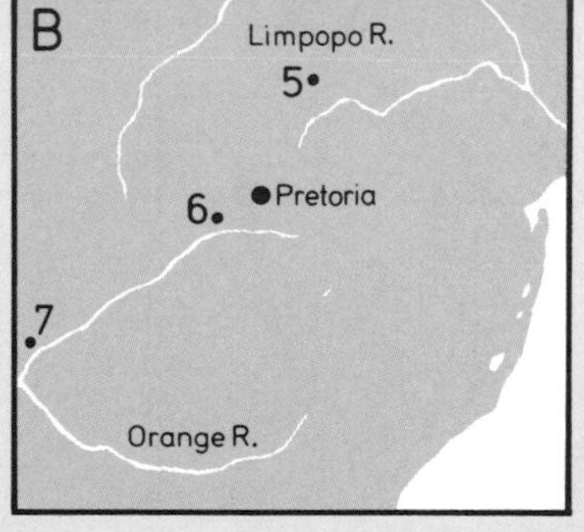

©DIAGRAM

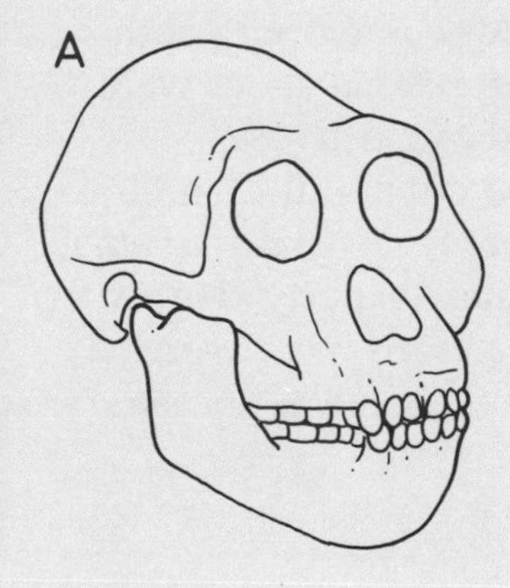

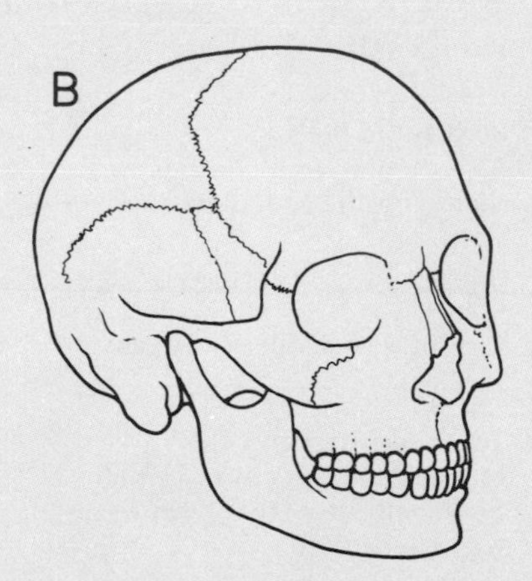

Skulls compared
A Skull of *Australopithecus africanus* in three-quarter view
B Three-quarter view skull of *Homo sapiens sapiens*

Australopithecus robustus

Once called *Paranthropus* ("beside man"), this "robust southern ape" was altogether larger and more strongly built than *Australopithecus africanus*. Some have argued that "robust" fossils were just males and *africanus* fossils females of a single species, but most experts reject that notion.

About 5ft 3in (1.6m) tall, *robustus* was the height of many Western women, but lighter than most at 110lb (50kg). Compared with *africanus* it had a larger, flatter skull which housed a bigger brain of about 500cc, and the face was considerably larger and broader in relation to the braincase. A tall central skull crest anchored powerful muscles that worked the massive jaws. Front teeth were no bigger than those of *africanus*, but cheek teeth were large and often worn despite a thick coat of wear-resistant enamel. All this suggests *robustus* must have eaten hard, tough foods, perhaps including seeds.

Robustus seemingly evolved by 2.5 million years ago, perhaps from *afarensis* or *africanus*. All undisputed *robustus* fossil finds come from South African caves, where the creatures' carcasses were evidently dragged or dropped by carnivores.

The species apparently died out about 1.5 million years ago. Before then it probably gave rise to *boisei* (see pp. 104–105), although many experts identify that with this species. Skull comparisons have even led some writers to suggest that *robustus* was ancestral to the gorilla.

Body build (left)
Australopithecus robustus is shown beside a modern man, for scale.
Height: 4ft 11in–5ft 7in (1.5–1.7m)
Weight: 110–154lb (50–70kg)

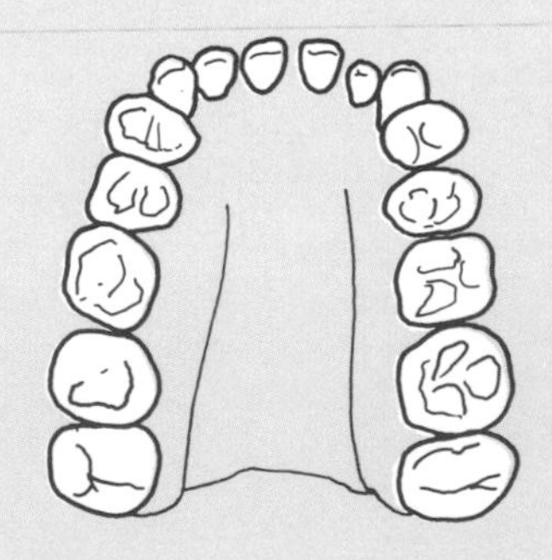

Tooth row (right)
Massive cheek teeth and small front teeth show that *robustus* was a herbivore.

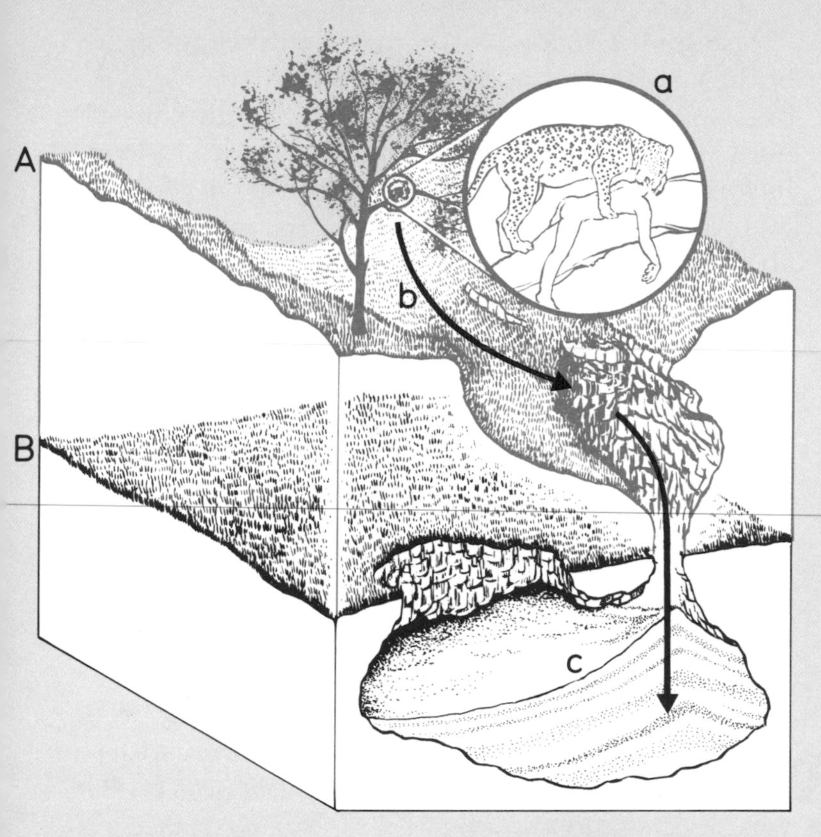

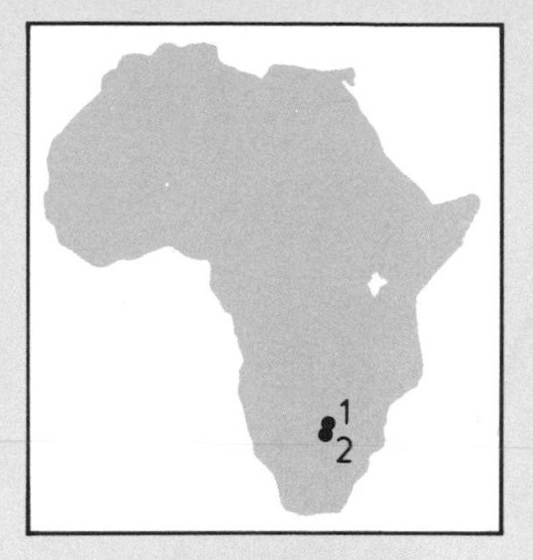

Where *robustus* lived (above)
1 Kromdraai
2 Swartkrans
Many experts would include sites named on page 105.

Tell-tale holes (below)
Damaged *robustus* bones hint at how their owners died.
a Holes in a *robustus* skull
b A leopard's lower canines match the holes.

Finds in fissures (above)
Robustus bones found in limestone fissures probably accumulated like this.
A Old land level
B Present land level
a Leopard drags *robustus* corpse up a tree out of reach of hyenas.
b *Robustus* bones fall into limestone fissure.
c Layers of sediment bury the bones.

Skulls compared (below)
A Three-quarter view of *A. robustus* skull, showing the skull crest and massive jaws. Brain capacity: 500cc.
B Three-quarter view of skull of *Homo sapiens sapiens*. Brain capacity: 1400cc.

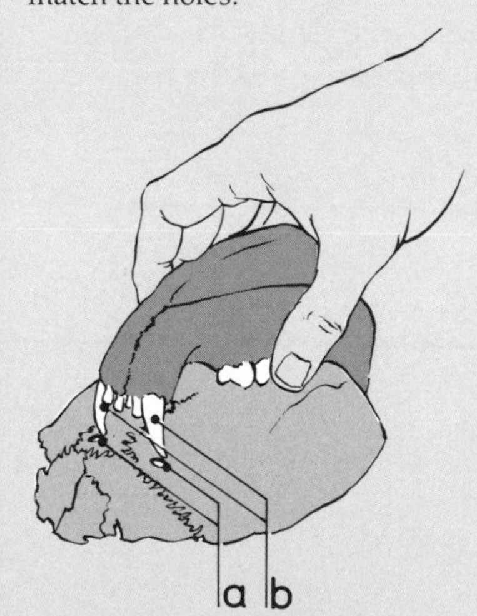

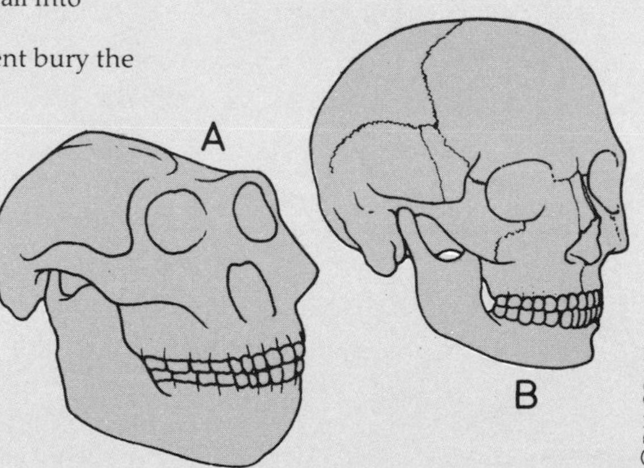

Australopithecus boisei

"Boise southern ape" was reputedly the biggest, burliest australopith, named after British businessman Charles Boise who helped fund fossil hunts that led to its discovery in East Africa in 1959. Its former name, *Zinjanthropus*, means "East Africa man," and this creature inhabited that region from about 2.5 million to 1 million years ago.

Boisei resembled a more massive version of *Australopithecus robustus*, from which it probably evolved, although many paleoanthropologists consider it just a regional variant of that species. Restorations based on finds of fossil skull and limb bones suggest individuals grew about the height of modern man, had a *robustus* sized brain, about one-third as big as ours, walked upright, yet were built on powerful lines recalling the gorilla. As with that ape, males seemingly grew far larger than females.

Like a gorilla's, *boisei*'s skull was large, with brow ridges, and a central crest for anchoring immense jaw muscles. But compared with a gorilla, *boisei*'s crest was slighter and located farther forward, its face was flatter, and the canine teeth were small.

Immense molars and premolars earned this animal the nickname Nutcracker Man, but biomechanical studies show that teeth exerted no more pressure than our own, one quarter their size. Instead of crunching hard-shelled foods, *boisei* seems to have

Body build (left)
Burly *Australopithecus boisei* is shown beside a modern man.
Height: 5ft 3in–5ft 10in (1.6–1.78m) or maybe less
Weight: 132–176lb (60–80kg)

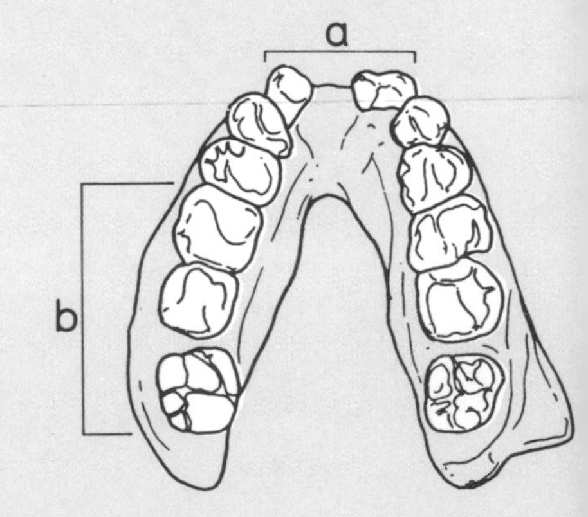

Tooth row (right)
This incomplete tooth row (reduced in size) shows features stressed in *Australopithecus boisei*.

a Small biting front teeth
b Immense grinding back teeth

chewed large quantities of leaves, a low-grade source of nourishment.

Fossil bones found with chipped pebbles about 1.8 million years old suggest that *boisei* might have made or used stone tools. More probably members of this species had fallen prey to its contemporary, *Homo habilis*, a likelier candidate as toolmaker (see pp. 106–107).

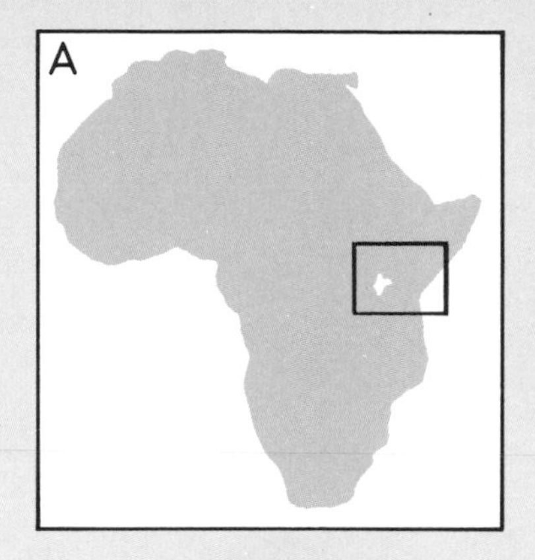

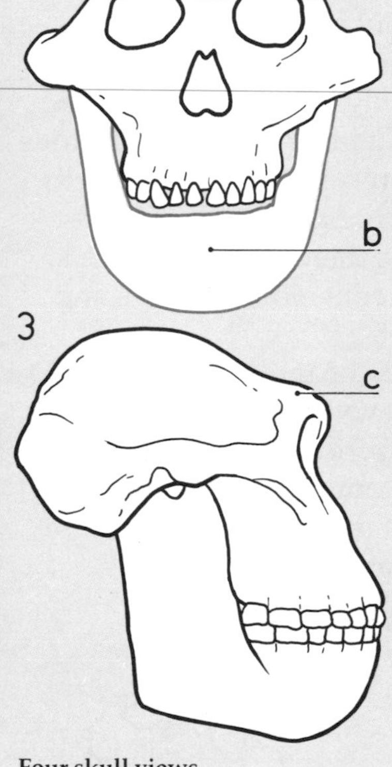

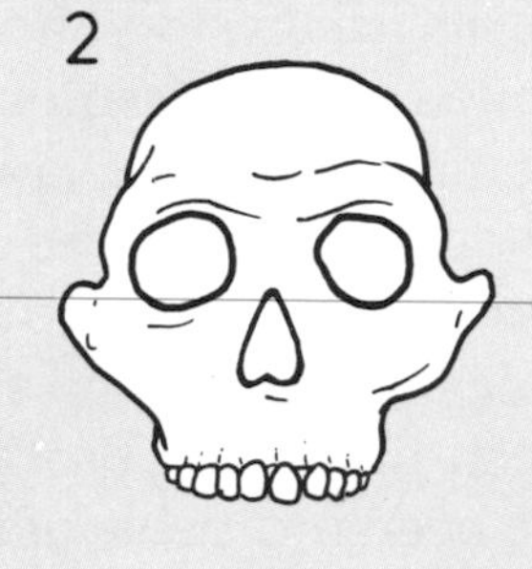

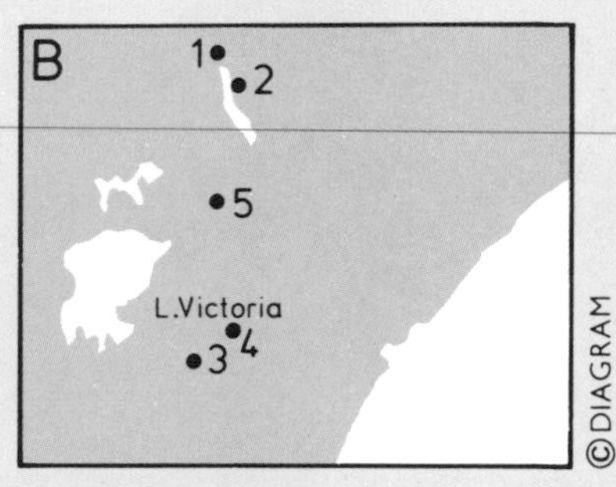

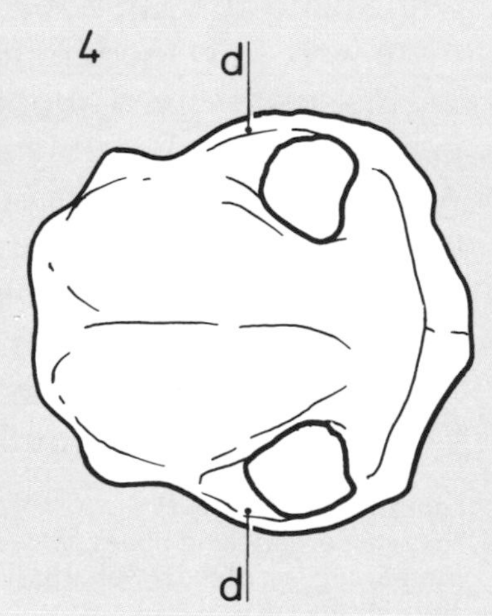

Where it lived (above)
A Africa. A rectangle shows the area involved.
B The area enlarged, with named localities.
1 Omo
2 Koobi Fora
3 Olduvai Gorge
4 Peninj
5 Chesowanja

Four skull views
1 Skull of male *Australopithecus boisei*
a Sagittal crest
b Reconstructed lower jaw
2 Skull of presumed female, a much smaller creature
3 Side view of male skull
c Brow ridge
4 Skull seen from above
d Zygomatic arches to take massive jaw muscles

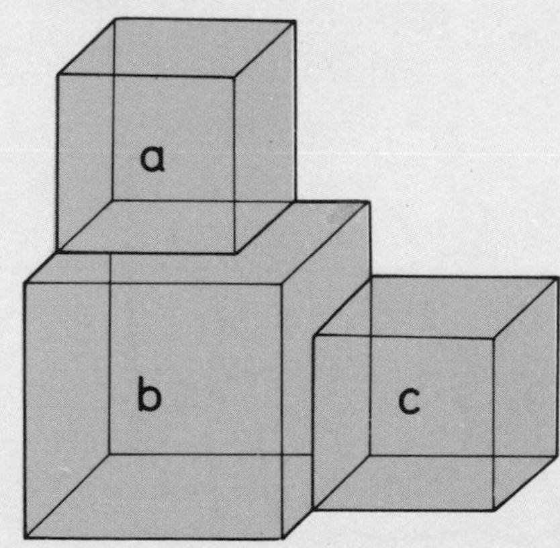

Brain capacities (right)
a *A. africanus:* 450cc
b Modern man: 1400cc
c *A. boisei:* 500cc

Homo habilis

Homo habilis ("handy man") was reputedly the first known species of our genus, *Homo*. *Homo* features a relatively bigger brain than *Australopithecus*, thus a bigger brain case, but smaller, less projecting face, and relatively smaller cheek teeth but larger front teeth, with an open-U-shaped tooth row. Arms are shorter in relation to legs, and hip bones permit both bipedal walking and giving birth to babies with large heads.

Homo habilis stood no more than 5ft (1.5m) tall. Its face was still old fashioned, with brow ridges, flat nose, and projecting jaws. But *habilis* had a more rounded head than the australopiths and a larger brain (650–800cc), yet still only half the size of ours. Inside the thin-walled skull a bulge shows the brain's speech-producing Broca's area, yet the larynx might have been incapable of making as many sounds as ours. Jaws were less massive than an australopith's, while hand and hip bones seem more modern, and there were fully modern feet.

This species lived about 2–1.5 million years ago, perhaps longer. Possibly it evolved from *Australopithecus afarensis* or *africanus*. But old-fashioned aspects of its anatomy persuade some experts to call all early *Homo* finds australopithecine.

Habilis lived in East Africa, and possibly (as "Telanthropus") in South Africa and (as "Meganthropus") in South-east Asia.

Foot and hand (top left)
Known foot and hand bones (tinted) suggest bipedal walking and a strong yet sensitive grip.

Body build (left)
Homo habilis is shown beside a modern man, for scale.
Height: 4–5ft (1.2–1.5m)
Weight: about 110 lb (50 kg)

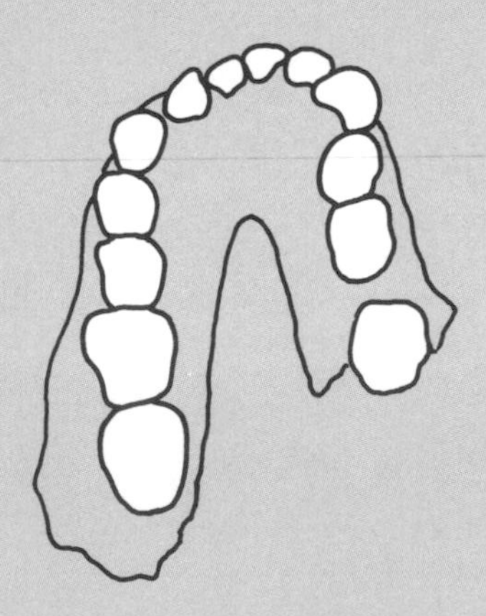

Tooth row (right)
This shows a more "modern" curve and narrower back teeth than in *Australopithecus*.

Artefacts found near its bones suggest it made basic stone tools, built simple shelters, gathered plant foods, scavenged big meaty limbs from carcasses of creatures killed by carnivores, and hunted small and maybe larger game.

Homo habilis probably gave rise to *Homo erectus*.

Where *habilis* lived (above)
1 Koobi Fora
2 Olduvai Gorge
3 Swartkrans (possibly)
There are claims for finds at other sites, mostly in East Africa.

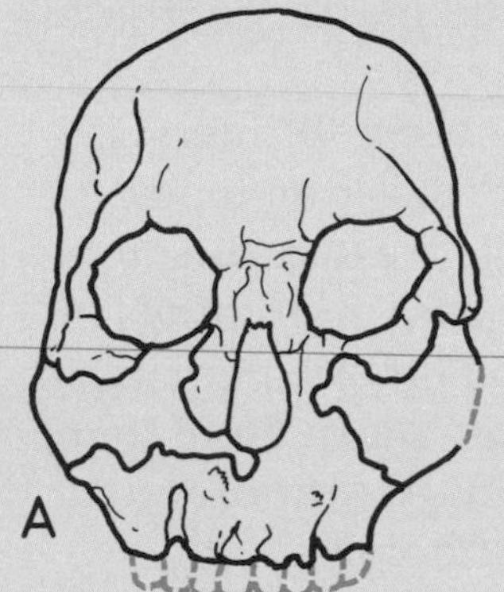

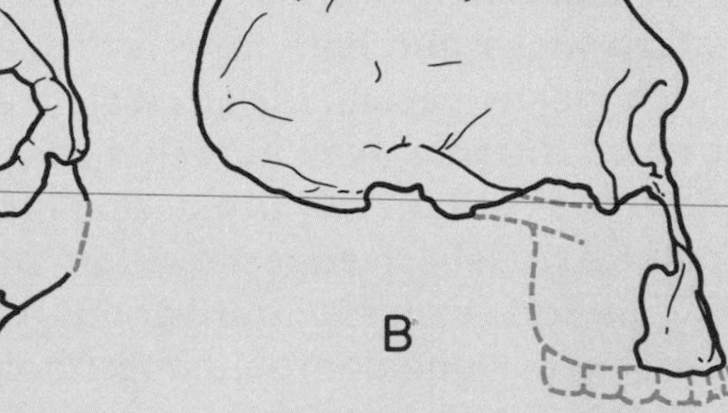

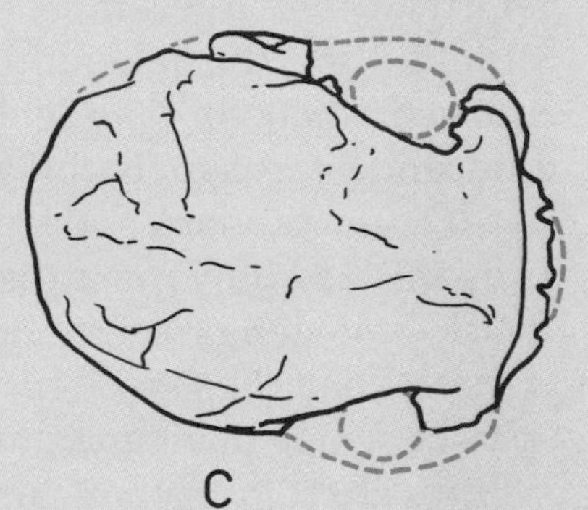

Three views of a skull
Homo habilis had a more rounded head, bigger brain, and smaller, narrower, longer, face than *Australopithecus*, yet with old-fashioned features. These views exclude the lower jaw. Broken lines are missing portions.
A Frontal view
B Side view
C View from above

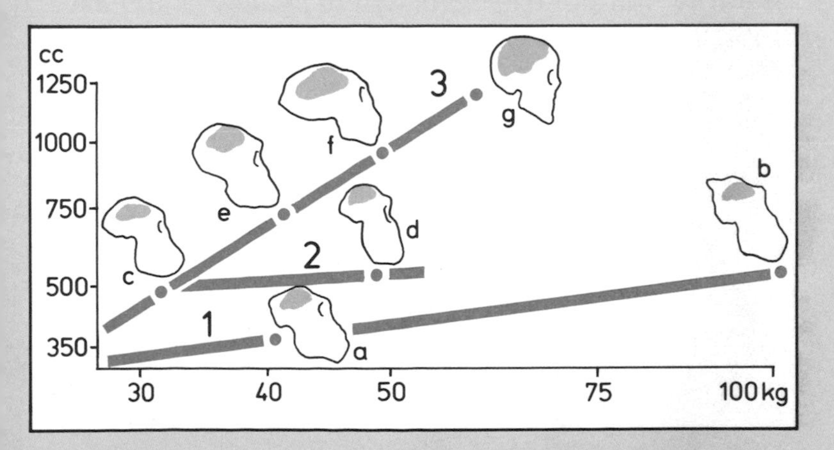

Brain size and body weight (left)
The steeper the line, the greater brain size increase compared to increase in body weight.
1 Apes
a Chimpanzee
b Gorilla
2 Australopiths
c *Australopithecus africanus*
d *Australopithecus robustus*
3 Man
e *Homo habilis*
f *Homo erectus*
g *Homo sapiens*

Early artefacts

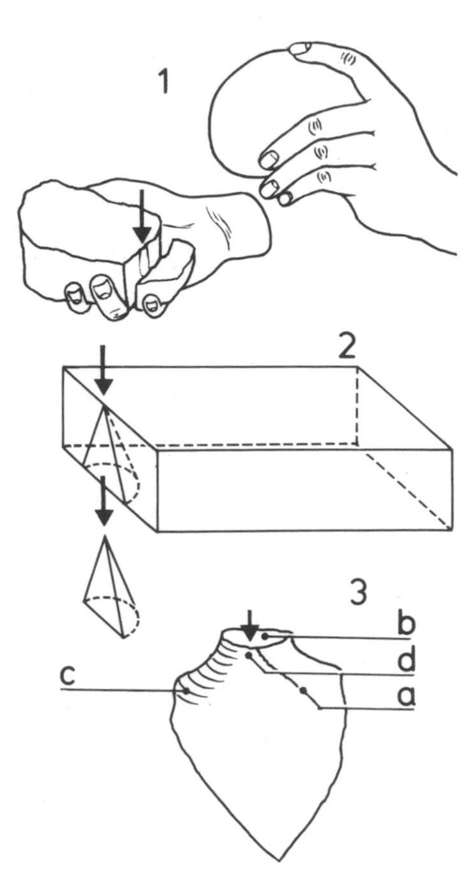

Making stone tools (above)
1 Bashing stone on stone to knock off flakes
2 Ideal cleavage of a flint flake
3 A man-made flint flake:
a bulbar scar
b striking platform
c concentric ripples
d bulb of percussion

Early man was slower and weaker than the big carnivores, and lacked their inbuilt weapons of fangs and claws. Yet early hominines learnt to make up for these disadvantages. They fashioned lumps of stone, bone, and wood to cut, scrape, or dig. Unlike fangs or claws, such tools could be picked up, put down, or interchanged at will. Tools in time gave man unprecedented mastery of the environment.

The first tools were probably bone splinters, sharp sticks, and bark trays for collecting food. Such mostly perishable artefacts are lost. But stone proved durable. We know that early hominines in Ethiopia deliberately broke small rocks, perhaps to make hard, sharp edges to cut up meat. From then, 2.5 million years ago, until about five thousand years ago, stone dominated technology.

Of the Stone Age's three subdivisions (Old, Middle, New) the Paleolithic or Old Stone Age endured wherever man set foot until about ten thousand years ago. Early Paleolithic toolmakers learnt to make sharp tools by bashing stone on stone. They selected only rocks that broke easily when struck, and along fracture lines the knappers could control. They chose rock hard enough to cut, grind, split, or scrape plant and animal materials.

Flint, chert, quartzite, and rock crystal all proved suitable, but none is everywhere available. In East

Oldowan tool kit (right)
Early stone tools from Olduvai Gorge in Tanzania:
A Lava chopper for cutting meat or cracking open bones
B Polyhedron, with 3 or more cutting edges
C Discoid, with a sharp rim
D Scraper, for working hides
E Lava hammerstone

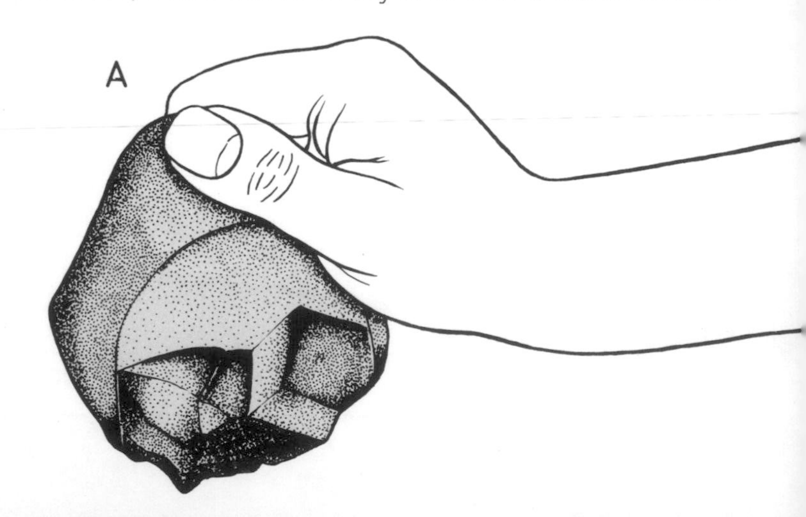

Africa, most of the oldest tools of all were made of lava which yields rougher surfaces than flint.

The best-known early toolkit is the Typical Oldowan from Tanzania's Olduvai Gorge. Here, 1.9 million years ago, *Homo habilis* (or some contemporary australopithecine) chipped basalt and/or quartzite pebbles into shapes identified as crude choppers, scrapers, burins, hammerstones and (by their shape) discoids, polyhedrons, and sub-spheroids. Some experts think the true tools were the "waste" flakes struck off from these artefacts. Certainly, near Kenya's Lake Turkana, hominines used small stone flakes to butcher antelopes 1.5 million years ago.

Tooth marks underlying cut marks made on bones imply that early hominines scavenged most of their meat from carnivores' kills. Yet early man quite likely made hunting weapons, too. Stone balls from Olduvai could have brought down antelopes if tied to thongs and thrown to wind around legs, as cowboys topple cattle with the bolas.

The Oldowan (alias chopper-and-flake or pebble tool) industry, with later variants, spread over much of early Stone Age Africa and Eurasia. In places it flourished until about 200,000 years ago – long after the invention of much more sophisticated Stone Age industries.

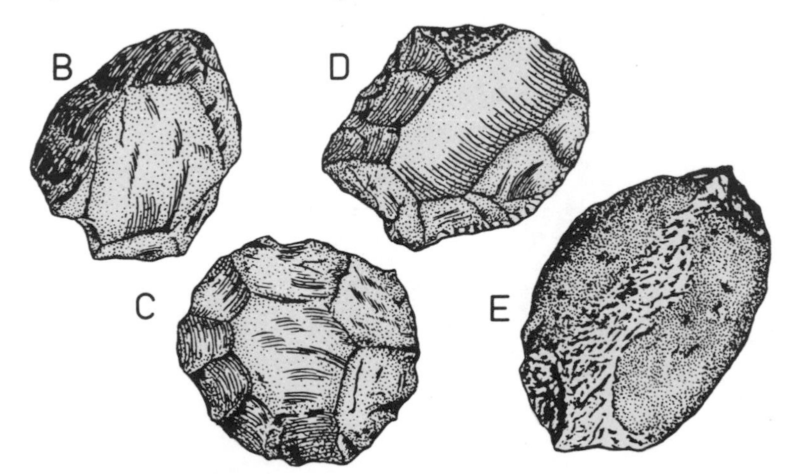

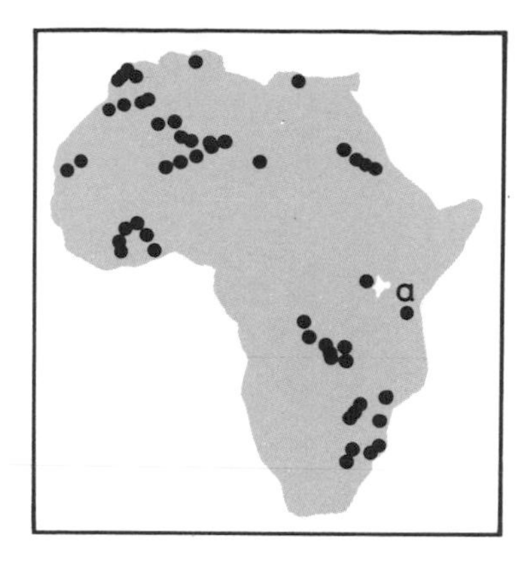

Oldowan Africa (above)
Map symbols show finds of stone tools supposedly like those from Olduvai (**a**).

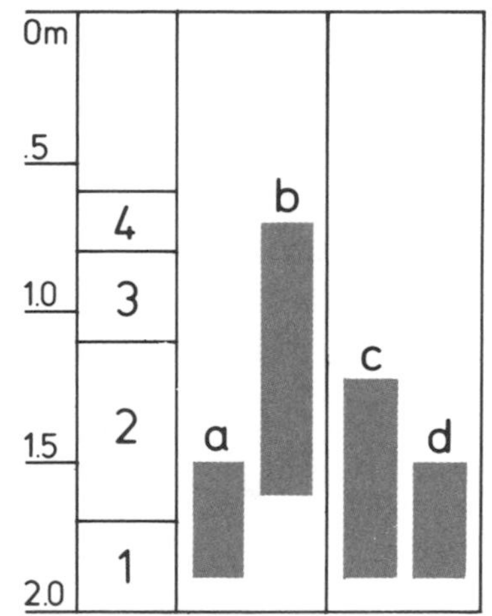

Oldowan time scale (above)
We show rock beds, tool kits and early homines at Olduvai.

1 Bed 1: 2.0–1.7 m yr ago
2 Bed 2: 1.7–1.1 m yr ago
3 Bed: 1.1–0.8 m yr ago
4 Bed 4; 0.8–0.6 m yr ago
a Typical Oldowan tools
b Developed Oldowan tools – more varied and overlapping *Homo erectus*
c *Australopithecus* species
d *Homo habilis*

The first camp sites

Clusters of stones and bones dug up at Olduvai Gorge in Tanzania suggest that by about 2 million years ago early humans already met in groups at centers where they butchered game, made tools, ate food, and built perhaps the world's first shelters.

Sharpened stones and the cut and broken bones of big grazing mammals reveal no chance assemblages produced by running water, but working and even living floors where families of *Homo habilis* or other hominines shared meat and marrow. They carried in

The oldest hut? (right)
This 1.8-million-year-old scattering of stones and fossils might represent the oldest known human habitation. Stones evidently anchored branches providing the structure of a lakeside hut at Olduvai.

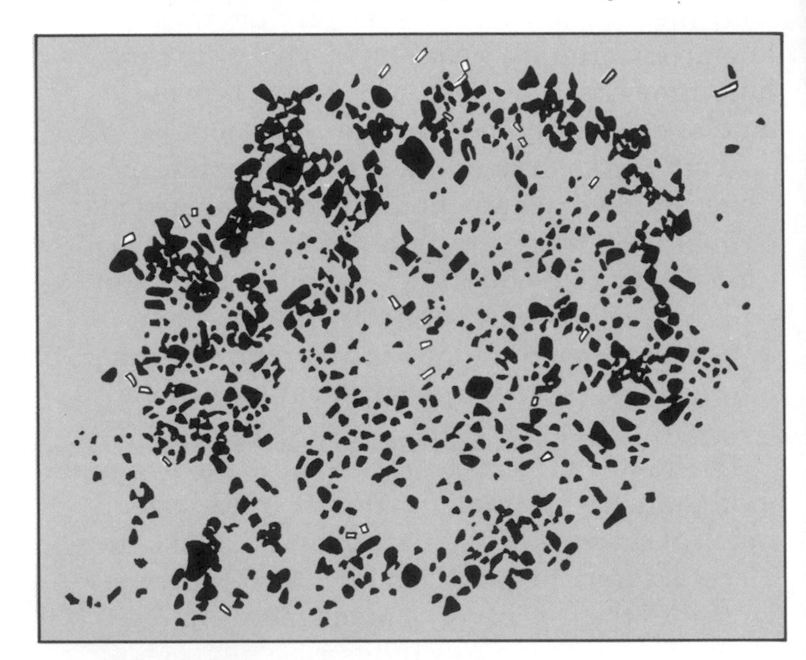

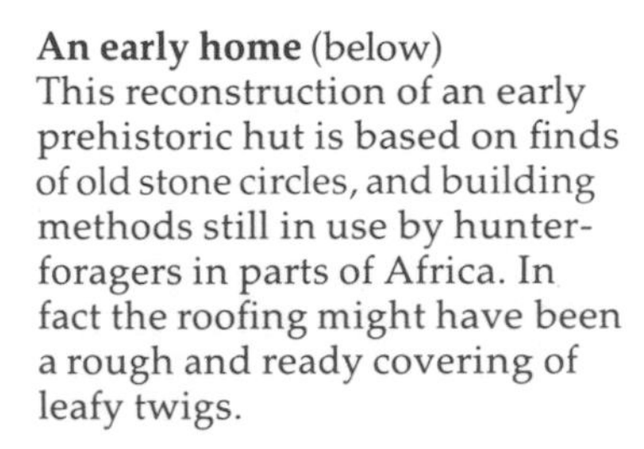

An early home (below)
This reconstruction of an early prehistoric hut is based on finds of old stone circles, and building methods still in use by hunter-foragers in parts of Africa. In fact the roofing might have been a rough and ready covering of leafy twigs.

bones scavenged from carnivores' kills, and lava lumps from nearby rocks, to shape them on the spot. About 1.7 million years ago they also seemingly used tools brought ready made from rocks 10 miles (16km) away, implying maybe even very early trade.

Close study suggests that most of the 20 or so very early sites at Olduvai were carefully located where drinkable freshwater streams flowed into a game-rich alkaline lake. Perhaps the most remarkable discovery of all is that these ancient hominines began to build. At one site, the scatter pattern of bones and stones suggests that a thorn fence or windbreak protected workers on their windward side. Another site features a stone circle about 13ft (4m) across – support for now long-vanished branches raised to form a hut like those still built in parts of Africa. About 1.8 million years old, this shelter is the first known human structure anywhere.

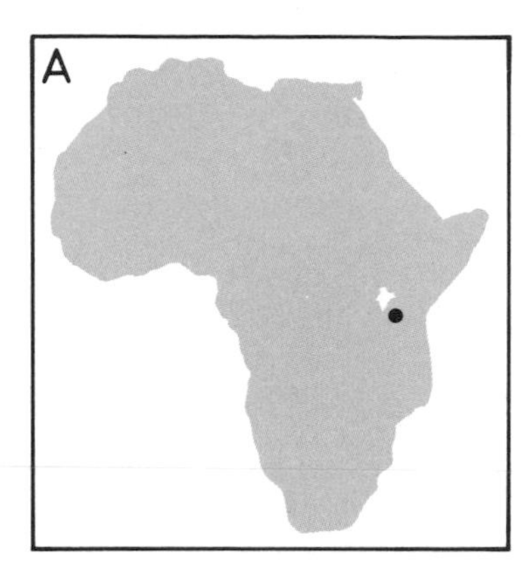

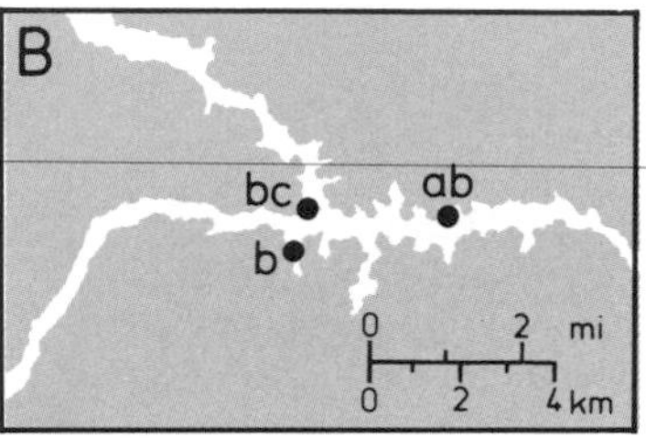

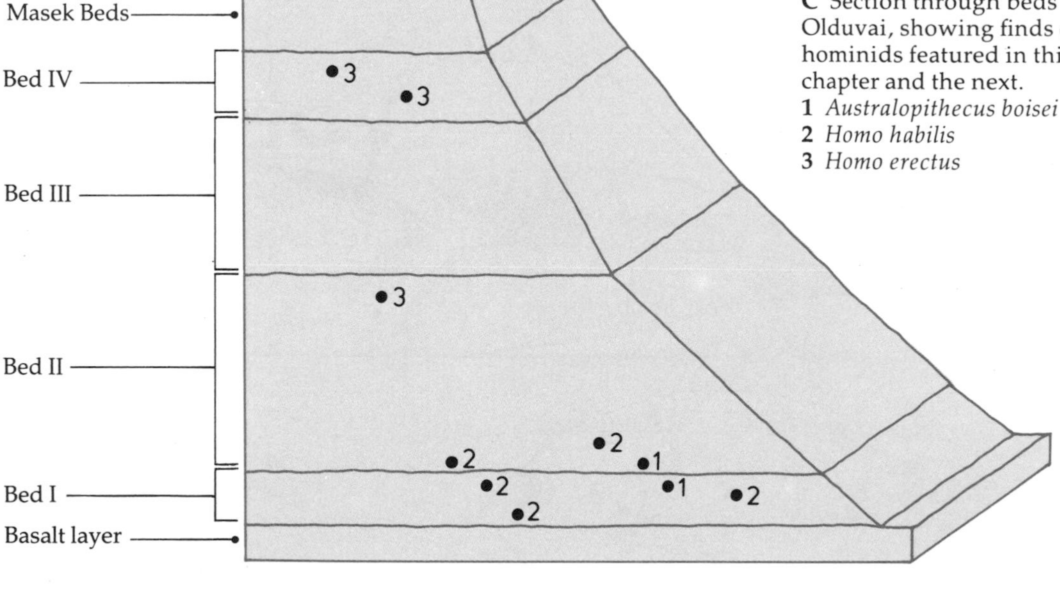

Finds from Olduvai
A Map of Africa showing the area including Olduvai Gorge.
B Olduvai Gorge showing a few sites of key early finds.
a Hut circle
b *Homo habilis*
c *Australopithecus boisei*
C Section through beds at Olduvai, showing finds of hominids featured in this chapter and the next.
1 *Australopithecus boisei*
2 *Homo habilis*
3 *Homo erectus*

Animals of ancient Africa

Thousands of broken animal bones litter East Africa's early butchery and living floors. Careful studies of sites from Tanzania, Kenya, and Ethiopia show scores of creatures large and small that shared savanna, lake, or forest with hominines 2–1.5 million years ago. Most detailed studies come from the former shoreline sites of a dwindling prehistoric lake in the area now pierced by Tanzania's Olduvai Gorge. Here are examples of these creatures, living and extinct, not shown to scale.

1 **Tilapia**, a cichlid, is a freshwater fish that tolerates brackish water. Length: up to 12in (36cm).

2 **Xenopus**, the clawed frog, is an aquatic gray frog with a pale belly. Length: up to 6in (15cm).

3 **Chamaeleo jacksoni**, is a slow-moving lizard catching flies on its long, sticky tongue. Males have three horns. Length: to 12in (30cm). Habitat: bush and open woodland.

4 **Phoeniconaias**, the lesser flamingo, is a long-necked wading bird with a bent beak. Length: 40in (1m). Habitat: brackish lakes.

5 **Megantereon** was a possibly lion-sized saber-toothed cat, stabbing big herbivores with its dagger-like upper canines.

Life at Olduvai
Below and right are eight creatures with numbers that correspond to their descriptions in the text. These animals are not all to the same scale.
1 *Tilapia*
2 *Xenopus*
3 *Chamaeleo jacksonii*
4 *Phoeniconaias*
5 *Megantereon*
6 *Deinotherium*
7 *Sivatherium*
8 *Pelorovis*

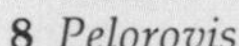

6 **Deinotherium** was an "elephant" with downcurved tusks in its lower jaw, perhaps for digging roots. Height: 13ft (4m).
7 **Sivatherium** was a short-necked giraffe with "antlers." Height: 7ft (2.2m).
8 **Pelorovis** was a giant African buffalo with a 6ft 7in (2m) horn span. It belonged to the Bovidae (cattle, antelopes, etc), which account for most bones found at Olduvai.

Chapter 6 UPRIGHT MAN

By 1.6 million years ago *Homo habilis* had most likely given rise to bigger, brainier *Homo erectus* – "upright man." Superior intelligence and technology helped this Early Stone Age hunter colonize new habitats, thinly populating Africa, Europe, and (mostly southern) Asia. Local populations apparently evolved in different ways. In Europe, by 400,000 years ago, individuals showed features found in early members of our species, *Homo sapiens*. By 200,000 years ago, *Homo erectus* was probably extinct – perhaps the victim of competition with its own descendants.

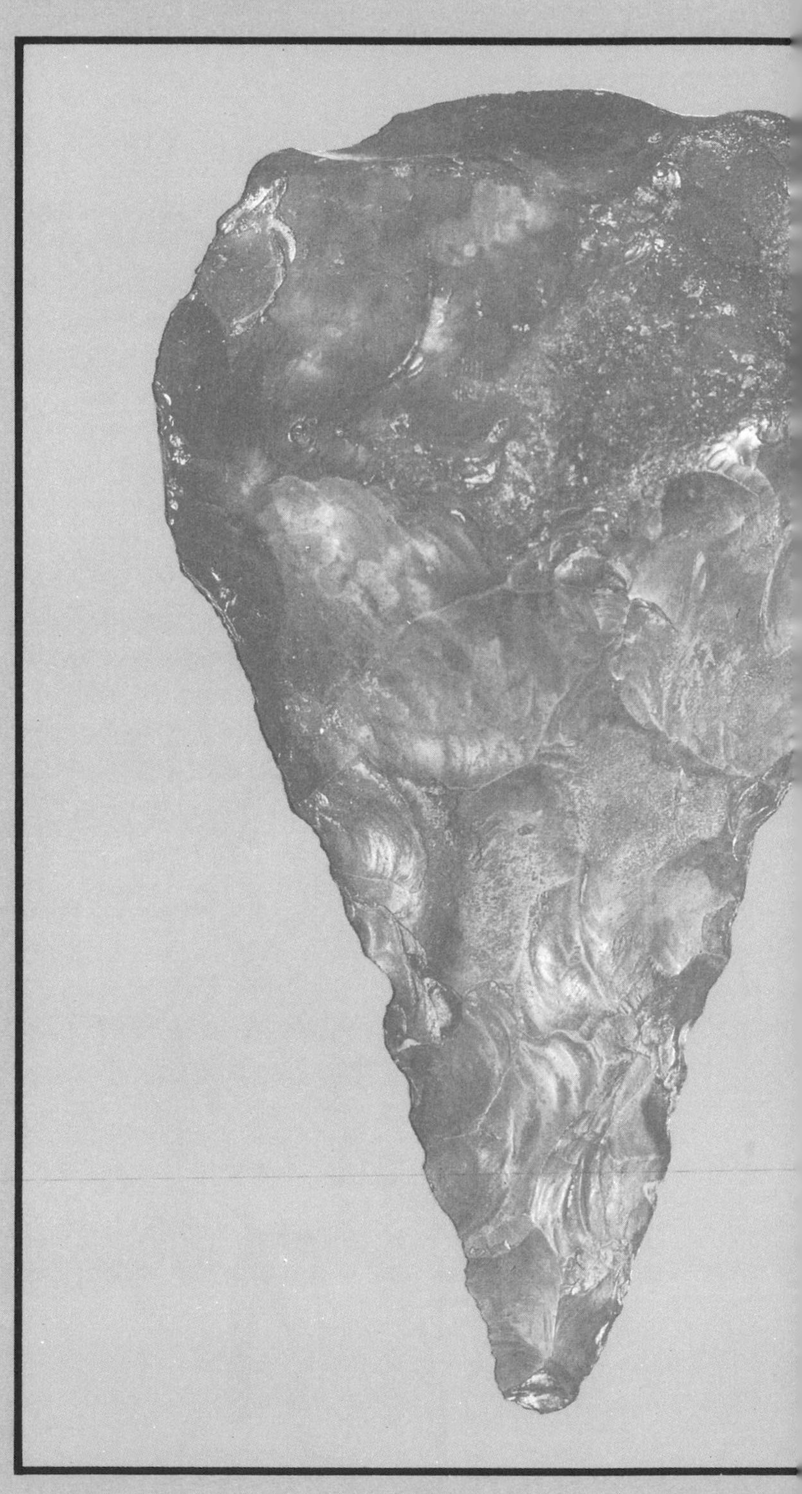

Two flint hand-axes – one more sophisticated than the other – represent the Stone Age artefacts produced by *Homo erectus*, and early *Homo sapiens*. Discovered in London about 1690, the nearer axe is the first stone implement known to have been collected as a piece of ancient human handiwork. Both are in the British Museum (Natural History), London.

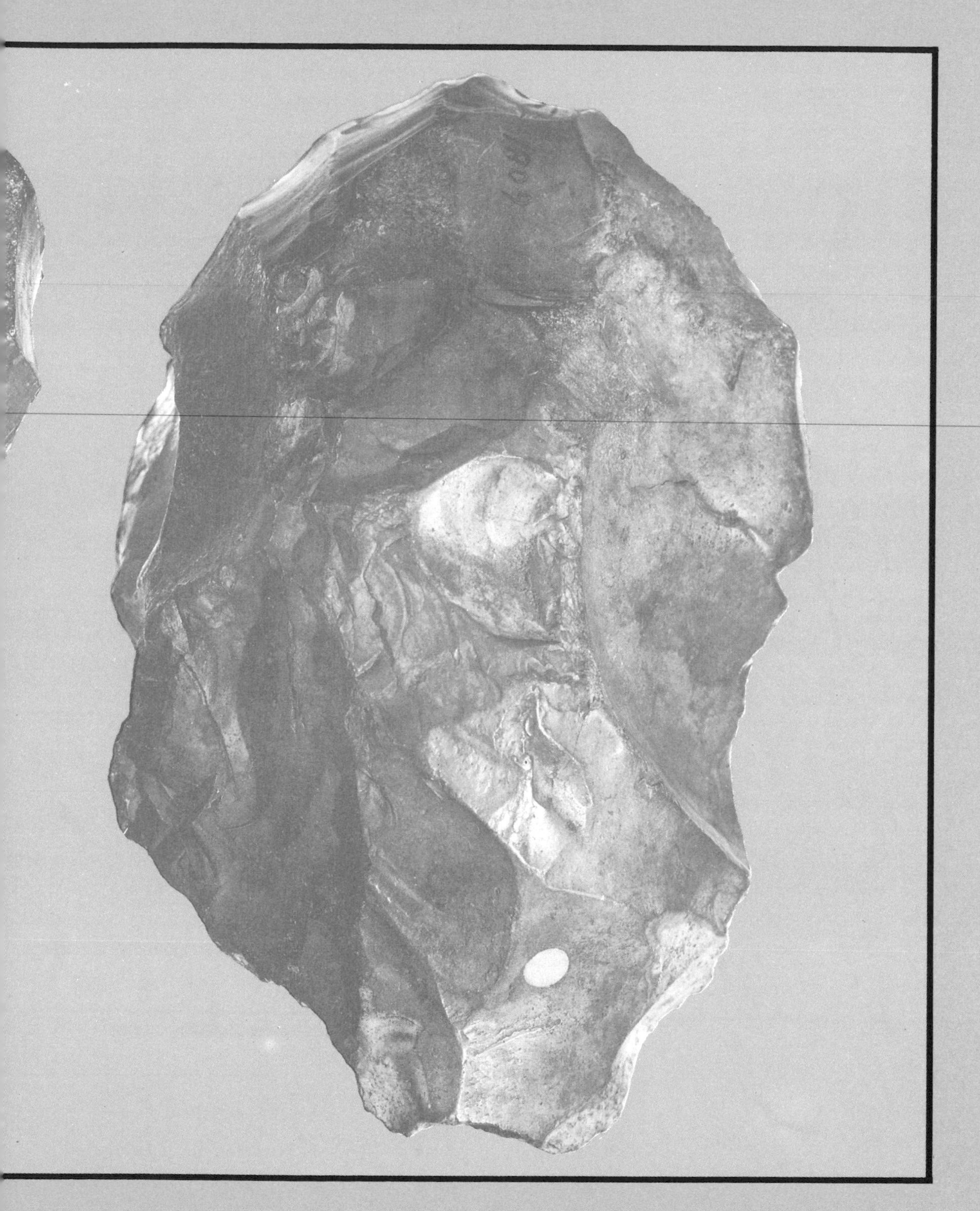

Homo erectus

Body build (below)
Homo erectus is shown beside a modern man, for scale.
Height: 5–6ft (1.5–1.8m)
Weight: 88–160lb (40–72.7kg)

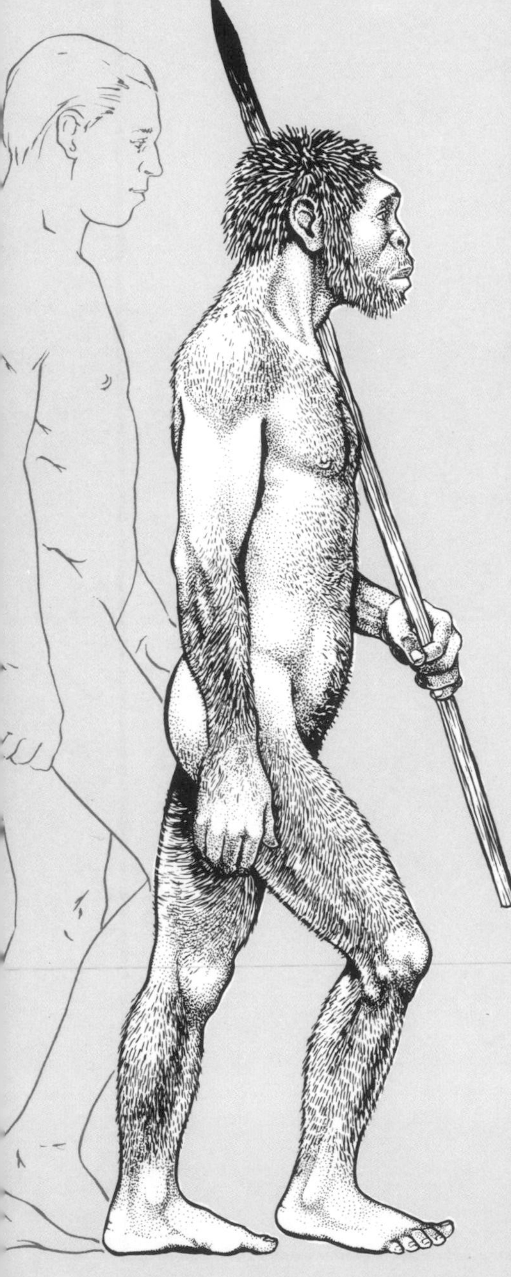

Homo erectus ("upright man") had a bigger brain and body than *Homo habilis*, its likely ancestor, and in many ways resembled a strongly built version of its direct descendant, modern man. The skull, though – the thickest in any member of the human tribe – retained old-fashioned features. It was long and low, with a bony bump behind, a shelving forehead, thick brow ridges, a flatter face than ours, big projecting jaws, teeth more massive than our own (yet slightly smaller than those of *Homo habilis*), and no chin. Strong muscles at the back of the neck joined the rear skull bump and stopped the front-heavy head sagging forward. Brain capacity averaged 880–1100cc (experts disagree) more than for *habilis*, though less than modern man's.

Some adults probably grew 6ft (1.8m) tall and were at least as heavy as ourselves.

Homo erectus lived about 1.6 million–200,000 years ago, arguably longer. Probably evolving in Africa groups spread to Europe, East Asia (including Peking man or "Sinanthropus"), and South-east Asia (Java man, or "Pithecanthropus"). Isolated populations seemingly evolved at different rates.

Improved technology including standard toolkits, big-game hunting, use of fire, and improved building methods put *erectus* far ahead of former hominids, enabling this species to invade new habitats and climates.

A
B
1
2
2
3
2 | 1.5 | 1 | .5 | 0

Rival family trees (above)
A *H. erectus* as our ancestor
B *H. erectus* as a dead end
1 *Homo habilis*
2 *Homo erectus*
3 *Homo sapiens*
(Figures given in millions of years ago)

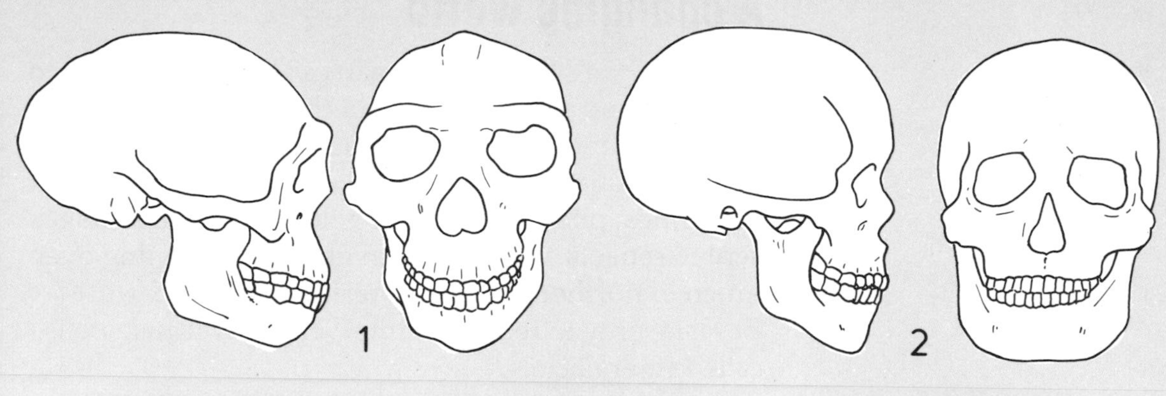

Skulls compared (above)
1 *Homo erectus* skull: long and low, with brow ridges, no chin, protruding jaws, but smaller teeth than *Homo habilis.*
2 *Homo sapiens sapiens* skull

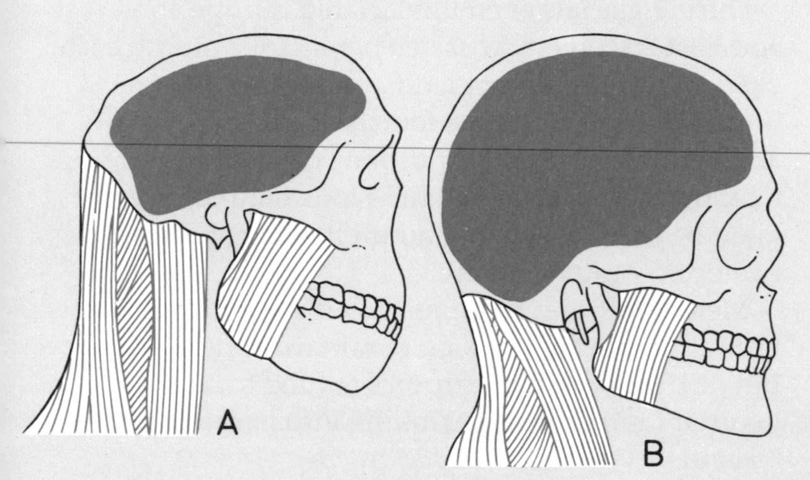

Brains and muscles (left)
Diagrams show skull shapes related to brain size and the size of muscles balancing the head and operating jaws.
A *Homo erectus* (small brain, big muscles)
B *Homo sapiens sapiens* (big brain, small muscles)

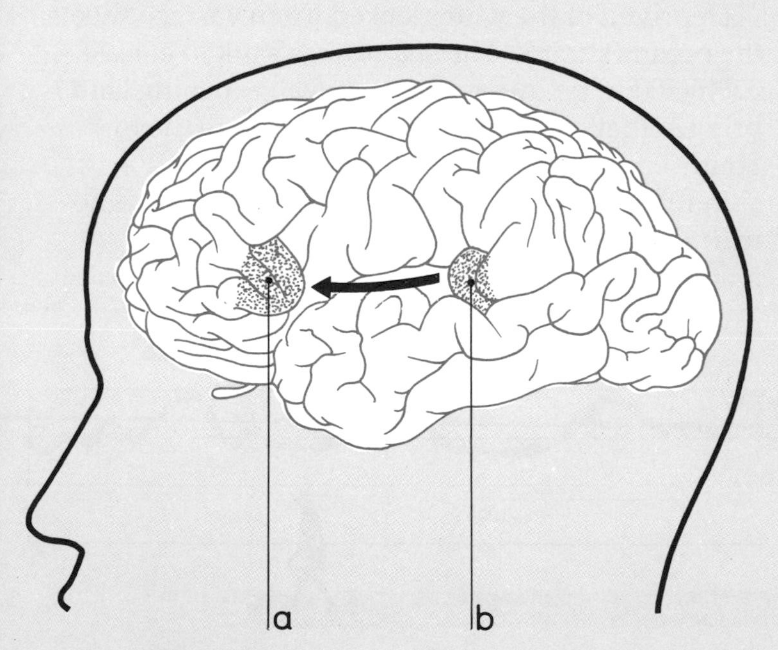

Speech centers (left)
Linked speech centers in the brain's left side produce swellings detectable in early fossil skulls of *Homo* (and, confusingly, but less pronounced, in apes).
a Broca's area, controlling speech production
b Wernicke's area, controlling understanding of speech

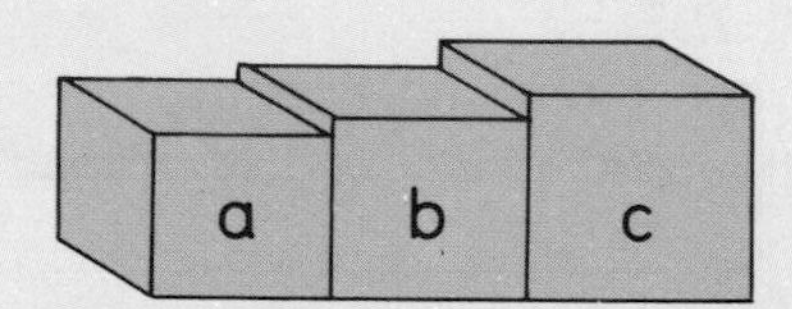

Increase in brain size (above)
a *Homo habilis:* 725cc
b Early *Homo erectus:* 850cc
c *Homo sapiens:* 1,400cc

A changing world

Homo erectus's likely time span of about 1.6 million to 200,000 years ago filled most of the early and middle sections of the Pleistocene – the geological epoch from about 2 million to 10,000 years ago. During these Ice Age times, phases of intense cold called glacial stages sent ice sheets and mountain glaciers sprawling over much of northern North America and North-west Eurasia, only to retreat in intervening warmer spells, called interglacials.

During glacials even unglaciated Europe and West and East Asia were frost free barely for a month each year. Accordingly their landscapes ranged from tundra to cool temperate forests of such trees as firs and beeches. But cool conditions favored large mammals including (in China for example) hyenas, giant beavers, red deer, and prehistoric species of rhinoceros and elephant.

Meanwhile, depressions forced south by ice sheets brought the subtropics more rain than they get now. But during glacials the tropics tended to be dry, their luxuriant rain forests shrinking into isolated "islands."

Deprived of the water locked up in vast ice sheets, the oceans shrank. The sea surface sank to at least 328ft (100m) below its present level, exposing land bridges that enabled humans to colonize the big South-east Asian islands.

In interglacials, some northern climates became warmer than they are today. Warmth-loving mammals such as hippo and Merck's rhinoceros

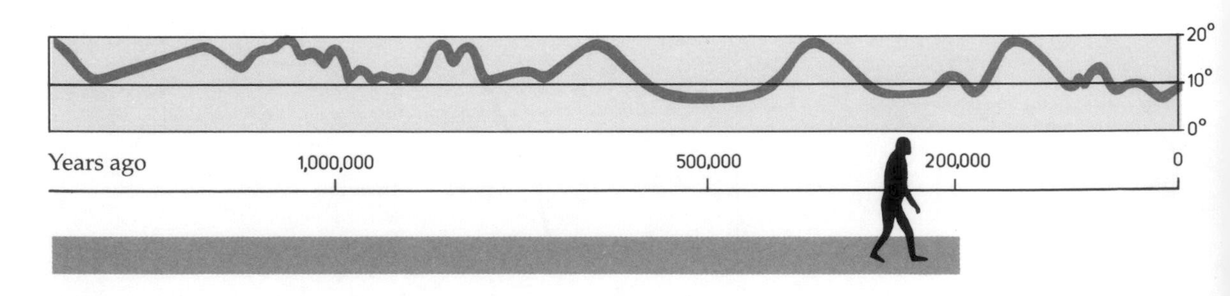

Climatic changes
A graph shows fluctuating July temperatures in degrees Centigrade for Central Europe for about the last 1,200,000 years. (Research now suggests that peaks and troughs occurred more frequently than this.) *Homo erectus* flourished throughout, perhaps until 200,000 years ago.

ranged north as far as southern England. At the same time the sea rose by up to 180ft (50m) above the level of today, isolating some offshore islands previously linked to land.

Any populations of *Homo erectus* isolated by climatic change were likely to evolve in slightly different ways, suited to conditions where they lived.

Ice-Age world
In glacial stages the Northern Hemisphere looked like this.
a Asia
b Europe
c North America

Extra land exposed in glacial stages

Sea

Glaciated areas

Summer extent of pack ice

Upright Man in Africa

Fossil finds point to Africa as the continent where complex feedback processes involving intensified use of hands, tools, and brain increased brain size in the *Homo* grade of man, to produce the bigger-brained, more intelligent, adaptable species *Homo erectus*.

The world's first known firmly dated *erectus* fossils come from the East Africa of about 1.6 million years ago. One skeleton represents by far the best preserved of any hominid predating deliberate burials, begun about 70,000 years ago.

Other fossils – mostly bits of skull or jaw – suggest that upright man eventually spread from East Africa to the farthest corners of the continent. But because so many fossils are so scrappy, and show no clear progressive evolution, and because the dividing line between this species and our own is blurred, African and European *erectus* specimens more recent than about 400,000 years ago are arguably *sapiens*.

Here are brief details of some of the major finds from Africa.

1 **Ternifine:** massive lower jaw with big teeth, found near two others and a skull bone. Age: perhaps 700,000 years. Place: Ternifine, Algeria.

2 **Koobi Fora:** cranium with heavy brow ridges, among the most complete and earliest *erectus* skulls discovered. Age: perhaps 1.6 million years. Place: Koobi Fora, east of Lake Turkana, Kenya.

3 **Swartkrans:** part of a lower jaw with five teeth, once called "Telanthropus" and later thought to be *Australopithecus* or *habilis*. Age: perhaps 1 million years. Place: Swartkrans, South Africa.

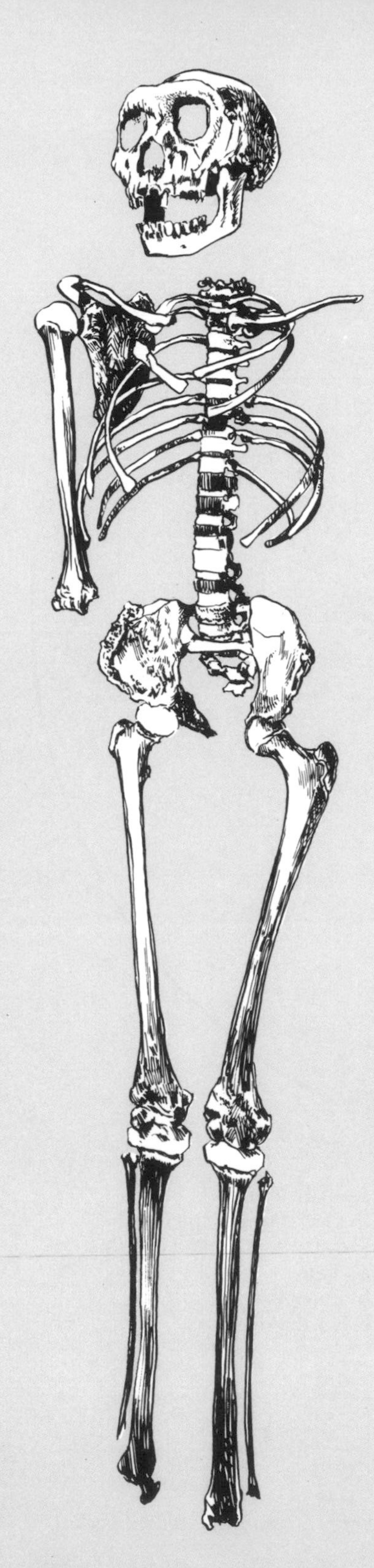

The oldest human skeleton
The 1.6-million-year-old skeleton shown left belonged to a *Homo erectus* boy. Aged under 13, he was already 5ft 4in (1.6m) tall, so might have grown to 6ft (1.8m) – taller than most modern men. Kenyan fossil hunter Kamoya Kimeu found these ancient bones in 1984, west of Lake Turkana.

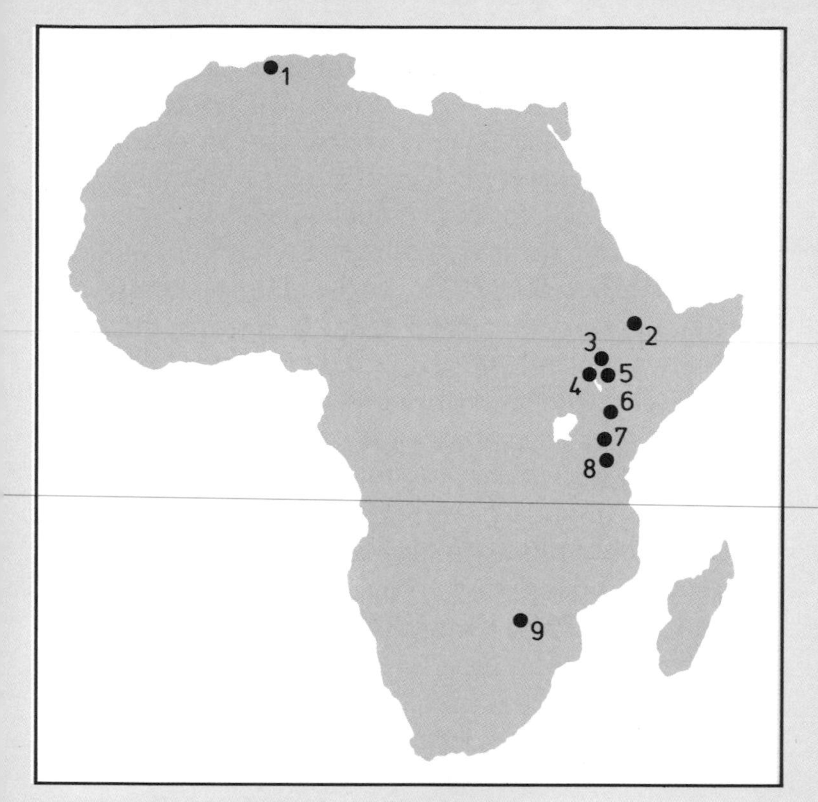

***Homo erectus* in Africa**
Major sites include:
1 Ternifine
2 Melka Kunture
3 Omo River
4 Nariokotome
5 Koobi Fora
6 Chesowanja
7 Olorgesailie
8 Olduvai Gorge
9 Swartkrans

Three fossil finds (below)
Numbers correspond to *Homo erectus* finds at African sites featured in the text.
1 Ternifine
2 Koobi Fora
3 Swartkrans

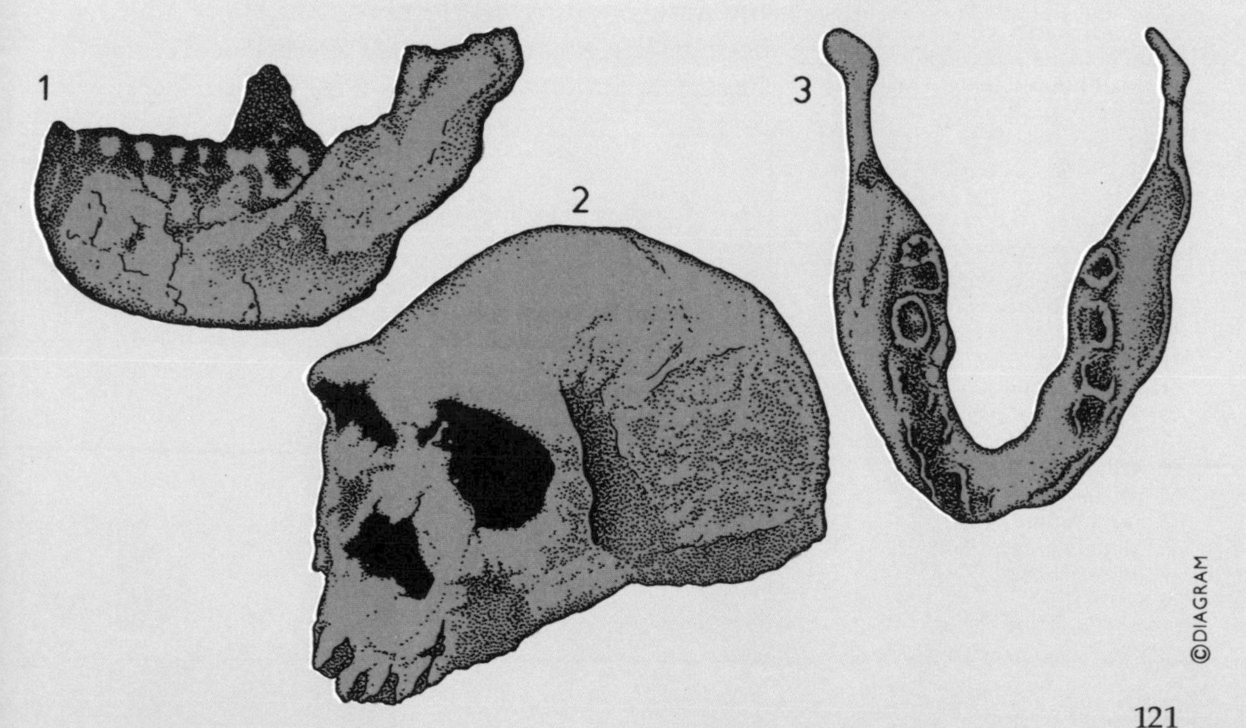

Upright Man in Europe

Old artefacts suggest that *Homo erectus* possibly reached Europe about 1.5 million years ago, but its supposed bones discovered there all seem to date from 500,000–200,000 years ago or less. Most are merely bits of jaw or skull. Confusingly, almost all show some features found in *Homo sapiens*. Some experts believe these fossils represent types transitional between both species. Perhaps their *erectus* ancestors invaded Europe in a warm phase when the ice sheets ebbed, but evolved toward our species while isolated from other human populations during phases of increasing Ice Age cold.

Deciding how to classify these early European humans is difficult. Certain experts lump all examples listed on these two pages with the archaic forms of modern man described on pp. 134–135.

1 **Heidelberg Man:** a massive, chinless lower jaw with teeth, designed to fit a broad, projecting face. Age: about 500,000 years. Place: Mauer, near Heidelberg, West Germany.

2 **Tautavel skull:** a skull with big brows, broad face and nasal opening, flat forehead, and long, narrow braincase. Age: about 400,000 years. Place: Arago Cave near Tautavel in South-west France.

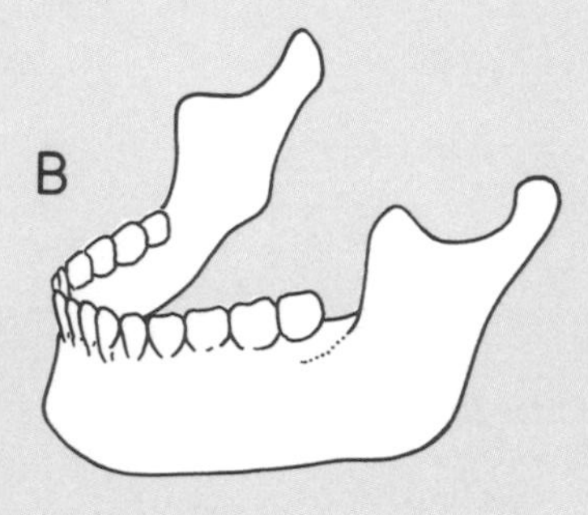

Jaw bones compared
A Mandible of Heidelberg Man. It is big and thick boned but structure and evenly proportioned teeth are largely similar to those of modern man.
B Mandible of modern man.

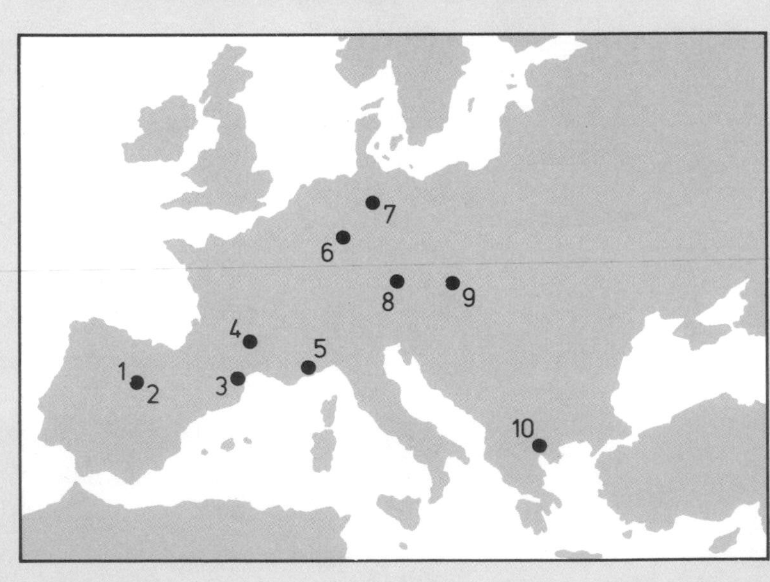

***Homo erectus* in Europe**
Selected sites show finds of bones or artefacts that have been attributed to *Homo erectus*. However, all possibly represent archaic *Homo sapiens*.
1 and **2** Ambrona and Torralba
3 Arago near Tautavel
4 Soleilhac
5 Terra Amata, Nice
6 Mauer near Heidelberg
7 Bilzingsleben
8 Prêzletice
9 Vértesszöllös
10 Petralona near Thessaloniki

3 **Vértesszöllös skull:** part of an occipital bone, from the back of the skull – thick, with a ridge for neck muscle attachment. Brain capacity might have been as great as ours. Age: about 400,000 years. Place: Vértesszöllös, west of Budapest, Hungary.

4 **Petralona skull:** a broad-based, broad-faced skull with beetling brows, sloping forehead, and angulated occipital bone but large cranial capacity – about 1230cc. Age: about 300,000 years. Place: Petralona, near Thessaloniki, Greece.

Four fossil finds
Numbers correspond to European fossils featured in the text.
1 Heidelberg Man (Mauer mandible)
2 Tautavel skull
3 Vértesszöllös skull
4 Petralona skull

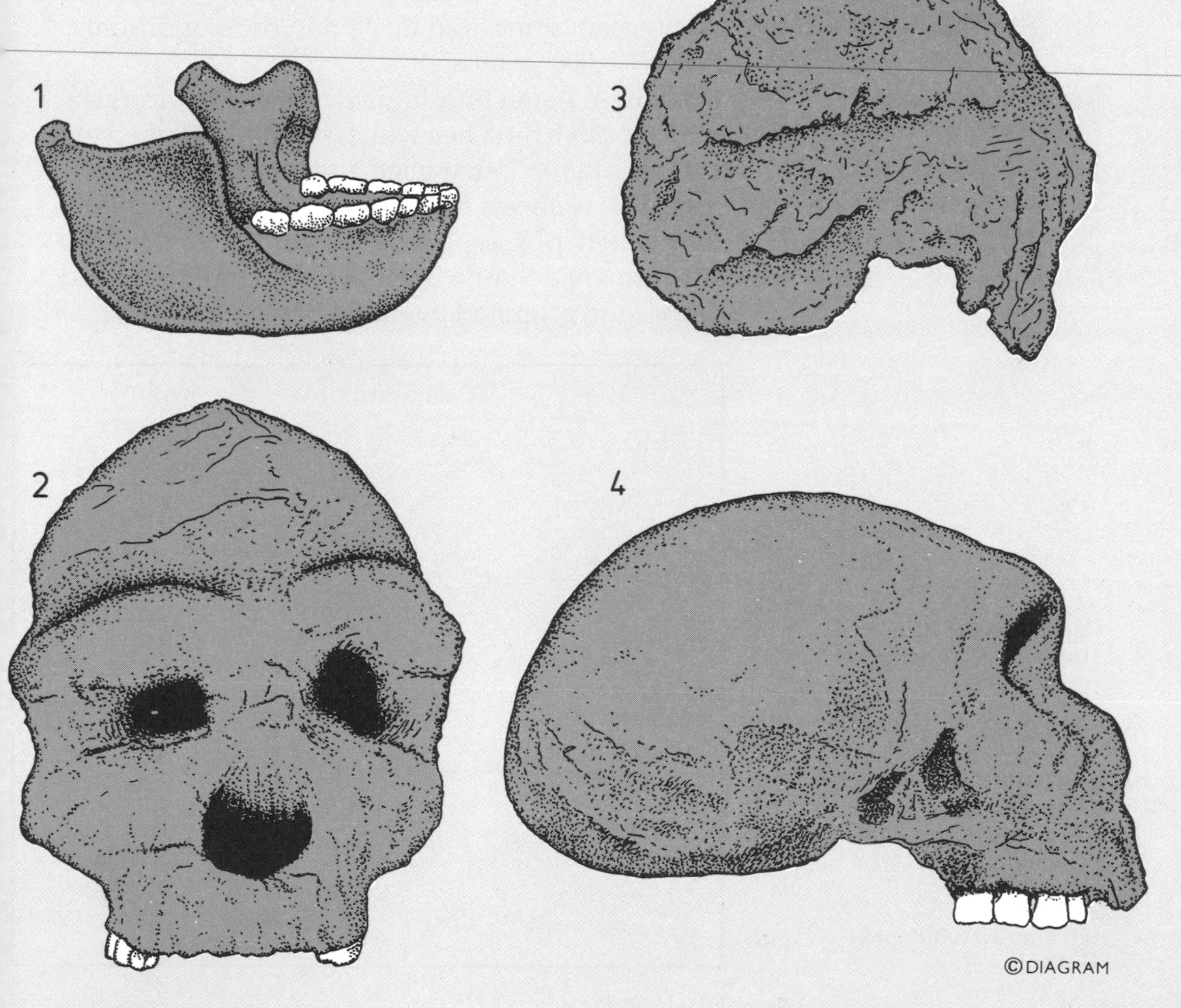

Upright Man in Asia

Most *Homo erectus* fossils come from Asia. Almost all were found in Java or China, with a possible *erectus* skull from India.

Very early specimens of "Pithecanthropus" from central Java's so-called Djetis Beds could date from more than 1.5 million years ago, while Java's Trinil Beds have yielded some possibly 700,000-year-old bones. China's best-known fossil human, "Sinanthropus," is known as Peking man, from the remains of over 40 individuals discovered near Peking; all disappeared in World War II, though casts survive. This Chinese form had a larger brain than older Asian forms, and thrived in cool conditions about 360,000 years ago.

All these Asian hominids lived near the fringes of the South China Sea which has been likened to a giant waterhole that filled and emptied as the northern ice sheets melted and advanced. In cold "low water" phases, *erectus* probably colonized the now drowned Sunda Shelf between Indonesia and China, and migrated overland between the two.

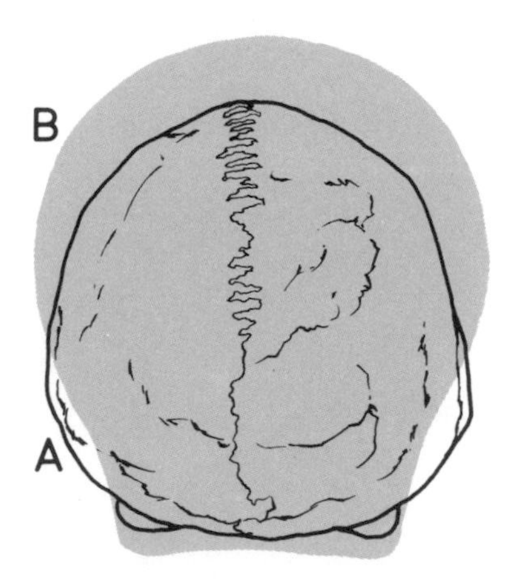

Back views compared
Above we compare back views of two skulls.
A Peking Man: skull broadest low down, but not as low as for *Australopithecus*
B Modern man: skull broadest high up

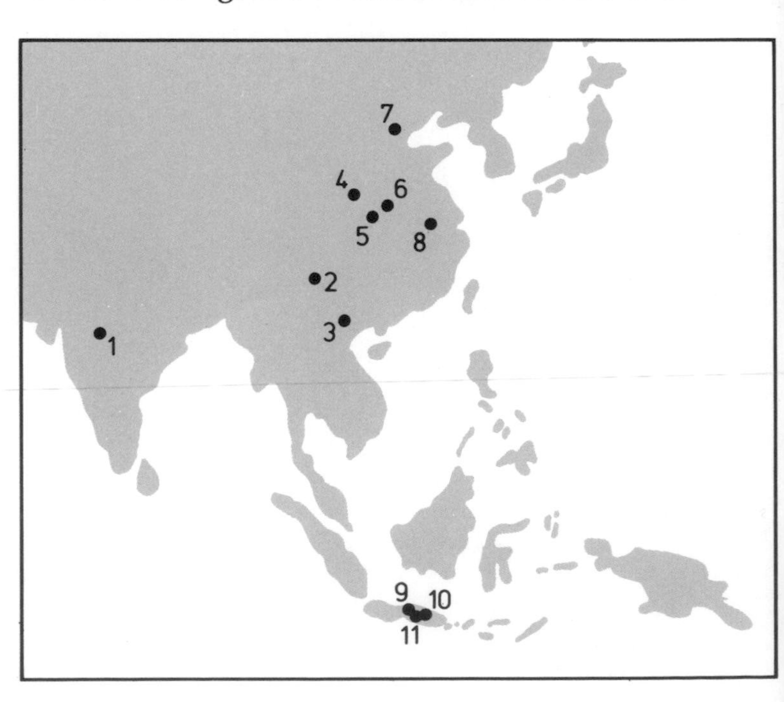

***Homo erectus* in Asia**
This map shows important selected sites.
1 Narmada
2 Yuanmou
3 Luc Yen
4 Lantian
5 Yunxi
6 Nanzhao
7 Beijing (Peking)
8 Hexian
9 Sangiran
10 Perning/Modjokerto
11 Trinil

Here are examples of specimens from Asia.

1 **"Pithecanthropus 4"** includes part of a big, thick skull, also a massive upper jaw with a gap between canine and incisor as if to take a large lower canine. Age: About 1 million years. Place: Sangiran, Java.

2 **Lantian skull:** a small, thick skull (cranial capacity 780cc), with strong, arched brow ridges. A separate chinless lower jaw at Lantian lacked third molars (a congenital condition seen in some people today). Age: about 600,000 years. Place: Lantian, Shensi province, China.

3 **"Sinanthropus"** had a low, wide skull (cranial capacity about 1075cc) but smaller teeth – without a gap – and a shorter jaw than older Asian forms. Age: 360,000 years. Place: Choukoutien, near Peking, China.

Three fossil finds
Numbers correspond to Asian fossils featured in the text.
1 "Pithecanthropus 4"
2 Lantian skull
3 "Sinanthropus"

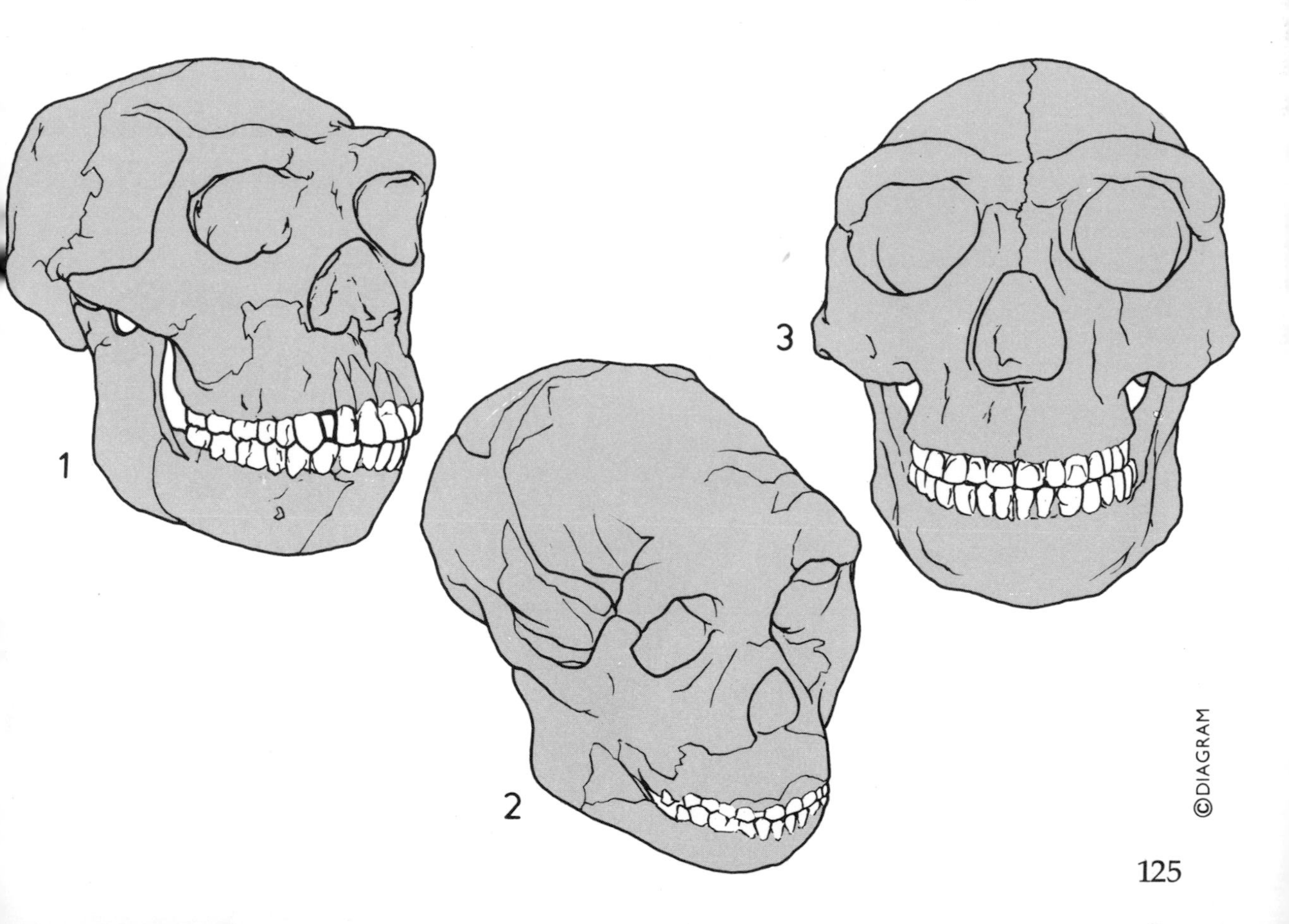

Hand-axes and choppers

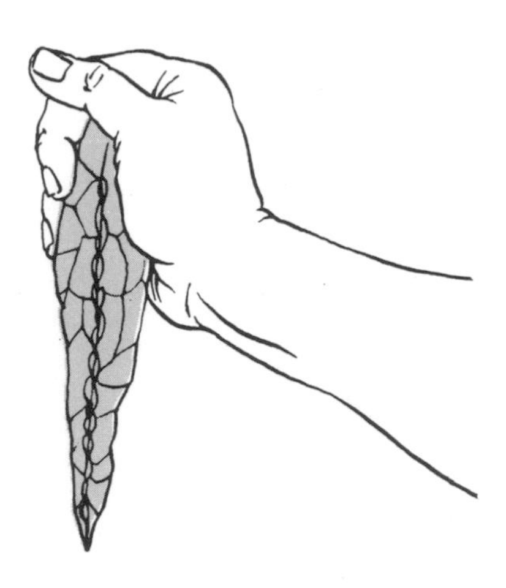

Hand-axe grip
The user could have gripped a hand-axe by its rounded butt, exerting pressure to cut meat or dig up edible roots.

About 1.6 million years ago a new, distinctive type of stone tool appeared in East Africa. This so-called hand-axe consisted of a fist-sized lump chipped into a shape resembling a hand or flattened pear, the sharpened edges formed by striking flakes from both sides. Experiments suggest this tool served largely as a butcher's knife to cut up carcasses already skinned by sharp stone flakes, some fashioned into long-edged cleavers.

The earliest hand-axes appeared about the same time as *Homo erectus*, and the foresight needed to produce such standard implements points to this advanced hominid as their probable inventor.

Old Stone Age toolkits featuring hand-axes, cleavers, scrapers, and flakes are called Acheulian, from 300,000-year-old finds at St. Acheul in northern France. From Africa, Acheulian toolmaking techniques reached India, and Europe where they persisted until about 100,000 years ago, but evidently never got to Indonesia or China.

Meanwhile cruder chopper-core cultures of the type called Oldowan at Olduvai in Tanzania spread to

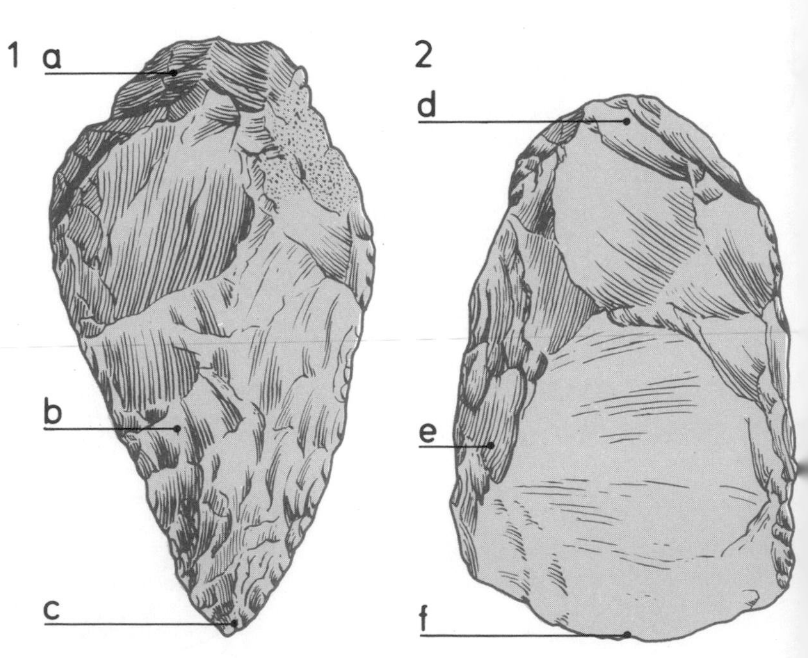

Acheulian tools
These Angolan examples are half actual size or less.
1 Hand-axe
a Butt
b Sharpened edge
c Point
2 Cleaver
d Butt
e Side
f Cutting edge

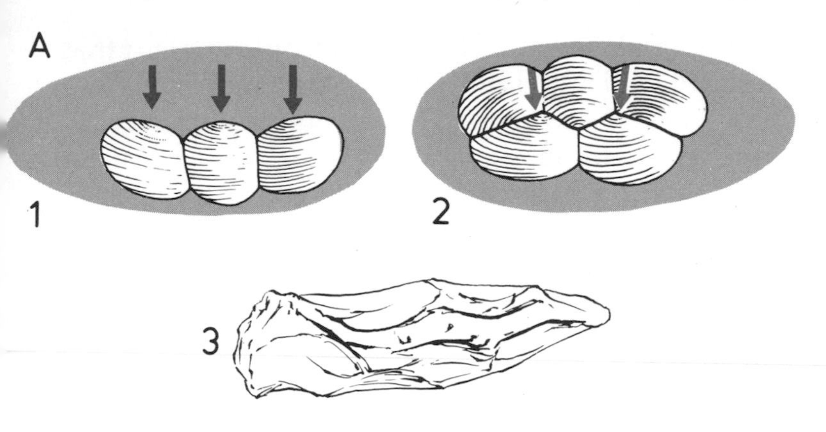

Primitive technique (A 1–3)
1 Blows with a hammerstone detach flakes from one side of a pebble, leaving deep, short, overlapping scars.
2 The pebble is turned over and struck again on ridges created by the scars. This forms another row of scars.
3 The result: A hand-axe with a strong, blunt, wavy cutting edge produced by deep scars meeting back to back.

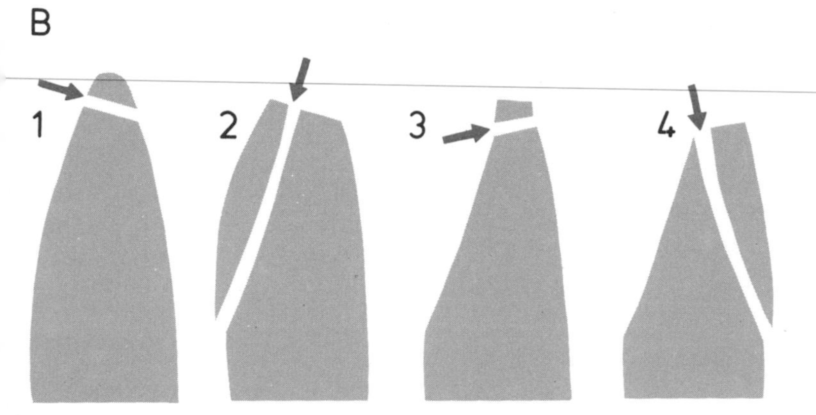

Advanced technique (B 1–4)
1 A blow removes part of a nodule's edge, leaving a flat "striking platform."
2 A blow removes a long shallow flake from one side of the nodule.
3 A blow prepares a new striking platform.
4 A blow removes a long shallow flake from the opposite side. The result will be a slimmer straighter cutting edge than that produced by the primitive technique.

Europe, and Asia from the Middle East to Java, the Philippines, and Choukoutien in northern China. Local versions include the Clactonian from Clacton, England (where its biconical stone cores, chopping tools, thick flakes, and notched flakes preceded an Acheulian industry), and the Tayacian from Tayac in the French Dordogne.

In places Acheulian and chopper-core techniques persisted side by side, and elsewhere which was used sometimes probably depended on locally available materials or the job in hand.

Traces of other *Homo erectus* artefacts include anvils and hammerstones; some of the first known borers, blades, and burins; and early evidence of bone and wooden tools – all from Ambrona or Torralba, Spain; also likely traces of a wooden bowl, from Nice in France.

Using a baton (C)
Blows described in (**B**) were probably delivered by a soft springy bone, horn, or wooden baton. This could hit a nodule near its edge without crushing, to remove long, thin, shallow flakes. This advanced technique shaped many hand-axes already roughed out by a hammerstone.

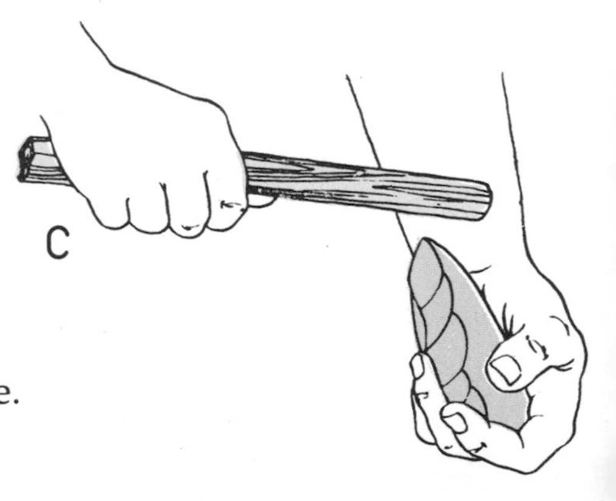

Hunting

A number of *Homo erectus* sites strongly suggest that these enterprising hominids were not just plant gatherers and scavengers, but active hunters of big game who combined in groups to plan and execute a chase or ambush. Finds from three continents give clues to hunting methods or the animals attacked. The three following examples all probably date from roughly 400,000 years ago.

Olorgesailie, Kenya, held one site with the remains of 50 *Simopithecus*. Perhaps early men had clubbed to death an entire sleeping troop of this large and now extinct baboon, as some Tanzanian tribesmen still kill its modern counterparts.

At Torralba, in North-central Spain, hunters seemingly used fire to drive dozens of migrating elephants, wild cattle, horses, deer, and rhinoceroses into a natural trap – a boggy gully in a steep-sided valley. Here died at least 30 elephants of an extinct straight-tusked species larger than the African, alive today. Many beasts may have been butchered at Torralba and nearby Ambrona.

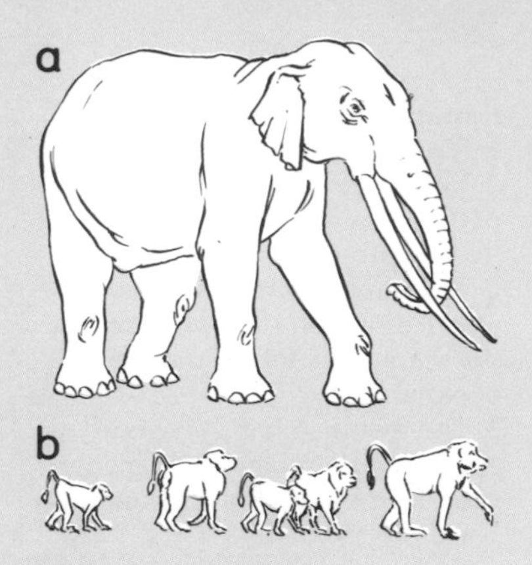

Hunters' prey (above)
a *Elephas antiquus*, a now extinct elephant, trapped and killed in South-west Europe.
b *Simopithecus*, a now extinct baboon, killed in East Africa.

An ancient feast (below)
This plan of an excavation at Ambrona, Spain reveals:
a Bones of fossil elephants and other animals
b Stone tools and waste flakes
c Burnt wood
d Possible hearth stones

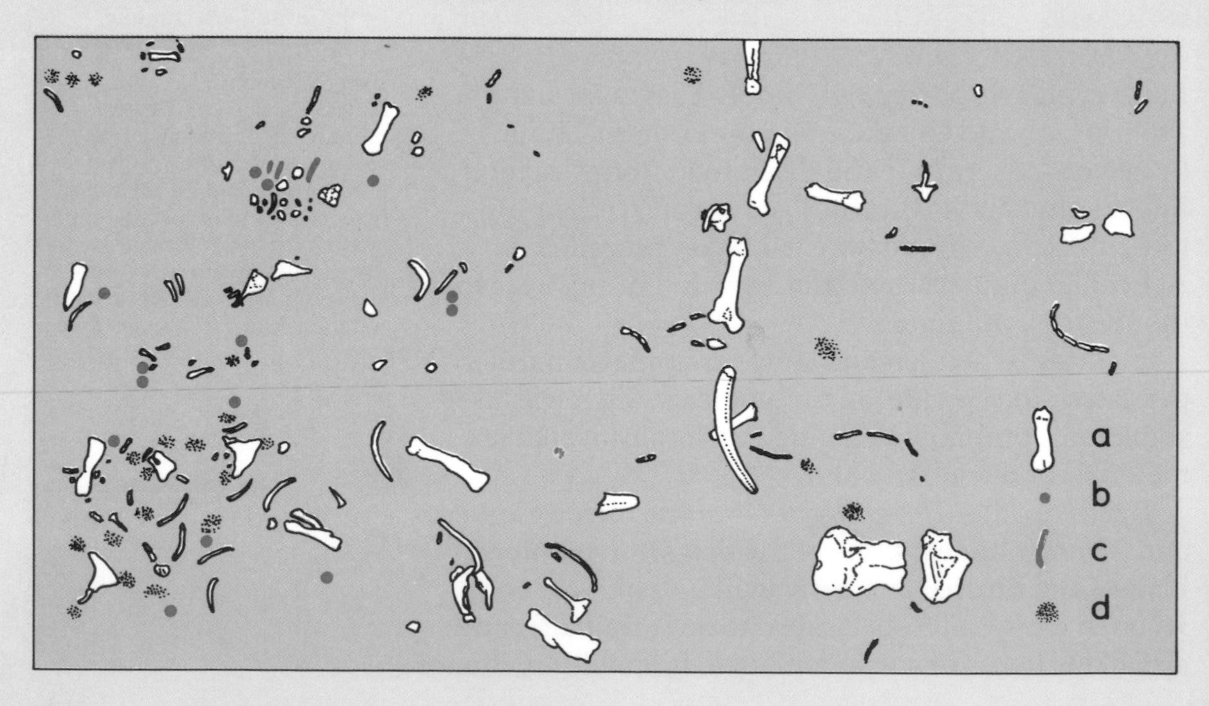

The most spectacular Asian evidence comes from Choukoutien, near Peking. Here, cave deposits hint that *Homo erectus* killed and ate animals including boar, bison, deer, gazelle, horse, and rhinoceros. Smashed human limb bones and human skulls all broken at the base suggest these hunters had been cannibals, who also ate their own species' brains and marrow.

Just how *erectus* may have killed large prey remains uncertain. Some evidence suggests the use of wooden spears and stone projectile points. Whatever methods served they would be hazardous, which might explain why most *erectus* skulls show signs of old, healed fractures.

Hunters in action (below)
Here are (numbered) four tools from Spain and (lettered) their likely uses.
1 Wooden spear
a Spearing a big meaty mammal
2 Denticulate: a stone tool with a notched, serrated edge.
b Sharpening a spear point
3 Quartzite cleaver, actual length 10in (25cm)
c Cutting up a large mammal
4 Double-edged side scraper made of jasper
d Scraping fat and flesh from a hide to clean it

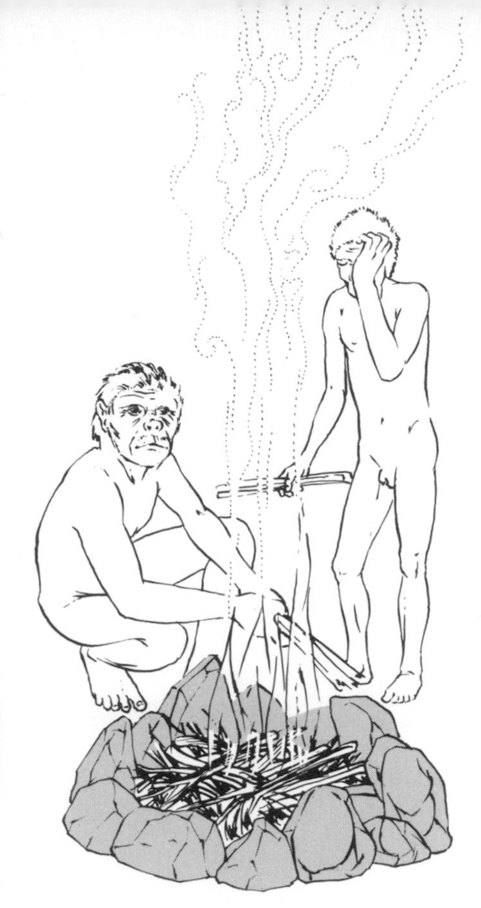

Ancient hearth
Stone windshields helped hunter-gatherers control open fires.

Homes and hearths

Collections of bones and stone tools show where *Homo erectus* family groups or larger bands made their camps. Most would have been home for just a few days while people planned hunts; butchered and shared out meat and edible plants; drank from a nearby spring, stream, or lake; refurbished their toolkits of stone, wood, and bone; and rested and slept. In the warm tropics, dry ground often must have sufficed. But *Homo erectus* also made shelters, best known from cool northern regions. Stone rings at Torralba and Ambrona in Spain recall weights used to hold down hides hung from the central pole of an Eskimo tepee. At Terra Amata in the French city of Nice, 90sq m (145sq km) of old living floors show where groups may have raised oval huts of interlocked branches braced by stones. Inside, fires burned in hearths sheltered by windshields of stones. The largest huts on this seasonal seaside site could have held 20 people.

Near Tautavel in the French Pyrenees, and at Choukoutien near Peking, hunters inhabited caves. They seemingly came to the French Cave of Arago on a seasonal basis, following migrating game. But at Choukoutien an alleged layer of ashes 19.7ft (6m) deep might imply prolonged habitation.

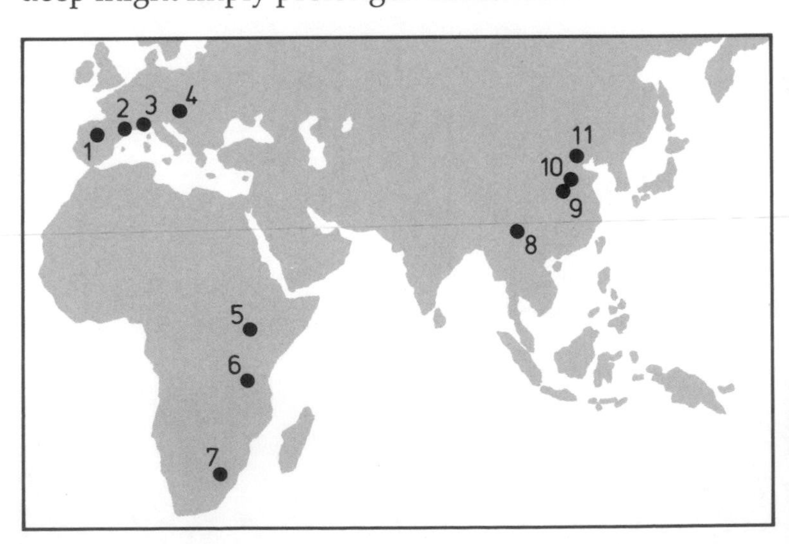

Fire and Early Man
A map shows sites where humans evidently made fires more than 100,000 years ago. One expert questions the evidence for fire at site 11 but 5, 6, 8, and 9 may date back a million years old or more.

1 Torralba
2 Escale
3 Terra Amata
4 Vértesszöllös
5 Chesowanja
6 Kalambo Falls
7 Cave of Hearths
8 Yuanmou
9 Xihoudu
10 Lantian
11 Choukoutien

A base in Spain
An excavated living floor at Torralba in North-central Spain reveals bones of big-game animals, stone tools used to butcher them, and other signs that hunters feasted here perhaps 400,000 years ago.

Homo erectus might not have discovered fire–charred ground 2.5 million years old is known from near Lake Turkana in Kenya. Also he (or she) perhaps only kept alight fires already started by lightning or volcanic eruptions. But this hominid was almost certainly the first to exploit fire systematically: for warmth, cooking, protection from predators, and as a weapon to scare and drive game.

In the Ice Age, building, burning, and a high-protein diet (plus perhaps the wearing of furs) opened up even cool northern lands to human colonization, while cooking made available once indigestible plant foods. For mankind such advances marked a significant switch in emphasis – cultural evolution now began to loom larger than biological change.

A Riviera home
Oval huts of interlocking branches may have housed hunters on France's Mediterranean shore. Such flimsy shelters vanished long ago, but archeologists can reconstruct them from their stones and postholes.

Chapter 7 NEANDERTAL MAN

By 300,000 years ago some archaic forms of *Homo sapiens* began assuming the rugged features of Neandertal (or Neanderthal) Man, best known from 70,000 to 40,000-year-old bones and artefacts.

This mainly European subspecies proved an innovative hunter-gatherer of the Middle Old Stone Age – able to endure quite cold climatic phases. Deliberate burials – the oldest known – hint at emergent human sensitivity. Yet Neandertals were probably a side branch off the evolutionary path to fully modern man. By 30,000 years ago that branch apparently became extinct.

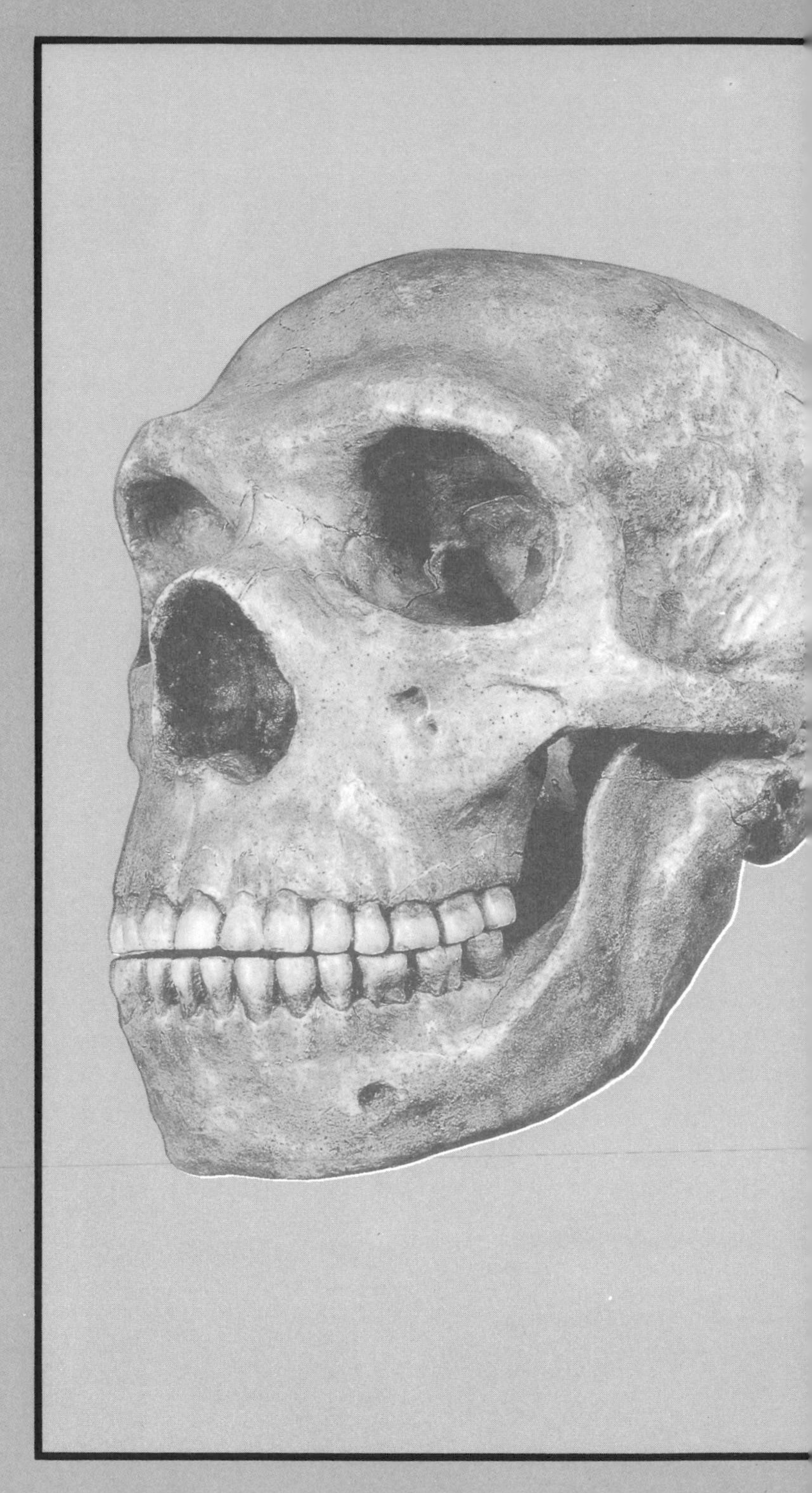

Neandertal skull (left) and reconstructed Neandertal head (right) in the British Museum (Natural History), London.

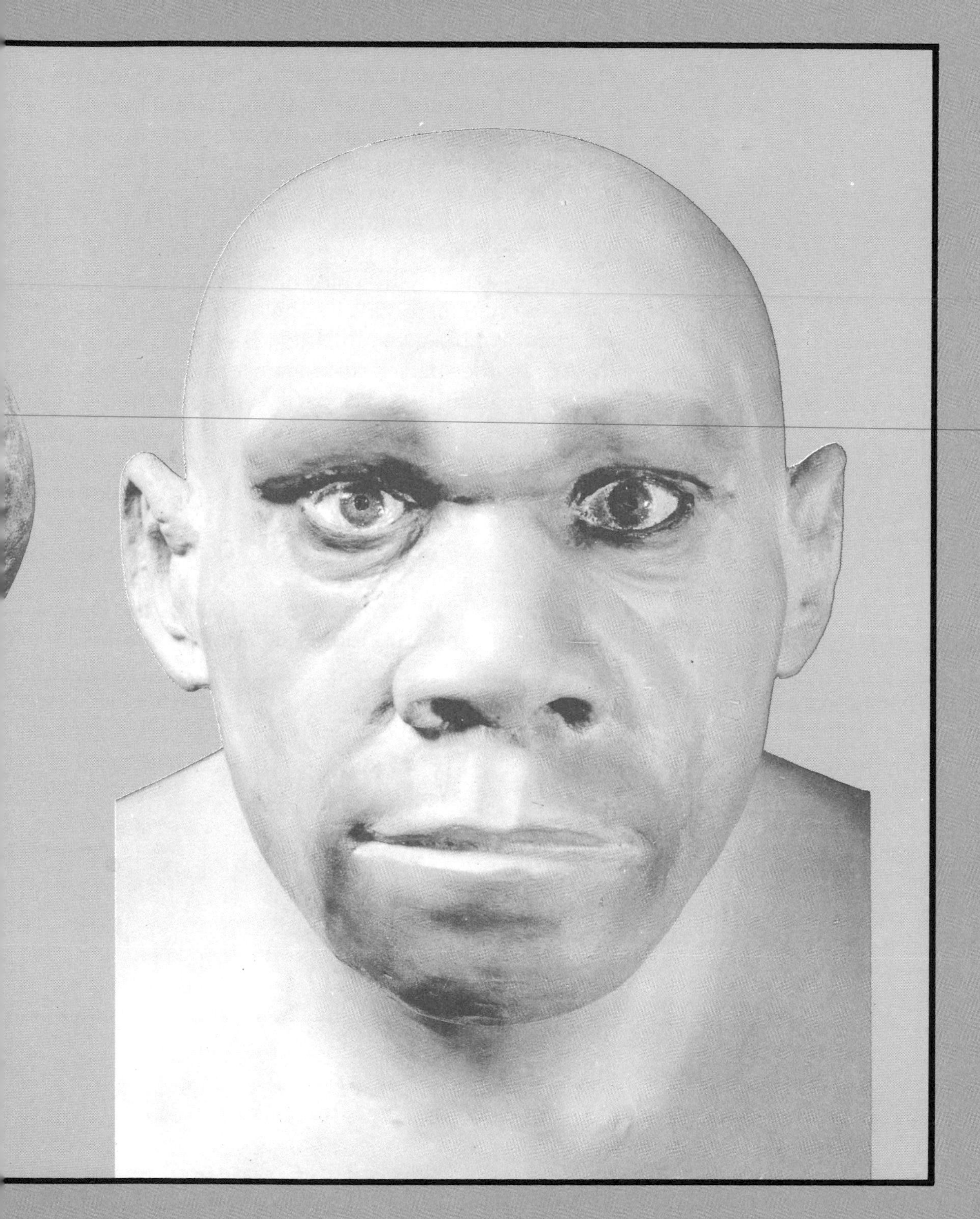

Trends in human evolution

By 300,000 years ago *Homo erectus* was evolving toward fully modern man, whose features figure on p. 156. But evolution advanced at different rates in different places – influenced by factors including mutations, increasing use of tools for heavy work once done by teeth or muscles, migration, and the isolation of some populations. Even in different parts of the body evolution progressed unevenly, producing skulls with a mosaic of modern and old-fashioned features. Despite these variations, paleoanthropologists rank most people from the later Ice Age in one species, as archaic forms of *Homo sapiens*, including its subspecies offshoot, the Neandertals described on pp. 138–139. Here, we give examples of archaic *sapiens*, including some so-called pre-Neandertals. (Elsewhere, most of our European *erectus* examples are also arguably *sapiens*.)

1 **Swanscombe Man:** parts of a thick female skull with an "advanced" rounded back and large brain capacity (1300cc). Age: perhaps 250,000 years. Place: Swanscombe, near London, England.

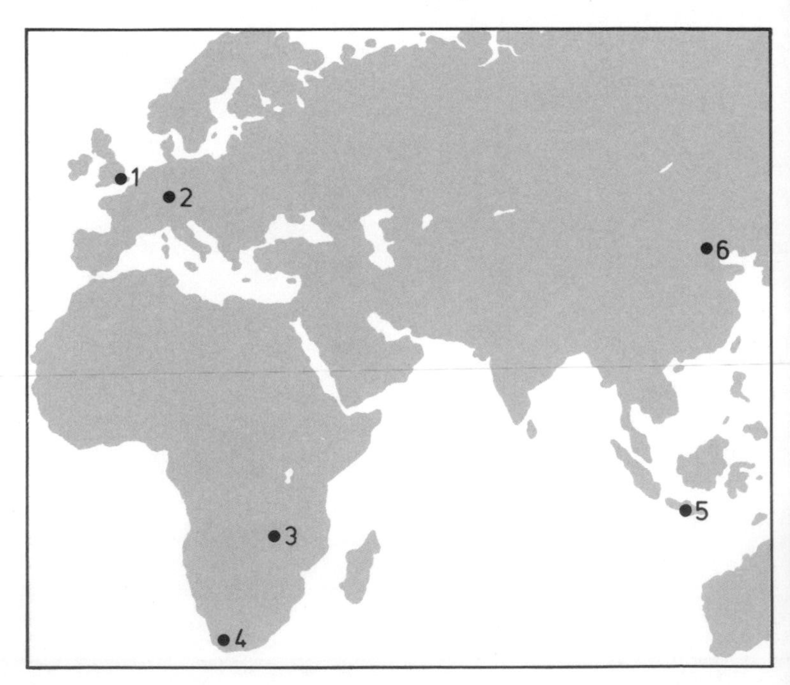

Where they lived
A world map shows approximate locations of six skulls of archaic *Homo sapiens*. Numbers correspond to items featured in the text.
1 Swanscombe, England
2 Steinheim, West Germany
3 Kabwe (Broken Hill), Zambia
4 Saldanha Bay, South Africa
5 Ngandong, Java
6 Hsuchiayao (Xujiayao) skull

2 **Steinheim Man:** low skull with brow ridges and brain capacity of 1100cc but with a rounded back, relatively small, straight face, and small teeth. Age: more than 300,000 years. Place: Steinheim, near Stuttgart, West Germany.

3 **Broken Hill Man:** (alias Rhodesian man or Kabwe skull): skull with sloping forehead, strong brow ridges, and angulated rear but with steep "modern" sides, and base, and brain capacity of 1300cc. Age: maybe 200,000 years. Place: Kabwe, Zambia.

4 **Saldanha skull:** similar to Broken Hill man. Age: at least 200,000 years. Place: Hopefield, Saldanha Bay, South Africa.

5 **Solo Man:** thick skullcaps with sloping forehead, flattened frontal bone, ridged back and brain capacity of 1035–1255cc; they recall much earlier Peking man. Age: perhaps more than 100,000 years. Place: Ngandong, Solo River, Java.

6 **Hsuchiayao Man:** one of a group of heavy Chinese skulls like *Homo erectus* but rounded at the back. Age: 250,000–100,000 for the group. Place: Datong, Shensi province, China.

Six early *"Homo sapiens"*
Numbers of these complete and partial skulls correspond to items featured in the text.
1 Swanscombe man (rear view of partial skull)
2 Steinheim man
3 Broken Hill man
4 Saldanha skull
5 Solo man
6 Hsuchiayao (Xujiyao) skull (occipital bone – from the back of the skull)

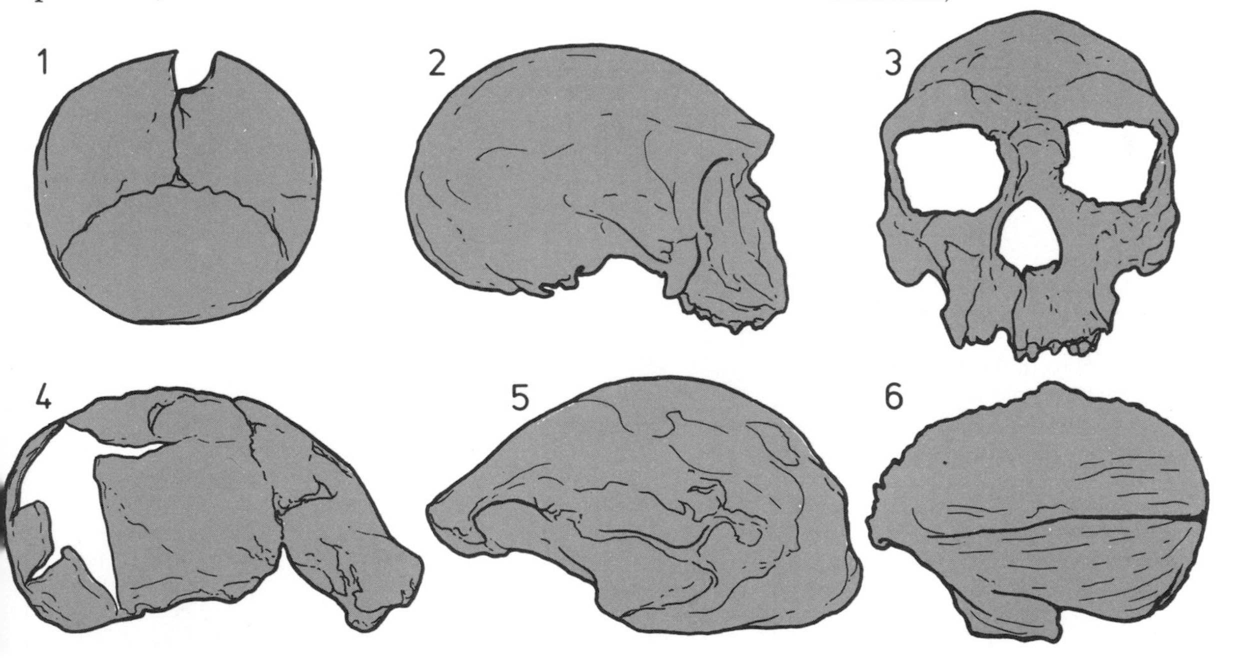

The changing world

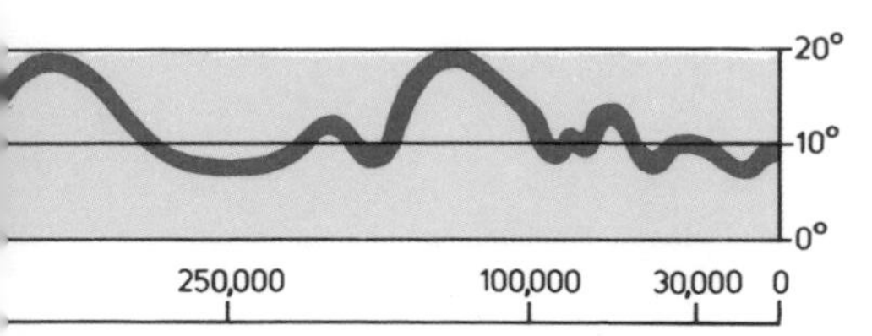

Climatic changes
A graph shows fluctuating July temperatures in degrees Centigrade for Central Europe for the last 250,000 years or so. (In fact peaks and troughs were probably more frequent than this.) During this time archaic *Homo sapiens* (indicated by the light-toned bar) gave rise to *Homo sapiens neanderthalensis* (the solid bar) who died out by 30,000 years ago.

Waves of cold with warmer intervening gaps continued dominating northern climates from 200,000 to 40,000 years ago – a time span embracing the Neandertals and their immediate precursors. Geologists traditionally divide this part of the Pleistocene epoch between two glacial stages (in Europe often called the Riss and Würm) separated by an interglacial (the Eemian); in fact research reveals more glacials and interglacials, each with climatic changes affecting the level of the sea.

At its most intense, the cold drove mammals south in North America and trapped some European species between advancing northern ice sheets and Alpine glaciers. In western and central Europe, temperate woodlands gave way to steppe or tundra. Here only cold-adapted beasts survived – including new species like the woolly mammoth, woolly rhinoceros, and musk ox. When ice sheets shrank, all moved north with the retreating steppe and tundra. Woodland spread in from the south with hippopotamus, straight-tusked elephant, giant elk, lion, and leopard. As cold returned, migration went into reverse.

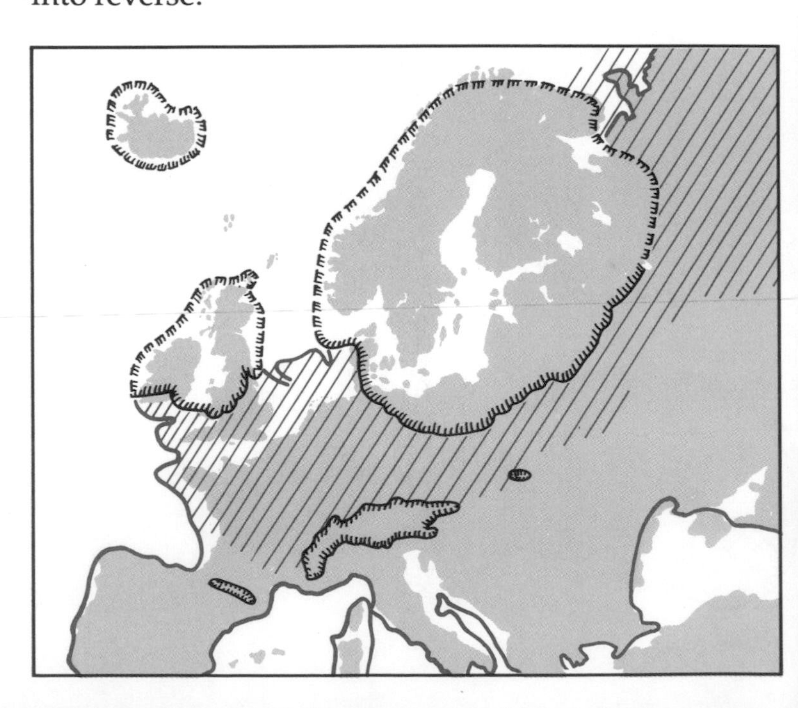

Ice-Age Europe (right)
This map shows harsh climatic conditions faced by Europe's Neandertals in the Würm glacial stage. The map includes both old and modern coastlines.

- Ice sheets and glaciers at their maximum
- Permafrost (ground with permanently frozen subsoil)
- Old coastline

Here we show four cold-tolerant mammals.

1 **Mammuthus primigenius,** the woolly mammoth, had long dark hair, woolly underfur, small ears, and huge curved tusks. Height: 9ft 6in (2.9m). Time: Mid-Late Pleistocene. Place: northern Eurasia and North America.

2 **Coelodonta antiquitatis,** the woolly rhinoceros, had a shaggy coat and two horns, the front one long. Height: 6ft 6in (2m). Time: Mid-Late Pleistocene. Place: Eurasia and North Africa.

3 **Ursus spelaeus,** the cave bear, had a great head with mighty jaws, yet was an omnivore. There were huge but also dwarf varieties. Length: up to 9ft (2.7m) from nose to tail. Time: Mid-Late Pleistocene. Place: Eurasia.

4 **Ovibos moschatus,** the musk-ox, has a large, horned head and stocky body with underwool and an outer coat of long, dark, almost ground-length hair. Height: up to 5ft (1.5m). Time: Mid Pleistocene to today. Place: northern Eurasia and North America.

Four Ice-Age mammals
Numbers correspond to items featured in the text.
1 *Mammuthus primigenius*
2 *Coelodonta antiquitatis*
3 *Ursus spelaeus*
4 *Ovibos moschatus*

About Neandertal Man

Body Build (below)
Muscular Neandertal man is shown beside a modern man for comparison.
Height: about 5ft 7in (1.7m)
Weight: about 154lb (70kg)

Homo sapiens neanderthalensis takes its name from fossils found in the Neander Valley, near Düsseldorf, West Germany. The so-called classic Neandertals from Europe had a large, long head, with a bigger brain inside a thicker skull than ours, yet the skull was thinner than that of *Homo erectus*. Somewhat like that species, Neandertals had heavy brow ridges and a sloping forehead. There was a distinctive bun-shaped swelling at the back of the skull with a big area below for tethering neck muscles. The broad face projected far forward, its backswept sides creating "streamlined" cheekbones. The large nose might have been either flat or bulbous. The powerful chinless jaw held larger front teeth than ours, and molars tended to contain big pulp cavities.

Classic Neandertals were short, extremely muscular, and stocky, with large joints and hands. Proportions recall the Eskimo, whose compact body helps conserve heat in a cold environment. But individuals and populations varied, as the following pages show.

The Neandertal subspecies evolved from an archaic form of *Homo sapiens* perhaps as much as 200,000 years ago. Their physique and improved technology made some of these Middle Paleolithic people probably the first hominids able to endure the rigors of winter in a cold climate. Also, Neandertal rituals seem to show a new high level of sensibility and human self awareness.

Rival family trees (right)
1 Neandertals as an extinct species
2 Neandertals as an extinct subspecies
3 Neandertals as part of the rootstock of modern man
a Early *Homo sapiens*
b Neandertals
c *Homo sapiens sapiens*

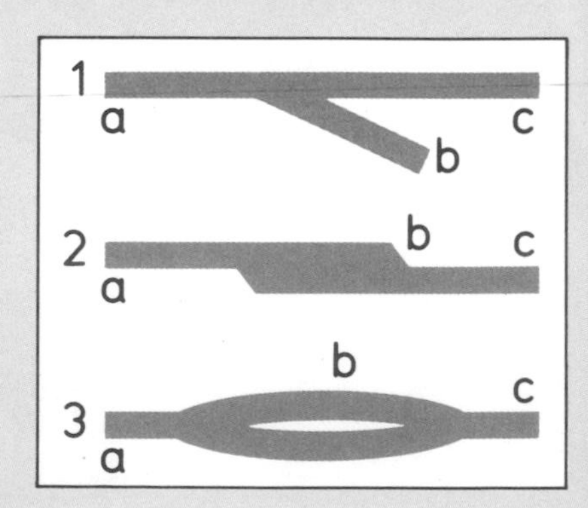

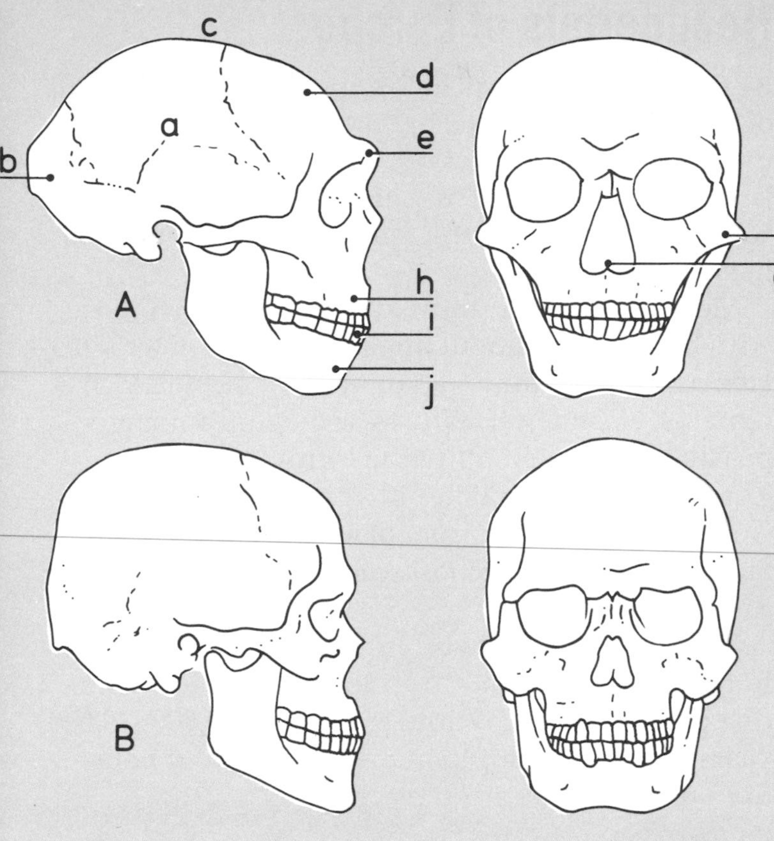

Skulls compared (left)
A Two views of a Neandertal skull show these features:
a Large cranial capacity
b Bun-like swelling
c Long, low cranium
d Sloping forehead
e Brow ridges
f "Streamlined" cheekbones
g Broad, long nasal opening
h Projecting midface
i Big teeth (but smaller than those of *Homo erectus*)
j Stout, usually chinless jaw
B Two views of a modern human skull for comparison

Yet about 30,000 years ago this group apparently died out. Some scientists suppose that the Neandertals were wiped out by emergent fully modern man who had evolved elsewhere. Rival explanations are that Neandertals interbred with or themselves evolved into our own subspecies. If so, their genes almost certainly survive in people now alive.

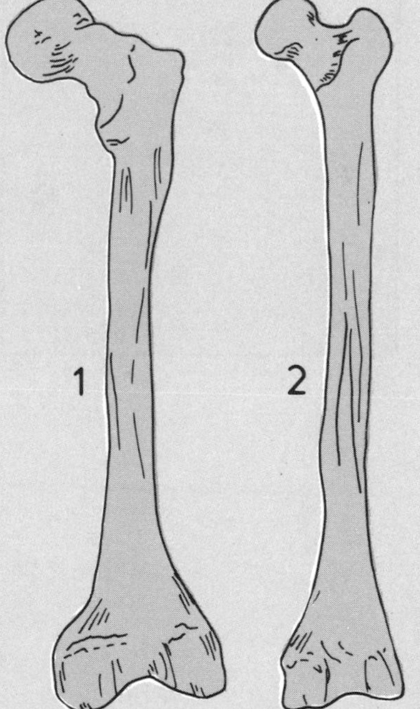

Femurs compared (right)
1 Neandertal femur: thick, strong, and noticeably curved
2 Modern human femur: slimmer, weaker, and straighter than 1.

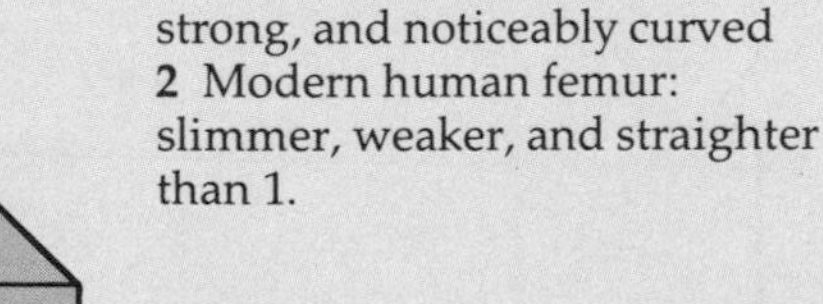

Brain capacities (left)
a Neandertal man: 1500cc
b Modern man: 1400cc

Neandertals of Europe

Where they lived
Selected European Neandertal and "proto-Neandertal" sites appear in maps below right and (an enlarged area) below.

1 Gibraltar
2 Pontnewydd
3 Spy
4 Neander Valley
5 Saccopastore
6 Monte Circeo
7 Krapina
8 Kůlna
9 Subalyuk
10 Staroselye
11 Kiik-Koba
12 Dzhruchula
13 Urtiaga
14 Mas d'Azil
15 Bañolas
16 Hortus
17 La Chapelle-aux-Saints
18 Le Moustier
19 La Ferrassie
20 La Quina
21 La Chaise
22 Fontéchevade
23 St.-Césaire

By the 1980s, scientists in Europe had found remains of about 200 Neandertal or "proto-Neandertal" individuals – most in caves. They include the oldest large samples of fossil man from any continent. Some show that even cohabiting groups varied considerably in skull and jaw design.

More than half the individuals came from France, with bits of 116 at about three dozen sites. Two sites – Hortus and La Quina – accounted for most French remains. Neandertal features also figure among possibly two dozen individuals from Krapina, Yugoslavia; 11 from Italy; 10 from Belgium; 8 from Germany; and others from places including Britain, Spain, Gibraltar, Czechoslovakia and Crimean Russia.

Finds date from maybe 250,000 to 30,000 years old but most "full-blown" Neandertals come from the first part of the last (Würm) glaciation: 70,000–30,000 years ago. Our examples represent different finds from different times and places.

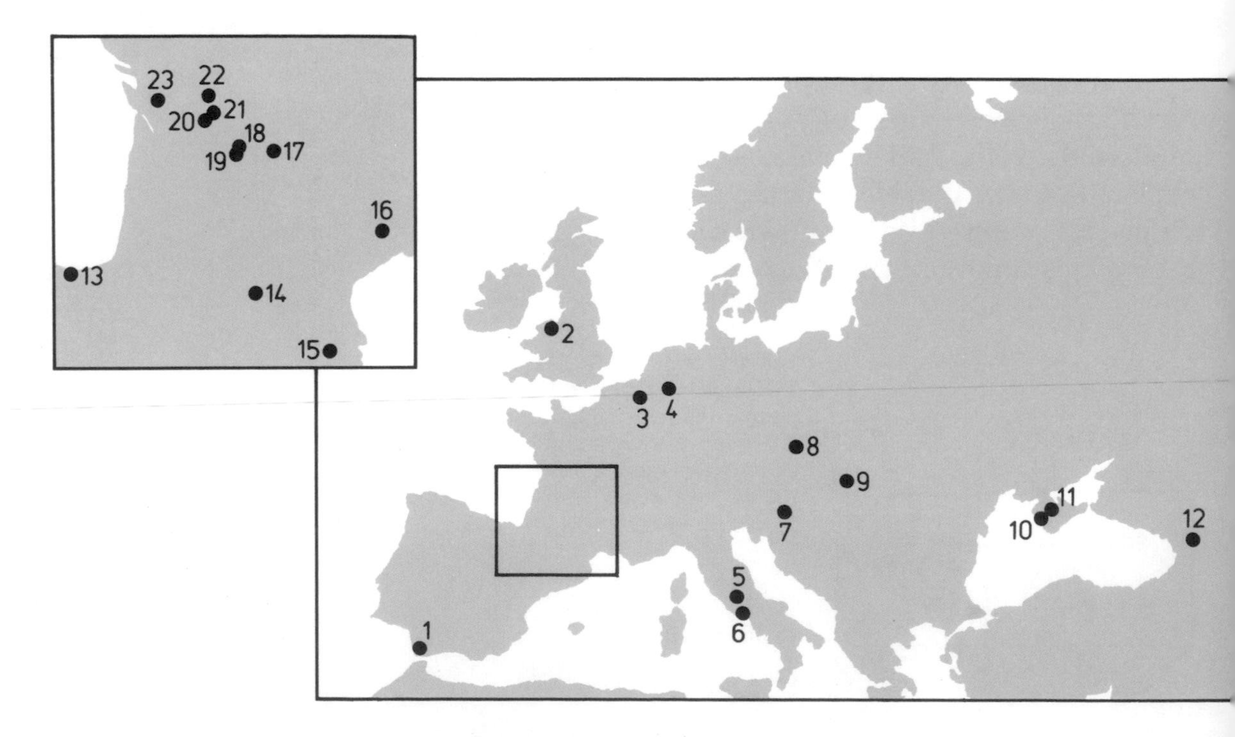

1 **Ehringsdorf skull** Remains include a Neandertal-type skull but with a high forehead, also a chinless jaw yet with small teeth. Age: perhaps 200,000 years. Place: Ehringsdorf, East Germany.
2 **Pontnewydd** Tooth and jaw remains discovered in the 1980s include features found among Neandertals. Age: about 250,000 years. Place: Pontnewydd Cave, near Rhyl, North Wales.
3 **Fontéchevade** This site produced skull fragments with and without brow ridges. Age: perhaps 150,000 years. Place: Fontéchevade, western France.
4 **La Chapelle-aux-Saints** is famous for the arthritic skeleton of an old male classic Neandertal. Age: perhaps 50,000 years. Place: La Chapelle-aux-Saints, south-central France.
5 **Gibraltar** This skull's discovery in fact predated that from the Neander Valley. Age: 50,000 years.
6 **Circeo** is noted for a man's mutilated skull. Age: about 45,000 years. Place: Monte Circeo, near Rome, Italy.
7 **Neander** The first-described Neandertal skeleton came from a cave beside this river. Age: maybe 50,000 years. Place: Feldhofer Cave, Neander Tal ("Neander Valley"), Düsseldorf, West Germany.
8 **Krapina** Skulls included broad, short forms with strong brow ridges, perhaps victims of a cannibal feast. Age: about 100,000 years. Place: Krapina, north Yugoslavia.

Fossil finds
Numbers correspond to items in the text.
1 Ehringsdorf
2 Pontnewydd
3 Fontéchevade
4 La Chapelle-aux-Saints
5 Gibraltar
6 Circeo
7 Neander Valley
8 Krapina

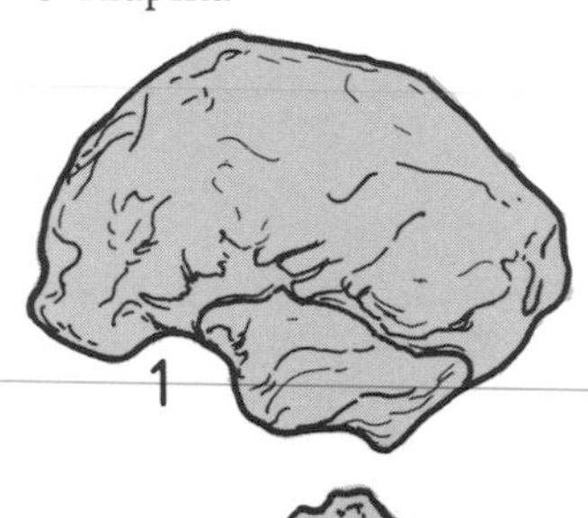

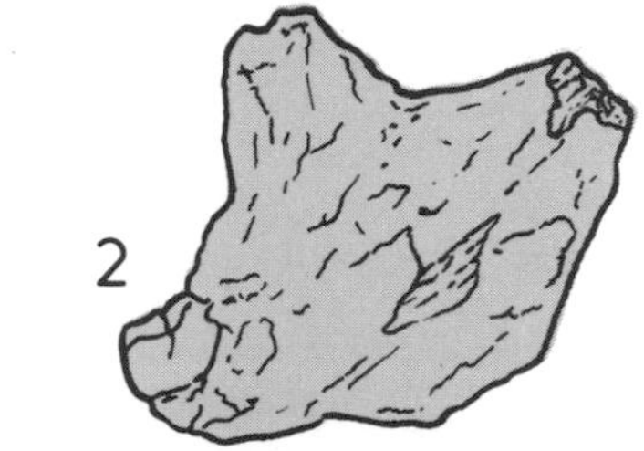

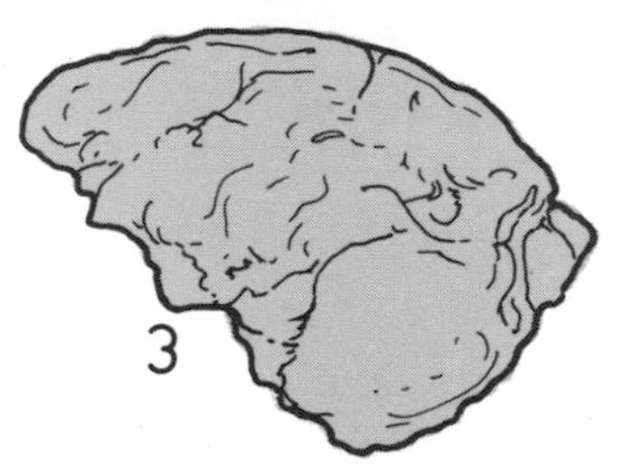

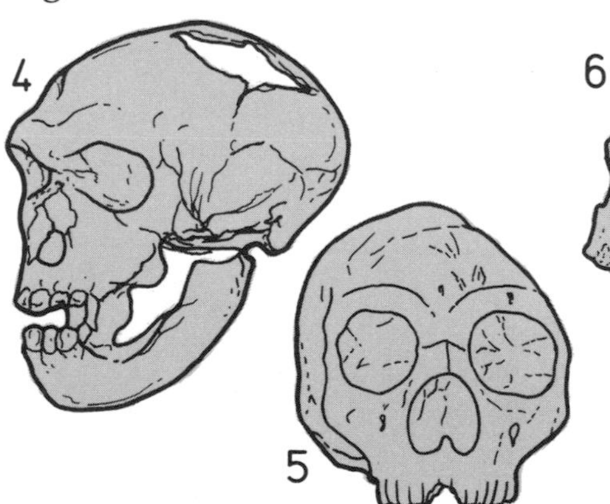

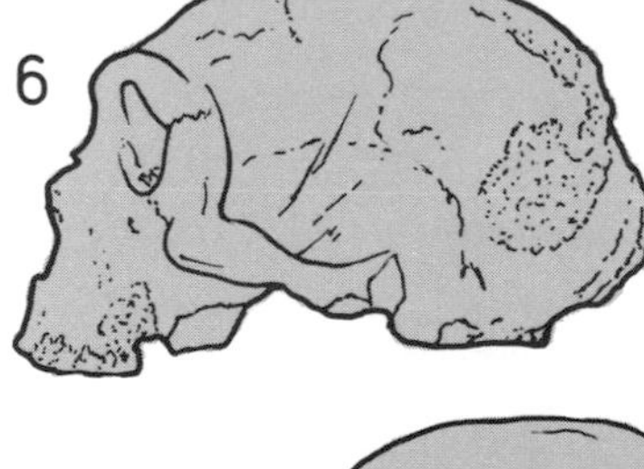

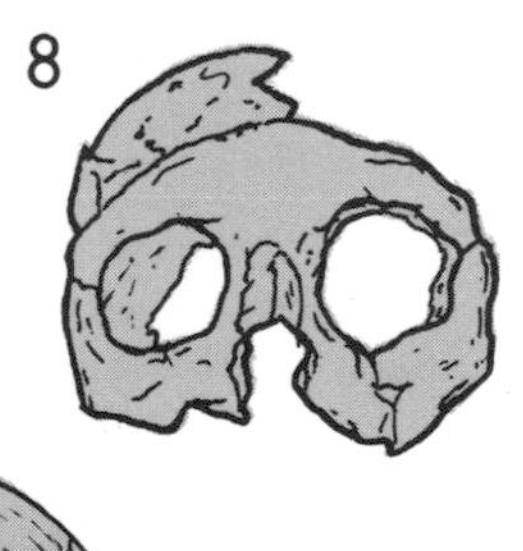

7

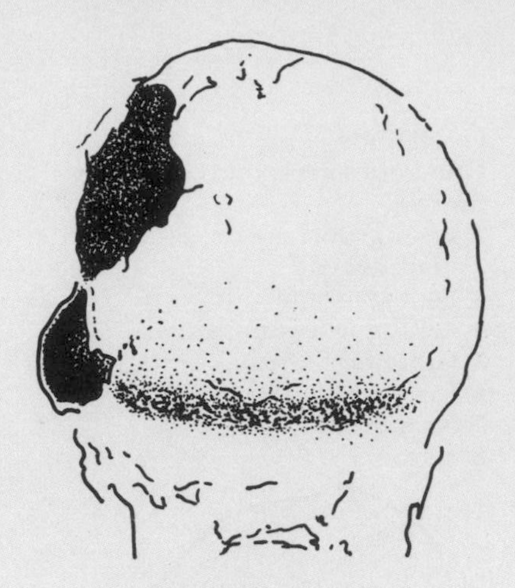

Neandertal skull
A rear view of Shanidar 1 shows a barrel-shaped Neandertal skull, broadest higher up than in skulls of *Australopithecus* or *Homo erectus*. Holes betray fatal crushing caused by a rockfall.

Where they lived
On this map numbered sites correspond to numbered items shown in the text and pictured on page 143.
1 Jebel Irhoud
2 Tabūn
3 Skhūl
4 Amud
5 Qafzeh
6 Shanidar
7 Teshik-Tash

Neandertals outside Europe

Neandertals lived in South-west Asia and maybe Africa, but some lacked much of the ruggedness of Europe's classic form, probably adapted to intense Ice Age cold. Some had straighter, slimmer limbs, less massive brow ridges, and a shorter, less sturdy, cranium. Also, brow ridges and a projecting face could occur in skulls with a high forehead, and high rounded cranium. No typically Neandertal fossils have been found outside Europe and South-west Asia. And by about 40,000 years ago, the last South-west Asian Neandertals evidently coexisted with people of entirely modern aspect. Some of the following skulls themselves might almost rank as fully modern man.

1 **Jebel Irhoud** skulls were long and low with large brow ridges but a modern face and slight "bun." Age: about 70,000 years. Place: Jebel Irhoud, Morocco.

2 **Tabūn** skull was low with a sloping forehead, brow ridges, and thick incisors, yet face and back of head were modern. A jaw had a chin. Curved limb bones resembled those of European Neandertals. Age: 50,000 years. Place: Tabūn Cave, Mount Carmel, Israel.

3 **Skhūl 5** combined a big brain, brow ridges, fairly high forehead, and modern face and back of head. Age: 40,000 years. Place: Skhūl rock shelter, a burial site on Mount Carmel, Israel.

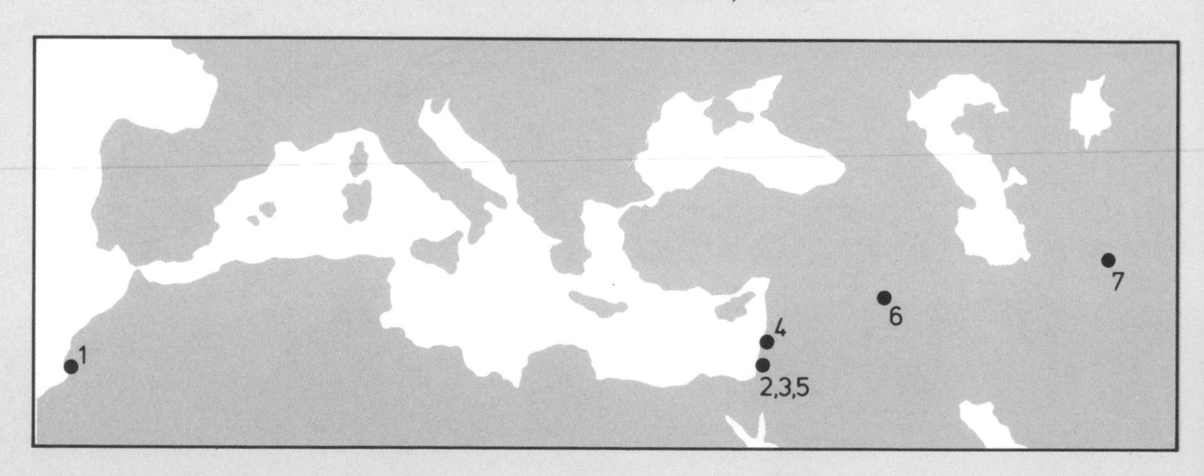

4 **Amud 1:** Neandertal-type skull with a brain capacity of 1740cc (among the largest known), plus long limb bones. Age: perhaps 45,000 years. Place: Amud Cave, near Lake Tiberias, Israel.
5 **Qafzeh** also included skulls with robust features. Age: perhaps 45,000 years. Place: Qafzeh Cave, Israel.
6 **Shanidar** revealed big-brained classic Neandertals, but their brow ridges were not continuous as in Europeans. Age: perhaps 70,000–45,000 years. Place: Shanidar Cave, north Iraq.
7 **Teshik-Tash** had a boy with undeveloped brow ridges, and other classic features yet more modern face and limbs. Age: perhaps 45,000 years. Place: Teshik-Tash Cave, Uzbekistan, USSR.

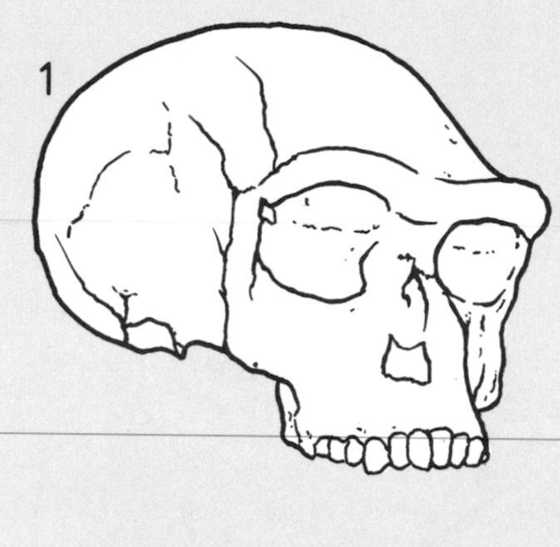

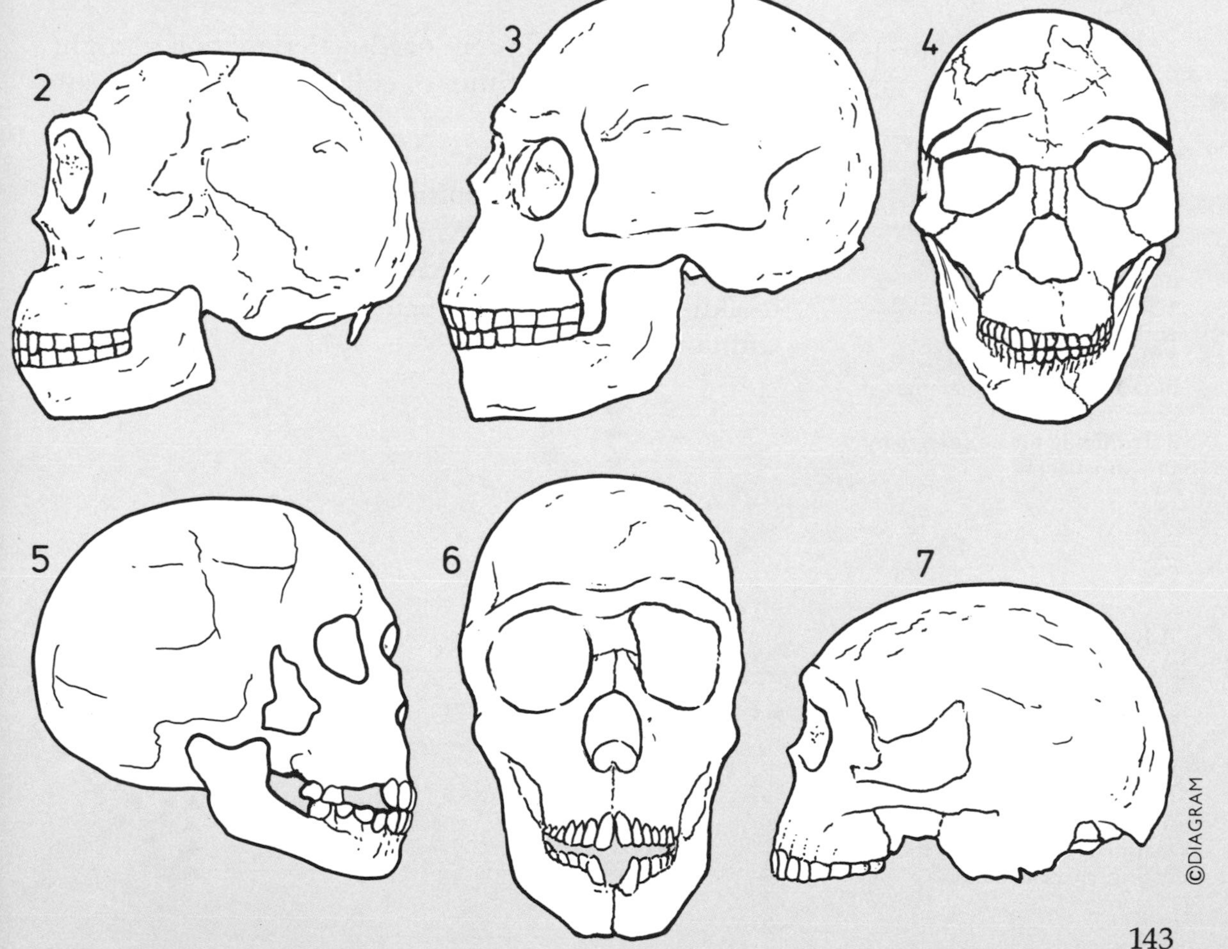

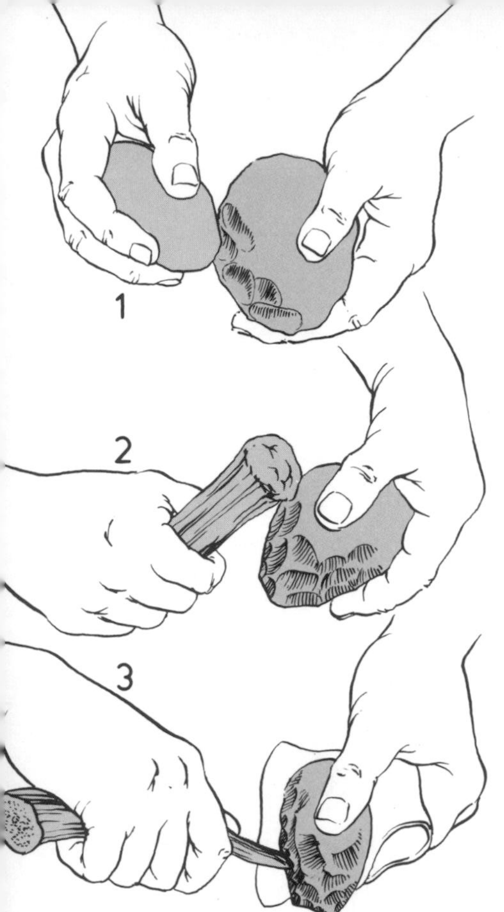

Making a flake tool (above)
Three pictures show stages in making a flint knife.
1 Roughly shaping a flint flake with a stone hammer
2 Refining the shaped flake with blows from a softer hammer of bone or antler
3 Trimming the knife edge by pressure flaking

Neandertal toolkits

Neandertal toolkits are termed Mousterian from finds at Le Moustier in France. They marked improvements over earlier chopper-core and hand-axe industries.

Key innovations included a variety of specialized, finely retouched stone tools, made from flakes not cores. Using fine-grained glassy stone like flint and obsidian, Neandertals improved on the already old-established Levallois technique for striking one or two big flakes of predetermined shape from a prepared core. They made each core yield many small, thin, sharp-edged flakes; then they trimmed the edges to produce side scrapers, points, backed knives, stick sharpeners, tiny saws, and borers. Between them these could have served for killing, cutting up, and skinning prey, and making wooden tools and clothing.

Europe's several types of Mousterian toolkit might represent different times or cultures, or just different tasks performed.

Mousterian toolkits evolved from the old Acheulian and chopper-core industries, and persisted from about 100,000 to 35,000 years ago. They appeared in Europe, North Africa, and South-west Asia. Related toolkits appeared as far afield as southern Africa and China.

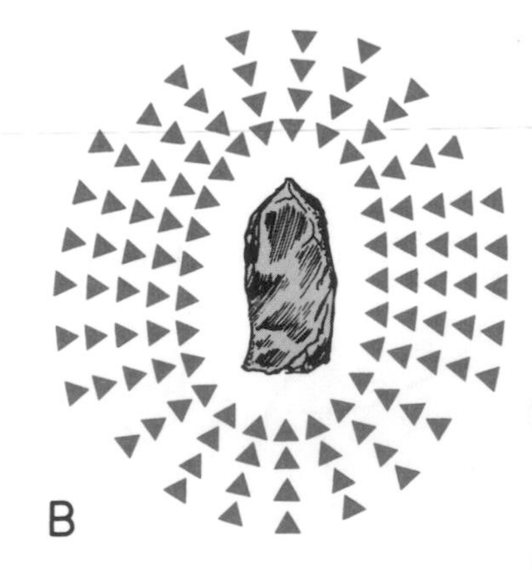

Advance in technique (right)
A A *Homo erectus* took 65 blows to produce this Acheulian hand-axe.
B A Neandertal took 111 blows to produce this Mousterian knife. Increased numbers of blows, and use of prepared cores conserved materials and widened the range of fine, specialized tools.

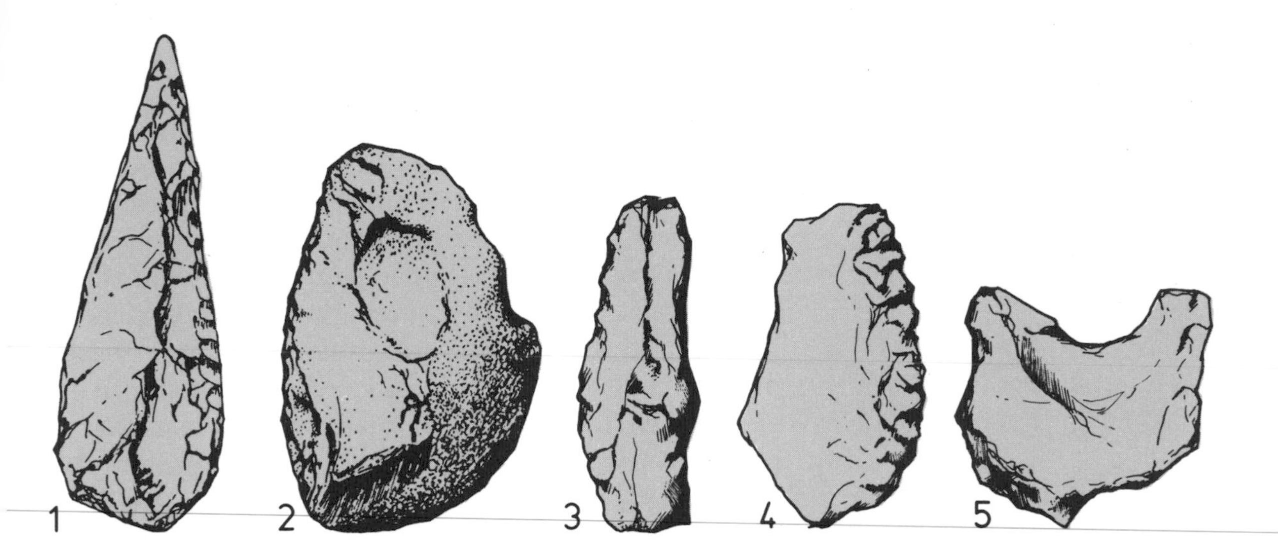

Our examples represent five typical tool types.
1 **Point**: a flake retouched to make a long, sharp, triangular point, perhaps lashed or wedged into a wooden shaft to form a dart head or spear head.
2 **Side scraper**: a convex scraper retouched to give a thickened working edge, maybe for dressing skins without tearing them.
3 **Backed knife**: a long sharp-bladed flake with a blunt back for exerting pressure; used for skinning, cutting meat, or trimming wood.
4 **Denticulate saw**: a flake with a retouched, saw-edged blade suitable for trimming wood.
5 **Notch tool**: a notched flake suitable for smoothing sticks used perhaps as spears.

Mousterian flake tools (above)
Numbers correspond to items featured in the text
1 Point
2 Side scraper
3 Backed knife
4 Denticulate saw
5 Notch tool

Working edge produced (below)
A Amount of cutting edge per pound of stone produced by *Homo erectus*
B Amount of cutting edge per pound produced by the Neandertals

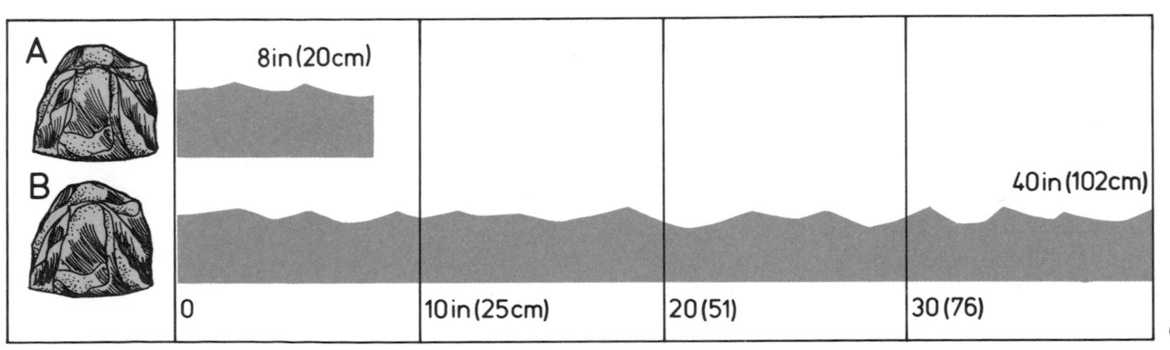

Hunting

Neandertals arguably included the then most effective hunters ever. They had to be to stay alive in harsh, cold Ice Age climates. In summer there were roots and berries to be had. But in winter groups relied on the concentrated nourishment in meat from tundra and cold-forest mammals.

Bones from caves and open camp sites tell us that the chief European prey were large creatures such as bison, cave bears, horses, reindeer, wild cattle, woolly mammoths, and woolly rhinoceroses. Some hunters specialized, for instance, in bison at Il'skaia in the northern Caucasus, and in reindeer at Salzgitter-Lebenstedt, north Germany. Smaller victims included foxes, hares, birds, and fishes. One Hungarian site alone has yielded more than 50,000 bones from 45 species of creatures large and small. (But many bear and mammoth bones at certain sites seem to have been scavenged from corpses of beasts that had already perished naturally.)

As well as for their meat, some animals were doubtless prized for fur, bones, and sinews – used respectively in clothes, tents, and snares.

We lack much evidence for how these people killed their prey. Likely methods include hurling spears or bolas, rolling boulders off a cliff, or setting snares and pitfall traps. Hunters would have picked off sick, old, young, or weakened animals, and even hibernating bears. With fire they could have driven whole herds of frightened horses over cliffs or into dead-end canyons for mass slaughter.

Spearing big game (above)
A hunting band jabs stone-tipped spears into a woolly rhinoceros. Neandertals lacked effective long-range weapons.

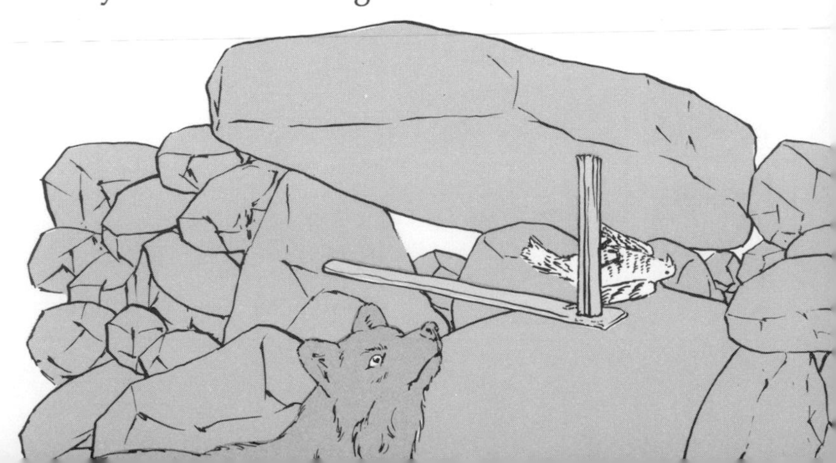

Deadfall trap (right)
In trying to seize meat bait this fox must dislodge a heavy stone that will fall and crush it. Such traps – still used by Eskimos – probably helped Neandertals secure fox pelts for clothing.

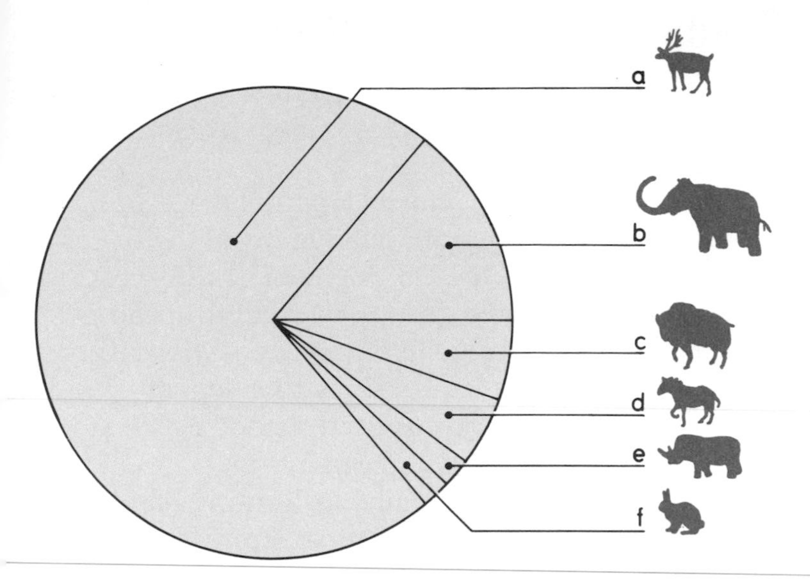

Prey percentages (left)
These are based on bones of beasts killed 55,000 years ago and found at a summer open camp site near Salzgitter-Lebenstedt in West Germany.
a Reindeer 72%
b Woolly mammoth 14%
c Bison 5.4%
d Horse 4.6%
e Woolly rhinoceros 2%
f Other animals 2%

Animals hunted (below)
Some cold-adapted, some warmth - loving, all these 15 kinds of creatures from a variety of habitats probably fell prey to the Neandertals.
1 Wild cattle
2 Cave bear
3 Brown bear
4 Ibex
5 Woolly rhinoceros
6 Reindeer
7 Woolly mammoth
8 Bison
9 Horse
10 Perch
11 Pike
12 Arctic fox
13 Arctic hare
14 Tortoise
15 Crane

Such strategems presuppose keen understanding of the victims' eating, drinking, or migration habits. They would also call for careful planning and cooperation.

In South-west Asia, outside the coldest zones and times, bands of men doubtless ranged far afield to track down and kill large creatures such as wild cattle, sheep, or goats. Meanwhile women and their children would have scoured the countryside around a home base – their targets: rodents, reptiles, insects, berries, gums, honey, roots, and tubers.

Caves, tents, and clothes

The Neandertals of Europe survived harsh Ice Age winters in warm microclimates formed by clothes and heated homes.

In some places caves afforded natural protection. Groups up to 40 strong inhabited many of the scores of limestone caves in the Dordogne of South-west France, and at least two dozen caves in European Russia. At Combe Grenal in France a post hole at a cave mouth hints at a wall of skins that kept out wind and snow or rain. There was even a stone wall inside the Cueva Morín, in northern Spain.

Where caves were not available, hunting bands built shelters in the open. The most impressive of these bastions against the elements were large tents or huts built in Ukrainian river valleys teeming with big game. Here, people raised branch frameworks up to 30ft (9m) long, 23ft (7m) wide, and 10ft (3m) high. They evidently covered these with skins weighed down with heavy mammoth bones found lying in the countryside around.

Tent in a cave?
Excavation of Le Lazaret cave near Nice in France suggests that 150,000 years ago archaic *Homo sapiens* made a partitioned skin tent inside a cave to keep out cold.
A Rear compartment
B Front compartment
a Shelter wall of stones perhaps including post supports
b Probable entrances
c Presumed partition
d Litter, perhaps including bedding
e Hearths

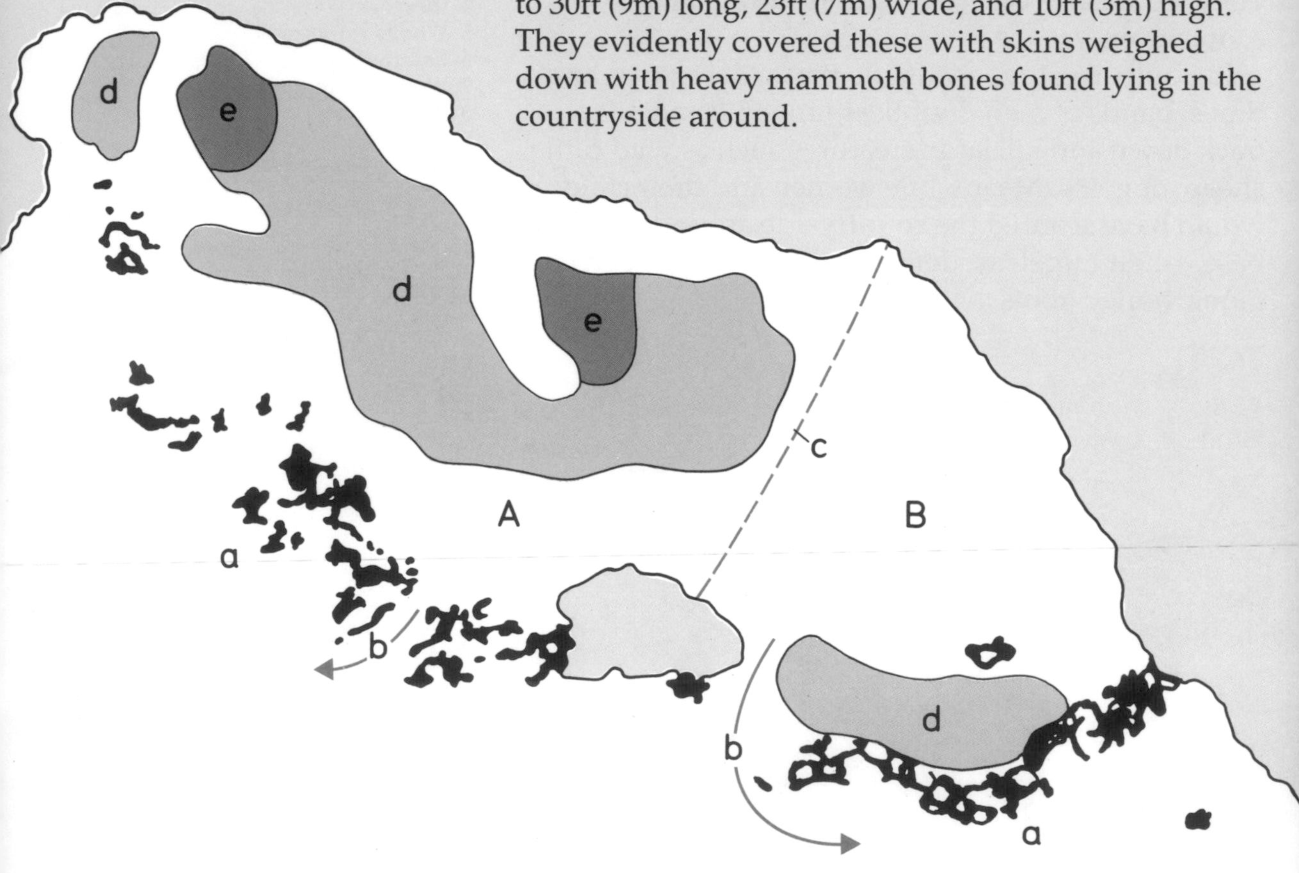

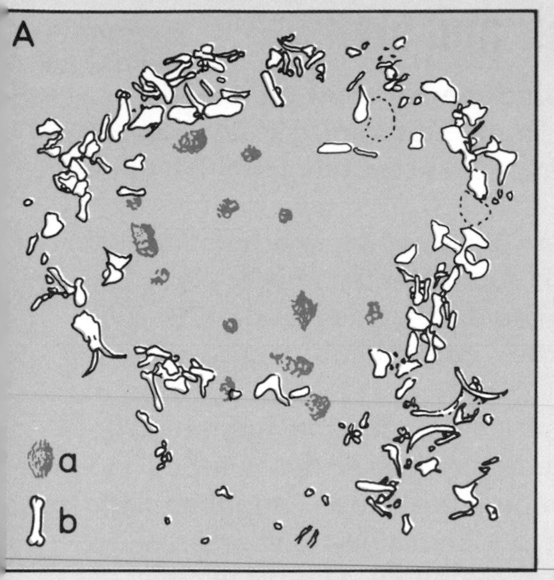

Evidence for huts (above)
A Excavated ground plan of a hut at Molodova in southern Russia.
a Hearths (15 were found)
b Mammoth bones
B The hut might have looked like this, but mammoth long bones would have served as supports if wood were scarce.

Making clothes (below)
1 Scraping flesh and fat from a hide to clean it
2 Lacing hides together with rawhide threaded through holes punched by an awl

Old hearths show where Neandertals had warmed their tents and caves by burning wood or bones. Discoveries suggest they knew how to start fires by striking sparks from iron pyrites and using dried bracket fungus as tinder.

Evidence for clothes is mostly indirect. People apparently used stone knives to cut furry skins to shape; bored holes in tailored skins with awls of stone or bone; then tied the skins with sinews. When skinning foxes, hares, and wolves, they evidently sometimes left the feet intact to serve as ties. The resulting clothes probably included unsophisticated trousers, tunics, cloaks, hoods, and wrappings for the feet.

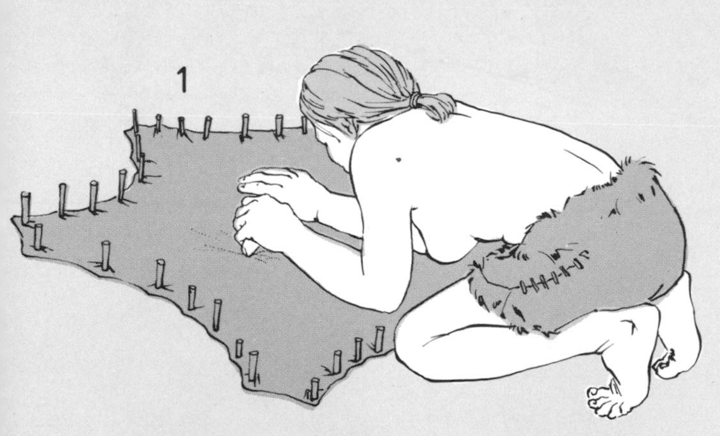

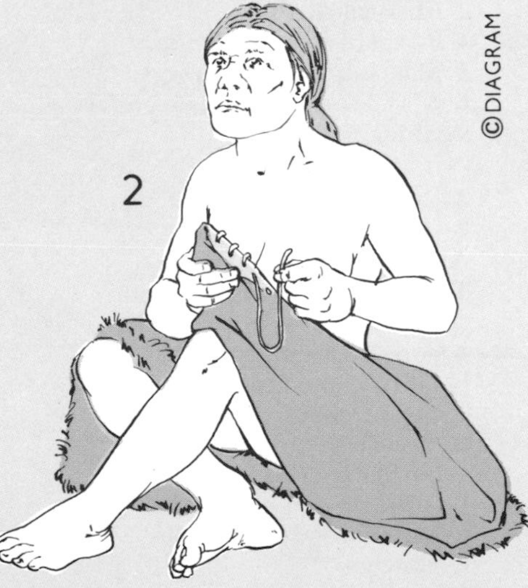

Burial, ritual, and art

Burial, ritual, and rudimentary art hint that the Neandertals were more self aware, socially caring, and generally capable of abstract thought than their ancestor *Homo erectus*.

Neandertals were the first people systematically to bury their dead. By soon after the middle of the 20th century scientists had excavated 68 burial sites, with more than 150 bodies, most in Europe, and almost all in caves.

There are clear proofs of deliberate interment. Skeletons lay in holes dug in cave floors. Many had been placed in sleeping posture and supplied with grave goods reputedly from stone tools and roasted joints of meat to a stone pillow, a bed of woody horsetail, and a scattering of late spring flowers (detected by their pollen grains).

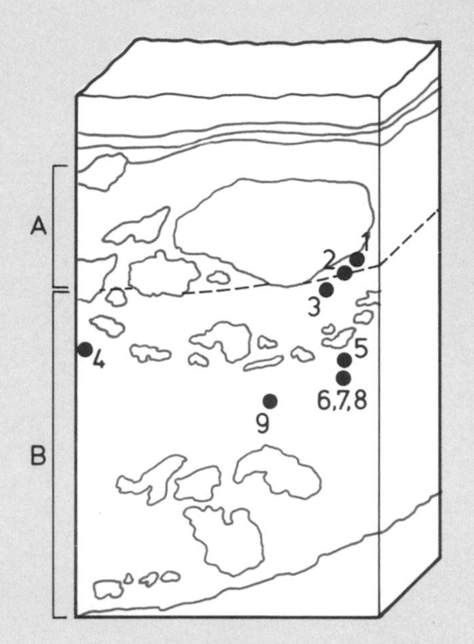

Shanidar burials (above)
This block diagram depicts Neandertal burials (some by rock fall) in the great cave of Shanidar in north Iraq. Excavated depth: 45ft (13.7m).
A Upper Paleolithic: 28,700–35,000 years ago
B Mousterian: 35,000–100,000 years ago
1 Disabled old man
2 Rock-fall victim
3 Rock-fall victim
4 Burial of rock-fall victim
5 Male buried with flowers
6–8 Male and two female burials
9 Child

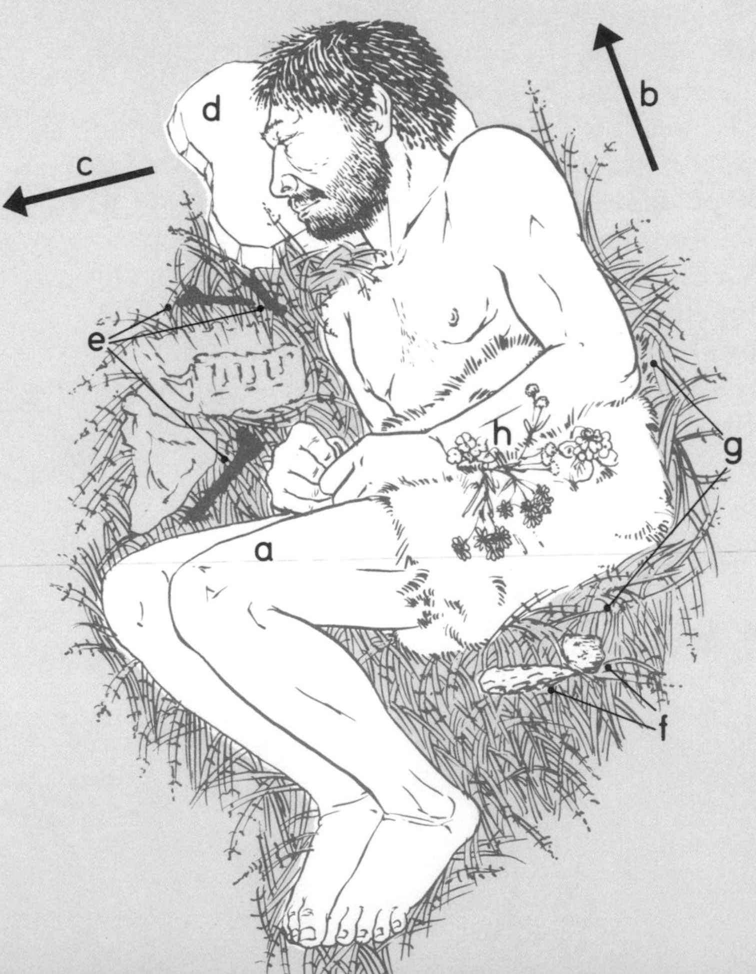

A Neandertal burial (right)
This composite reconstruction shows:
a Body in sleeping posture
b Body aligned east-west
c Head facing south
d Stone pillow
e Burnt bones
f Flint implements
g Bed of woody horsetail
h Flowers

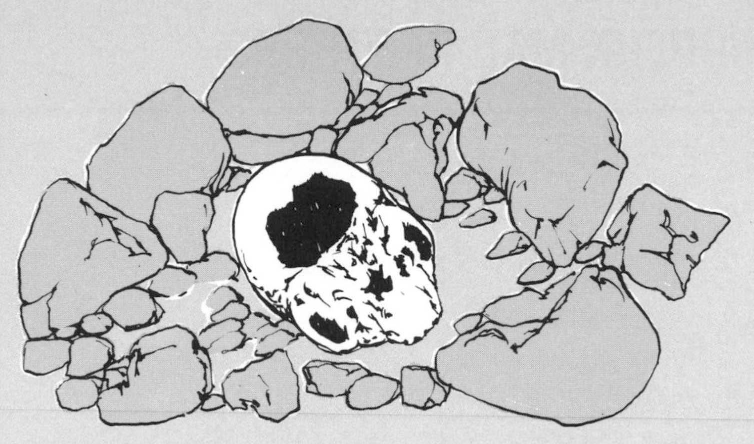

A cannibal's feast?
The big hole in the base of this human skull was made maybe by enlarging the foramen magnum. Someone possibly scooped out and ate the brains, then placed the skull within this ring of stones found in a cave at Monte Circeo, Italy.

All this suggests that Neandertals attached importance to an individual's life and death, and perhaps looked forward to an afterlife. Maybe they showed compassion, too. At Shanidar, Iraq, only caring companions could have kept alive an old man, half blinded, and crippled by arthritis and a withered arm long before he died.

Yet, like their predecessors, these people seemingly ate each other; mutilating skulls and limb bones to extract brains and marrow. Probably the eaters hoped to gain the strength of those they ate.

Cannibalism was evidently one of several rituals. In one cave Neandertals placed mountain goats' horns in a ring; elsewhere they seemingly stacked cave bears' skulls. Perhaps they aimed to please the spirits of the animals they killed.

Ritual perhaps infused their art, known only from such items as a bone amulet, scratched pebbles, and lumps of red iron oxide and rubbed manganese – these last evidently used to paint the body. (Red maybe symbolized blood and therefore life.)

Bear cult? (below)
This bear's skull lies in a crevice walled off from the cave beyond by stones. This might suggest religious ritual in the making. The bones were found near Veternica in Yugoslavia.

Chapter 8 MODERN MAN IN EUROPE

Evidence of fully modern man – the subspecies *Homo sapiens sapiens* – first emerged with bones discovered at Cro-Magnon in South-west France. Cro-Magnon Man left skeletons and artefacts in many parts of Europe. Here, about 40,000 to 10,000 years ago, lived people probably ancestral to Europe's modern so-called Whites or Caucasoids. Late Old Stone Age hunters survived the harshest rigors of the Ice Age by making new, sophisticated implements of stone and bone, and using them to kill big meaty animals that ranged across the continent.

But new, Mesolithic (Middle Stone Age), hunting and gathering cultures flourished in post-Ice-Age Europe 10,000 to 5000 years ago.

A prehistoric cave painting shows archers shooting arrows at running deer. This Late Paleolithic or Epipaleolithic scene comes from the Cueva de los Caballos near Castellón, in eastern Spain. (Illustration taken from *A History of Technology*, edited by Singer, Holmyard, Hall and Williams)

Body build (below)
Cro-Magnon man was probably an early Caucasoid ("White") ancestor of today's Europeans.
Height: 5ft 6in–5ft 8in (1.69–1.77m)
Weight: about 150lb (68kg)

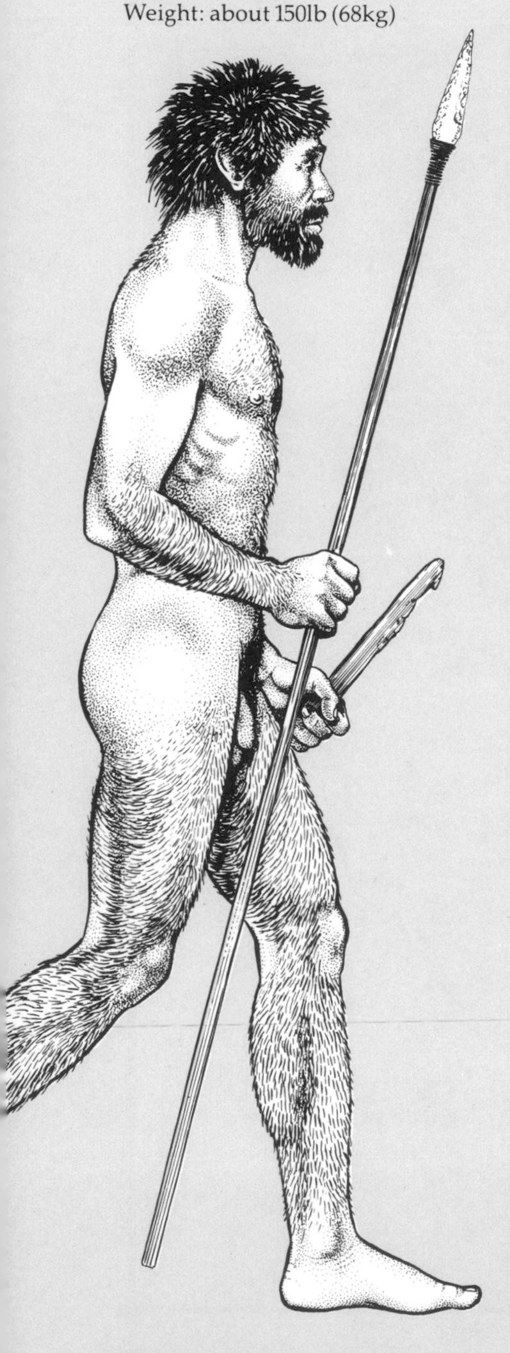

About early modern man

Fully modern man – the subspecies *Homo sapiens sapiens* – crops up widely in the fossil record in 40,000-year-old sites as far apart as Borneo and Europe. Some early skeletons even show similarities to modern Whites, Blacks, Orientals, and aboriginal Australians (see also Chapter Nine).

Best known are the bones of early modern Europeans, called collectively Cro-Magnons, from skeletons discovered at Cro-Magnon in South-west France. Cro-Magnons were taller and less rugged than Neandertals, with thinner bones than these or *Homo erectus*. Compared with the Neandertals, the head was relatively tall but short, with a more rounded braincase containing a slightly smaller brain of 1400cc average capacity. Other innovations were an upright forehead; a straight not forward-jutting face; absent or only slight brow ridges; a smaller nose; smaller jaws; more crowded teeth; and a well-developed chin.

Some paleoanthropologists think fully modern man evolved in one continent (most likely Africa) and spread to all the rest, replacing local archaic forms of *Homo sapiens*. Other experts have argued that local archaic forms evolved into our own subspecies independently of one another. A compromise suggests that modern man arose in one place but interbred with older, local forms to help produce the so-called races of today.

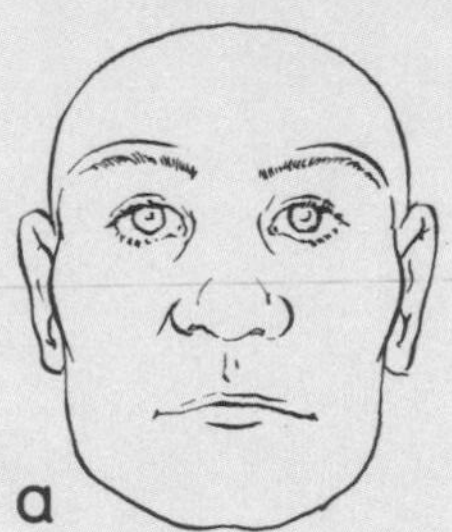

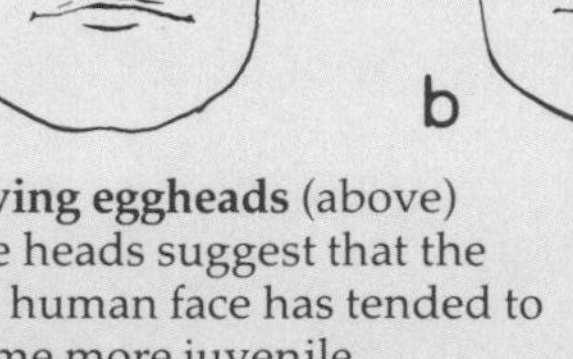

Evolving eggheads (above)
Three heads suggest that the adult human face has tended to become more juvenile.
a Neandertal face: large jaw, large nose, low cranium
b Cro-Magnon face: smaller jaw, smaller nose, higher cranium
c Some humans today: still smaller jaw and nose, still higher cranium

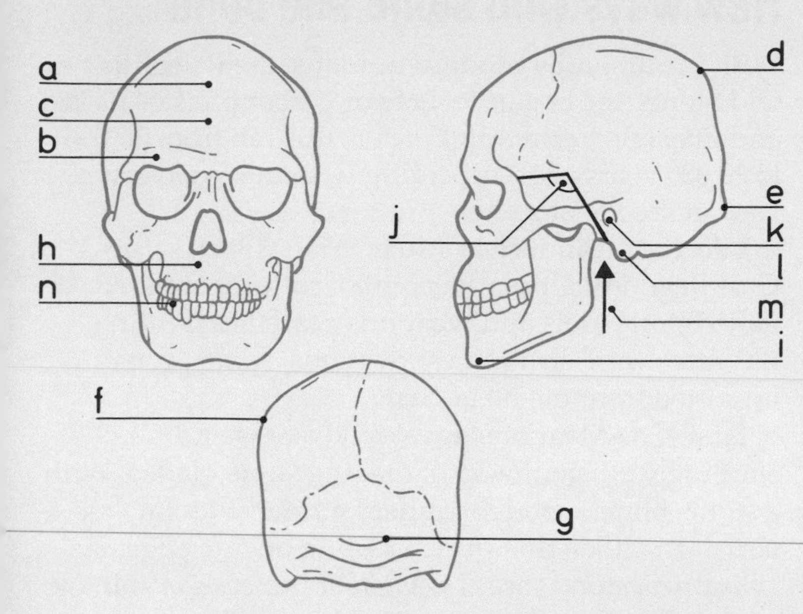

Skull features (left)
Three views of a modern human skull (left). Most listed items distinguish it from older *Homo*.

a Large cranial capacity
b No overhanging brows
c High, vertical forehead
d Convex cranial vault
e Low, rounded occiput
f Skull widest high up
g Small arched torus
h Naso-maxillary region not inflated
i Projecting chin
j Relatively acute sphenoidal angle
k Round ear hole
l Sizable mastoid process
m Central foramen magnum (hole in the skull base)
n Pulp cavities of teeth seldom enlarged

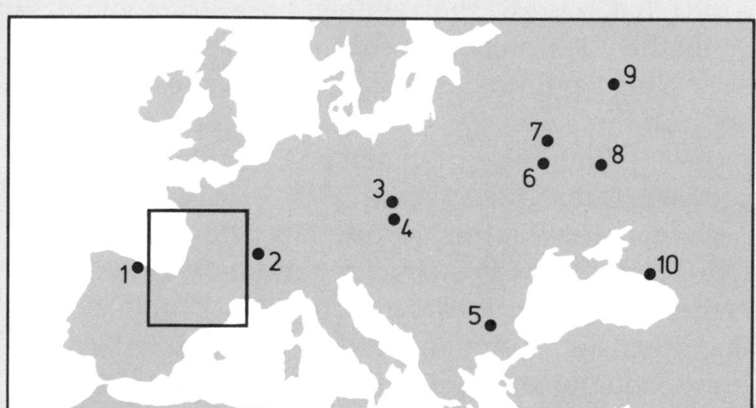

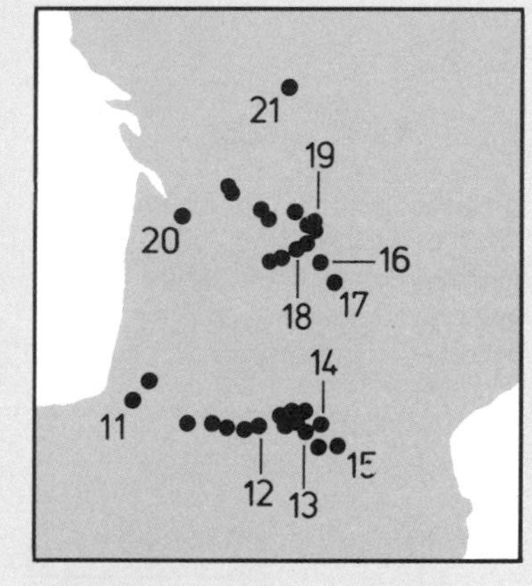

Cro-Magnon Europe (above)
Maps above left and above (an enlargement of part of France) show selected European Cro-Magnon sites, with bones or artefacts, or both.

1 Altamira	
2 Solutré	**12** Gargas
3 Mladeč	**13** Enlène
4 Předmost	**14** Mas d'Azil
5 Bacho Kiro	**15** Niaux
6 Chulatovo	**16** Cougnac
7 Eliseevichi	**17** Pech-Merle
8 Kostenki	**18** Cro-Magnon
9 Sungir	**19** Lascaux
10 Akhshtyr	**20** Chancelade
11 Isturits	**21** Les Roches

Fossils fail to show why our own subspecies has proved so successful. Indeed, until about 10,000 years ago, our forebears still lived in largely mobile bands of Old Stone Age hunter-gatherers. Yet these colonized all continents except Antarctica, and developed tools, techniques, and new behavior patterns that would transform the human way of life and bring explosive population growth. The next pages trace major trends in later prehistoric Europe.

New ways with stone and bone

With the first fully modern humans, the Paleolithic or Old Stone Age began its Late or Upper phase – its last and finest flowering. In Europe, this ran from 35,000 to 10,000 years ago, coinciding with the frigid last gasp of the Pleistocene.

Like the Neandertals' Mousterian cultures, the Cro-Magnons' Upper Paleolithic cultures stressed flaked stone tools and weapons geared to hunting. But there were striking innovations, some probably imported from the Near East.

First, Cro-Magnons developed a new and startlingly efficient way of making stone blades. With a stone, bone, wood, or antler hammer a toolmaker struck an antler-tine punch resting on one edge of a cylindrical stone core. This *indirect percussion* split off a long, flat, narrow, sharp-edged flake. The toolmaker could then delicately trim this flake by *pressure flaking* – pressing a pointed tool against the blade edge to snap off tiny subflakes. These methods produced more and finer tools from just one stone than any previous technique.

Compared with the Neandertals before them, Cro-Magnons made a far wider and more finely crafted range of knives, scrapers, saws, points, borers, and other stone implements. But half their tools were bone – tougher and more durable than wood, and shaped by chisel-edged stone burins, which also fashioned artefacts of antler, wood, and ivory.

From these materials Cro-Magnons made such novel implements as eyed needles, handles, fishhooks, harpoons, and spear throwers. Between them, these simple-seeming artefacts enormously increased mankind's control of the environment.

Indirect percussion (above)
This produced long, slim, parallel-sided flint blades.
a Hammerstone
b Bone or antler punch
c Core
d Anvil stone
e Blades and waste flakes

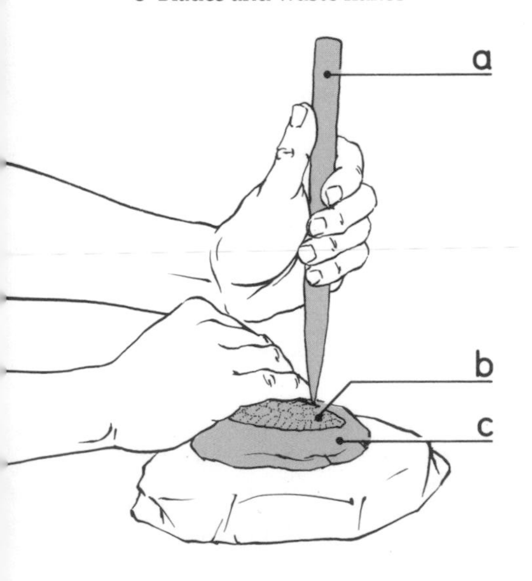

Pressure flaking (left)
Pressing a pointed tool down on a flint flake's outer edge snaps subflakes from its underside.
a Pointed stick or bone
b Flint flake artefact
c Bark cushion on stone anvil

Late Paleolithic tools
Shown right are four tools of flint and one of bone (not all to scale). (More figure on the following pages.)
a Flint knife, with a back blunted by pressure flaking
b Flint end scraper, rounded at one end by pressure-flaking
c Burin chisel, for working antler, bone, or wood
d Microburin drill, for piercing holes in skin, wood, bone, or antler
e Bone needle, with an eye pierced by a microburin drill

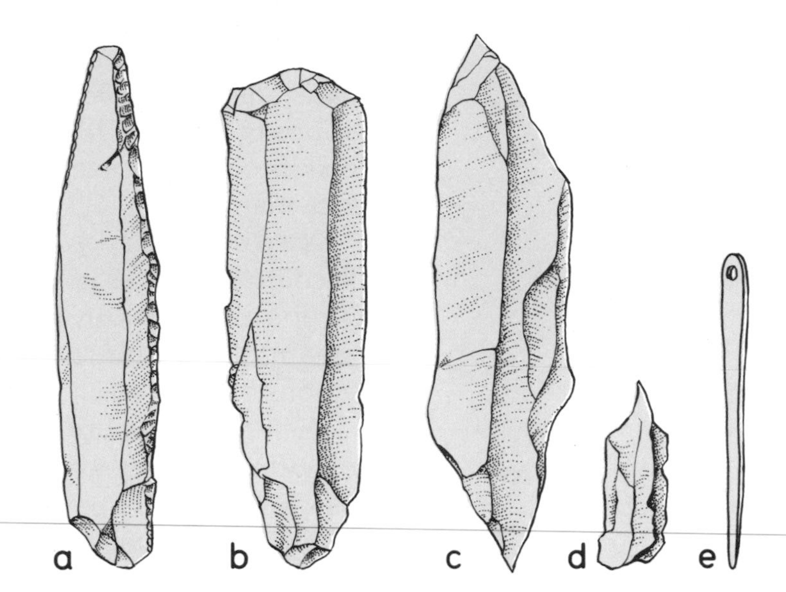

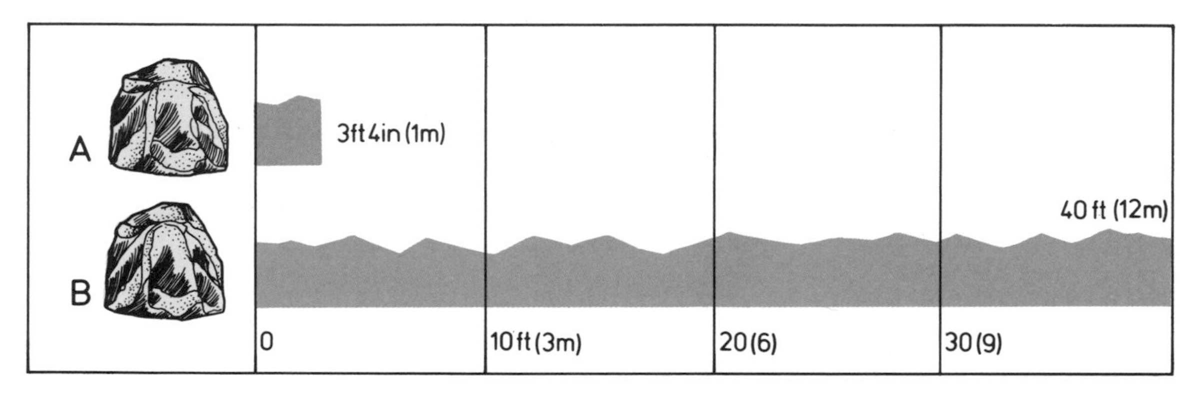

Economy with stone (above)
A Length of cutting edge per pound of stone produced by the Neandertals' technique
B Length of cutting edge per pound produced by the Cro-Magnons' technique

Advance in technique (right)
More than 250 blows (including pressure flaking) went to make a finely crafted Cro-Magnon flint flake knife. Compare this with the fewer blows shown on page 144 for implements produced by Neandertal Man and *Homo erectus*.

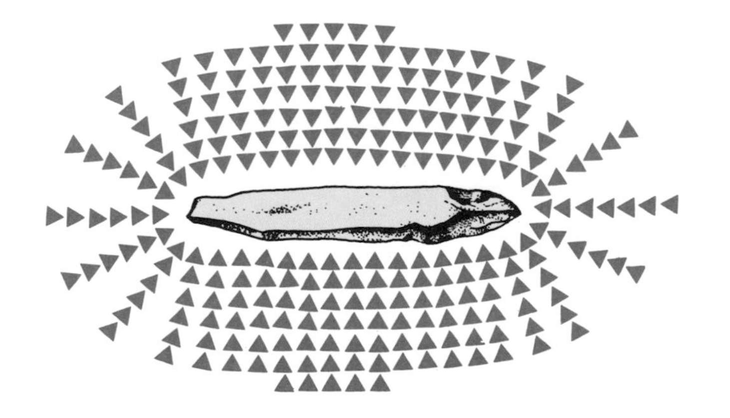

Cro-Magnon cultures

Europe's Cro-Magnon toolkits varied with time and place. Some archeologists think this reflects different groups of people. Other experts argue that different toolkits only represented different schools of craftsmanship.

Based on tools and other artefacts, four main Upper Paleolithic cultures have been named for Western Europe, where South-west France is rich in their remains. (Many more cultures come from other regions of the world.)

The Périgordian, of about 35,000–32,000 years ago in France, featured side scrapers, saws, and curved-back knives, perhaps deriving from the Mousterian tradition, and overlapped the Châtelperronian Neandertal culture. The later Gravettian, with long, tapering gravette points, many burins and bone tools, had a geographic range from France to Russia where it lasted until 10,000 years ago.

The Aurignacian, of about 35,000–29,000 years ago, had many scrapers, borers, and burins, but it was relatively poor in knife-like blades. The Near East had similar and even older industries.

The short-lived Solutrean (about 22,000–18,500 years ago) stressed fine workmanship in such articles as slender, leaf-shaped blades tapered at each end and superbly crafted on both sides.

The Magdalenian (about 18,500–11,000 years ago) is noted for hooked rods used as spear throwers, and fishing implements from barbed hooks to harpoons.

Cultural complexes
Four lines show time spans of four Late Paleolithic West European technocomplexes. (We omit the short-lived post-glacial Azilian which began about 10,000 BC: arguably a Mesolithic – post-Paleolithic – culture.)
1 Périgordian, here extended to include the Upper Périgordian or Gravettian, which began at least 29,000 years ago.
2 Aurignacian
3 Solutrean
4 Magdalenian

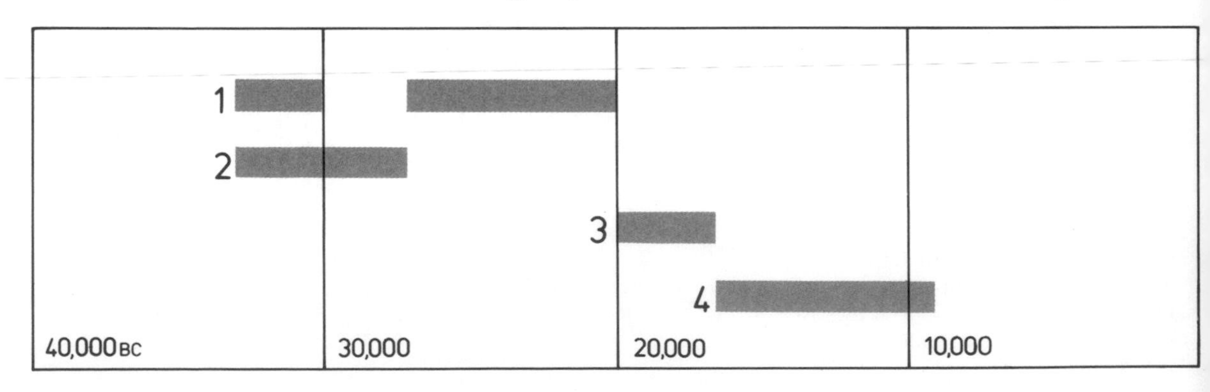

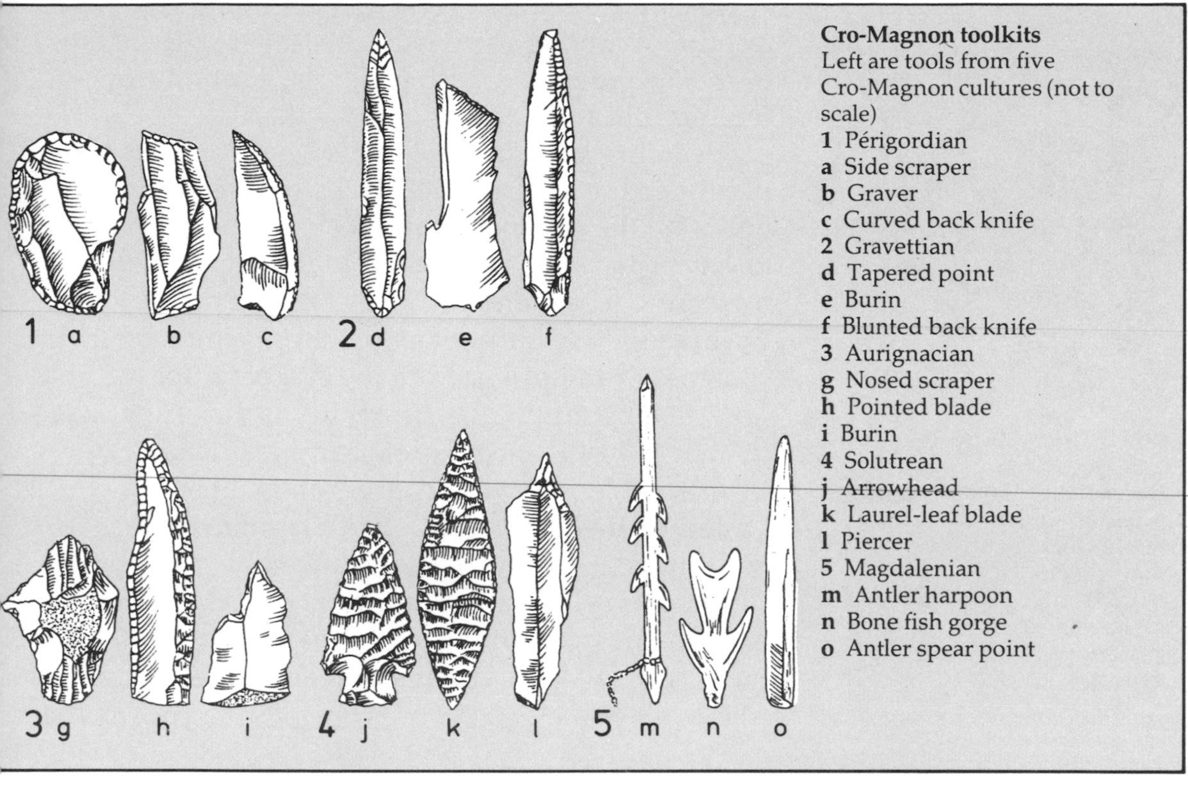

Cro-Magnon toolkits
Left are tools from five Cro-Magnon cultures (not to scale)
1 Périgordian
a Side scraper
b Graver
c Curved back knife
2 Gravettian
d Tapered point
e Burin
f Blunted back knife
3 Aurignacian
g Nosed scraper
h Pointed blade
i Burin
4 Solutrean
j Arrowhead
k Laurel-leaf blade
l Piercer
5 Magdalenian
m Antler harpoon
n Bone fish gorge
o Antler spear point

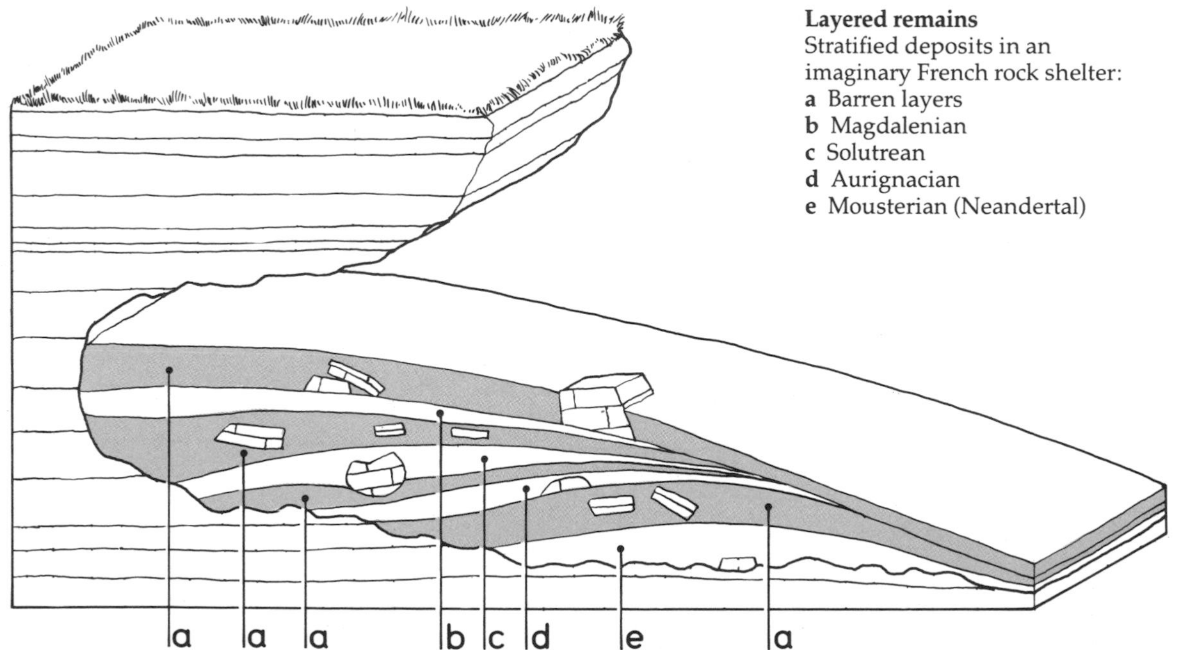

Layered remains
Stratified deposits in an imaginary French rock shelter:
a Barren layers
b Magdalenian
c Solutrean
d Aurignacian
e Mousterian (Neandertal)

Hunting methods

Archeological finds suggest that Cro-Magnon hunting weapons and techniques far surpassed those of the Neandertals – with major implications for food supply and population growth.

Spear throwers that looked like giant crochet hooks improved the leverage exerted by the human arm, doubling the distance over which a hunter could hurl his spear. Now he could kill at long range, before his prey grew scared and ran away. Barbed points led to the invention of harpoons for killing salmon on their spawning run upriver from the sea. Now, for the first time, fish became a major item in the diet, especially in South-west France. Runnels in spearheads increased blood flow from stricken animals, hastening their end. Cro-Magnons snared birds and almost certainly developed deadfall traps for birds, wolves, foxes, and much larger game. Some experts think that pitfall traps had caught the 100 mammoths whose remains were found near Pavlov in Czechoslovakia.

Harpoon head and shaft (above)
a Barbed bone or antler head which stuck in a stabbed fish or other prey.
b Cord linking head and shaft
c Shaft socketed to take the head and gripped by the hunter before and after stabbing.

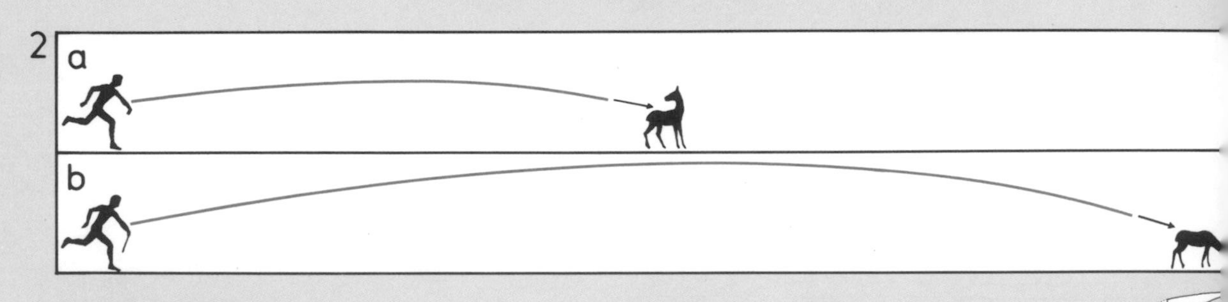

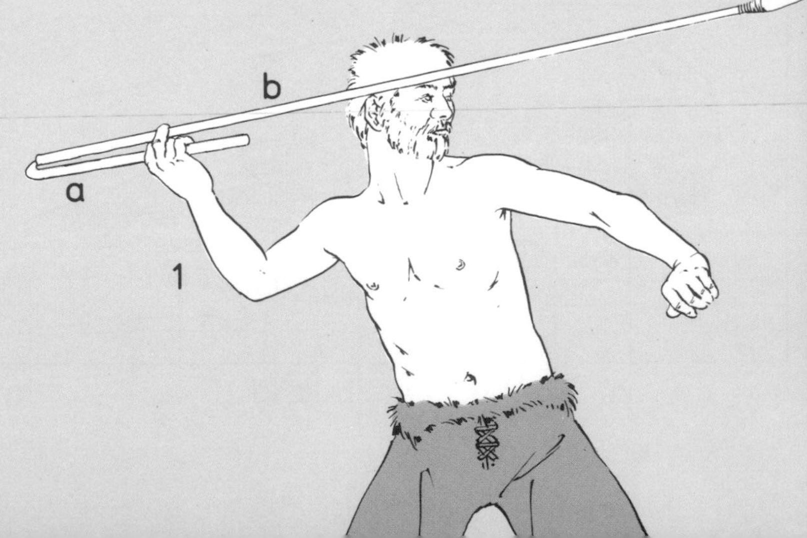

Spear thrower in action
1 A Cro-Magnon hunter used a spear thrower like this.
a Spear thrower
b Spear
2 Diagrams above show how this increased his weapon's range.
a Unaided arm power hurls a long spear 70yd (64m); actual killing range might be 15yd (13.7m)
b A spear thrower helps hurl a spear 150yd (137m); killing range increases to 30yd (27.4m).

Wounded bison (right)
Arrows seem to pierce a bison in this Late Paleolithic cave painting from Niaux, in South-west France. Despite such images and a few finds of arrowheads, there is little evidence that hunters used bows and arrows before the Ice Age ended.

Cro-Magnons were arguably the most skillful ever hunters of big game. Seasonal occupation of selected sites implies that they followed reindeer and ibex on seasonal migrations to or from new feeding grounds. Great piles of bones imply that hunters learned the knack of driving herds to easy killing grounds. This might explain the 1000 bison evidently killed in a ravine in southern Russia. Even more spectacular are the skeletons of 10,000 wild horses found below an inland cliff near Solutré in East-central France. About 17,000 years ago, hunters had seemingly stampeded whole herds over the cliff-top, or trapped and killed them at the bottom.

Using techniques and tools like these, Cro-Magnons won an almost inexhaustible supply of highly nourishing food. This evidently helped them multiply, and people even harsh, cold regions of Siberia.

Migrating herds (below)
Arrows show likely spring migration routes of reindeer from mild winter quarters to coastal and mountain pastures.
a Périgord
b Atlantic coast (now submerged)
c Cantabrian Mountains
d Pyrenees
e Massif Central
f Mediterranean coast (now submerged)
g Alps

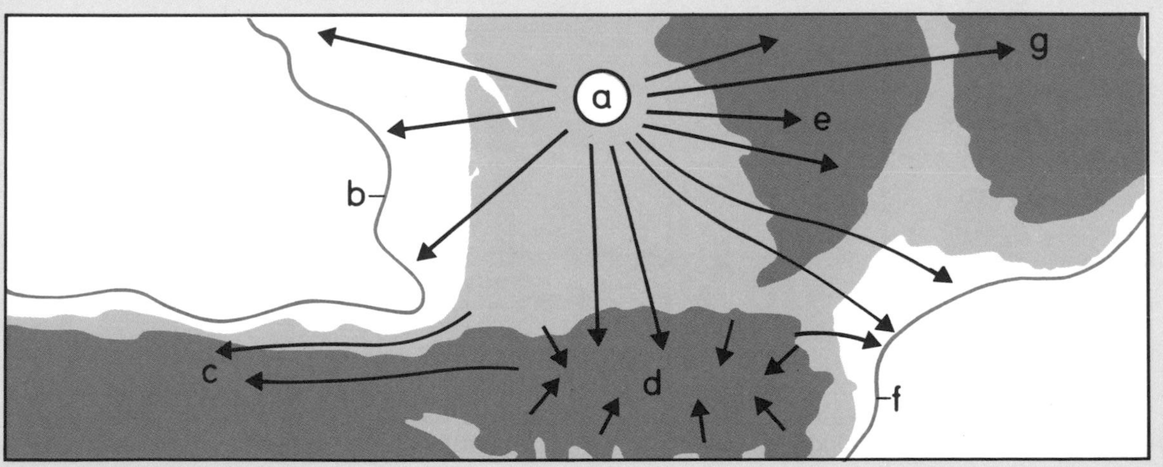

The hunters' prey

Late Old Stone Age Europe teemed with big, wild, meaty or fur-bearing mammals – never since surpassed in numbers or variety. Most species would have been familiar to Neandertal man, but Cro-Magnon hunters' abundant, widespread leavings shed new light on which lived where.

Wild horses, reindeer, bison, mammoths, saiga antelopes, and woolly rhinoceroses grazed open steppes. Wooded slopes and valleys sheltered red deer, wild boars, cave bears and cave wolves. Chamois, ibex, and wild goats climbed to high mountain summer pastures. Wild cattle thrived on steppes and in the forest edge.

Local differences suggest that in the early Late Stone Age reindeer and horses predominated in north Germany, mammoth and rhinoceros in southern Germany, with cattle and horses farther east, around the Black Sea. But there was change through time. In one Italian site, the chief prey animals – cattle, horse, red deer, and boar – gave way to horse and ass, in turn replaced by boar, red deer, and cattle. Probably climatic changes had repeatedly affected vegetation.

Species wax and wane
Diagrams below show how numbers of various animals and plants waxed and waned in South-west France as temperatures fluctuated in the last part of the Pleistocene.
a Rise and fall in temperature 30,000–10,000 BC.
b Red deer
c Arctic fox
d Deciduous trees
e Conifers

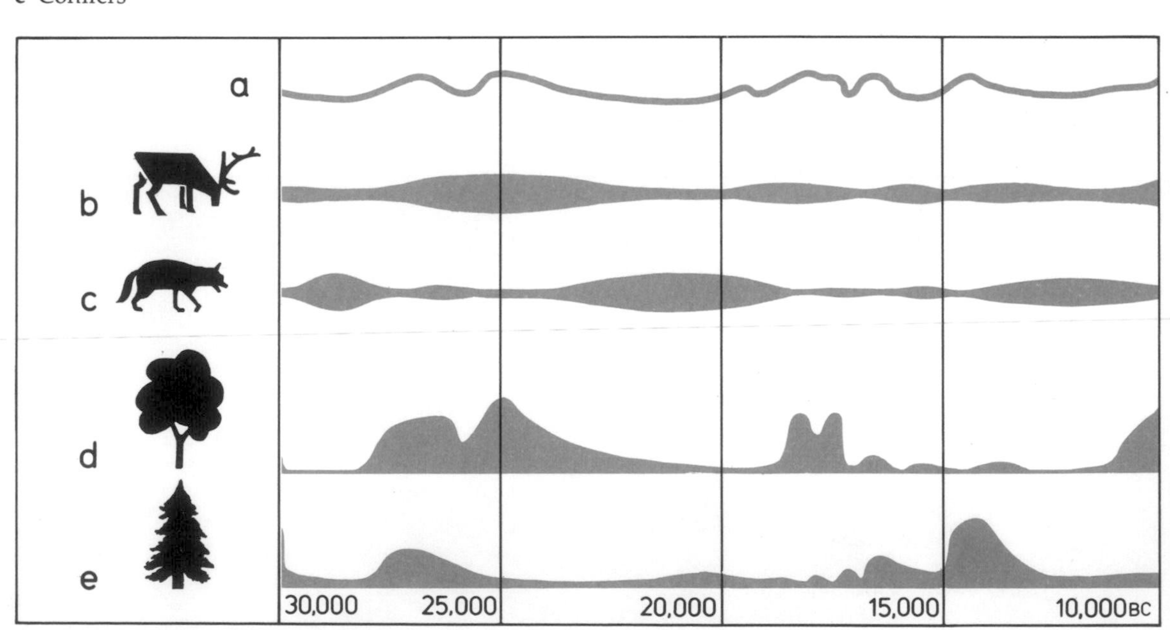

Here are five species that formed staple items in Cro-Magnon diets.

1 **Rangifer tarandus,** the reindeer, is the only deer with antlers in both sexes. Shoulder height: to 4.6ft (1.4m). Time: Pleistocene-modern. Place: Circumpolar.

2 **Cervus elephas,** the red deer, is the largest deer in most of Europe. Shoulder height: to 5ft (1.5m). Time: Pleistocene-modern. Place: Eurasia and North Africa.

3 **Bos primigenius,** the aurochs, was a huge ox. Shoulder height: 6ft (1.8m) Time: Pleistocene–AD1627. Place: Eurasia and North Africa.

4 **Equus caballus,** the horse, was a small, wild subspecies. Shoulder height: to 4.6ft (1.4m). Time: Pleistocene-modern. Place: northern continents.

5 **Capra pyrenaica,** the Spanish ibex, is a mountain goat. Males have long horns curved out and up. Shoulder height: to 30in (76cm). Time: Pleistocene-modern. Place: Spain, and once South-west France.

Five prey species
Numbers correspond to items featured in the text.
1 Reindeer
2 Red deer
3 Aurochs
4 Horse
5 Spanish ibex

Homes and clothes

Dressed for warmth
Carved figurines from Russia show how people kept warm through Ice Age winters.
a Figure wearing a close-fitting fur costume, with hood and trousers. This mammoth-tusk carving comes from Buret in Siberia.
b Headless female figure wearing a decorative belted costume. This carving comes from Kostenki, south of Moscow.

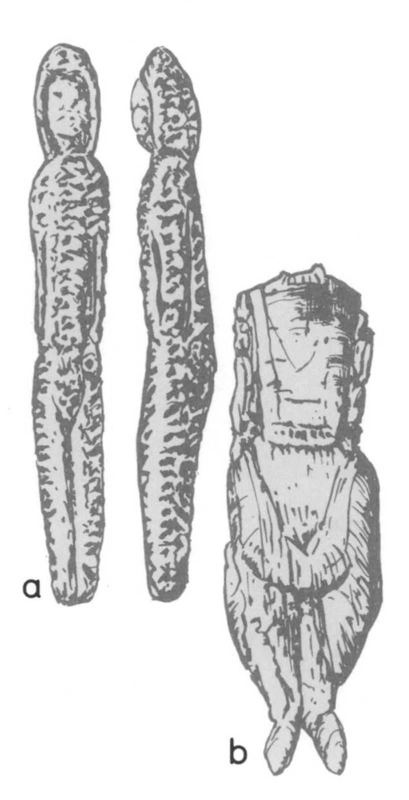

Cro-Magnon homes largely followed old, Neandertal, traditions, but some homes and probably all clothing incorporated innovations that improved survival prospects in the last cold millenia of the Pleistocene.

Like the Neandertals before them, Europe's cave-dwelling Cro-Magnons made use of the limestone river cliffs of South-west France – especially in the Dordogne, but also on the northern slopes of the Pyrenees. Many caves faced south, warmed by sunshine and sheltered from cold northern winds. The caves lay near plentiful supplies of water, and commanded views of pastures grazed by big meaty herbivores. Where food abounded all the year, up to several dozen individuals might live permanently in one large cave. But certain caves show signs of only seasonal activity.

Again like the Neandertals, Russia's Gravettians built winter homes in river valleys. Some were paved with stone, some sunk in the ground, and many walled and roofed by tented skins propped up by mammoth thigh bones and weighed down around the rim by other heavy bones and tusks. Among the largest structures was an 89ft (27m) long longhouse from Kostenki in the Don Valley; its central row of hearths suggests that several families wintered here beneath one roof consisting of a row of linked skin tepees.

Other finds and old drawings in French caves reveal that nomadic hunters also put up flimsy summer huts like huts still built by certain modern primitives.

Increased ability to live and work in Ice Age cold owed as much to novel clothing as new building methods or the use of fire. Bone needles and carvings of fur-suited people speak of closely fitting pants, hooded parkas, boots, and mittens – with seams closely sewn to keep in body heat.

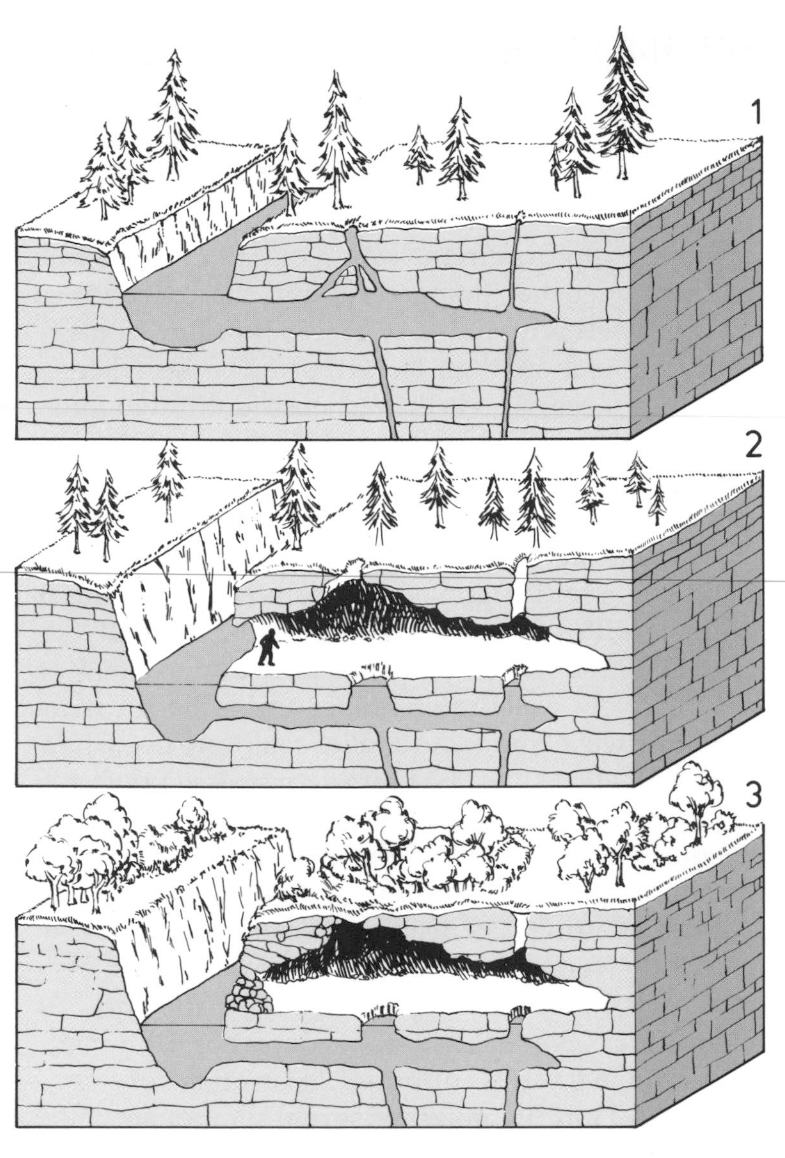

Story of a cave (left)
1 Water seepage dissolves limestone in a plateau, eating out an underground cave that opens into a river.
2 A drop in water level has left the cave high and dry. Cro-Magnons now inhabit it.
3 Today a rock fall has shut the cave mouth and vegetation blocks the sinkholes above, preserving Cro-Magnon relics in cave-floor sediments.

A Russian longhouse (below)
This is a reconstruction of a communal longhouse sunk slightly in the ground, and roofed by tented skins propped up by poles, with mammoth bones and tusks to reinforce the bottom of the walls. Found at Pushkari, north-east of Kiev, the site might really represent three tents ranged close together in a row.

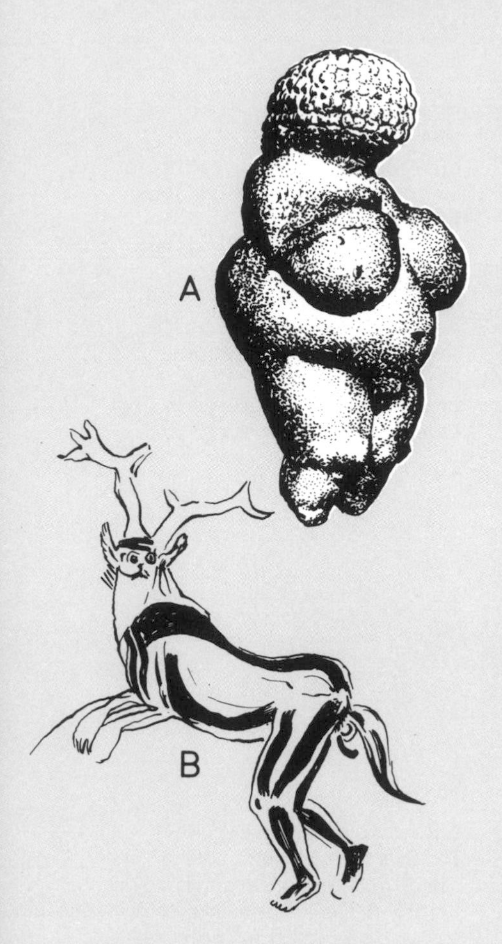

Art and ritual

From 35,000 to 10,000 years ago Europe passed through its great age of prehistoric art. Works ranged from engravings of animals and people done on portable bits of stone, bone, ivory, and antler, to clay or stone sculptures in the round or half round, and paintings in ocher, manganese, and charcoal dabbed on cave walls with moss, or blown through a straw.

Studies of more than 100 caves and rock shelters convince some experts that art passed through four stages of development. Period I (32,000–25,000 years ago) featured animals and other forms mostly poorly drawn on small, portable objects. Period II (25,000–19,000 years ago) produced early cave art including hand prints and engraved and painted silhouettes of animals with sinuously curved backs. Period III (19,000–15,000 years ago) marked cave art's climax as seen in lively, well drawn horses and cattle at Lascaux in South-west France, and relief sculpture elsewhere. Period IV (15,000–10,000 years ago) stressed portable art, symbolic marks, and the superbly life-like creatures in caves such as those of Altamira in northern Spain and Font-de-Gaume in France.

Most paintings lay deep in caves where artists evidently worked in the light of burning wood, or moss-wick lamps. Perhaps their skills had ritual significance. Some illustrations depict part-human, part-animal figures, supposed "scorcerers" or shamans. Hunting magic or sexual symbolism probably explain the many scenes of meaty creatures of the chase. Fertility symbolism could account for human figurines with exaggerated female features. And geometric patterns suggest notation systems, one supposedly depicting phases of the Moon. But all interpretations are open to dispute.

Two problematic figures
A "Venus" of Willendorf in Austria, a mammoth-ivory figurine with exaggerated female features
B "Sorcerer" from Les Trois Frères cave in France – painting of a figure that seems half stag, half man.

Phases of the moon? (below)
a Bone plaque from Sergeac, France (somewhat reduced)
b Engraved symbols on the plaque, one per night, may trace phases of the moon.

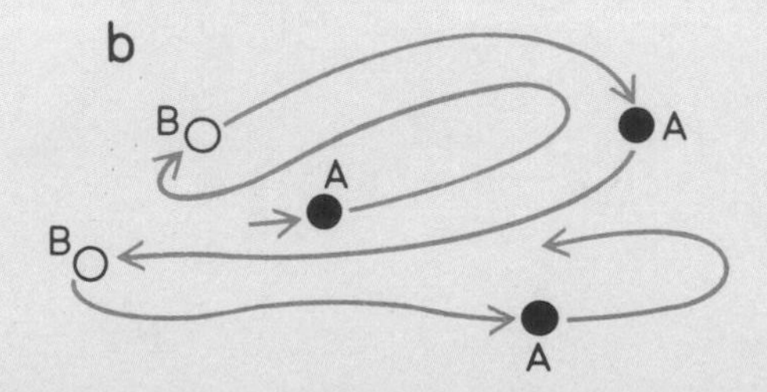

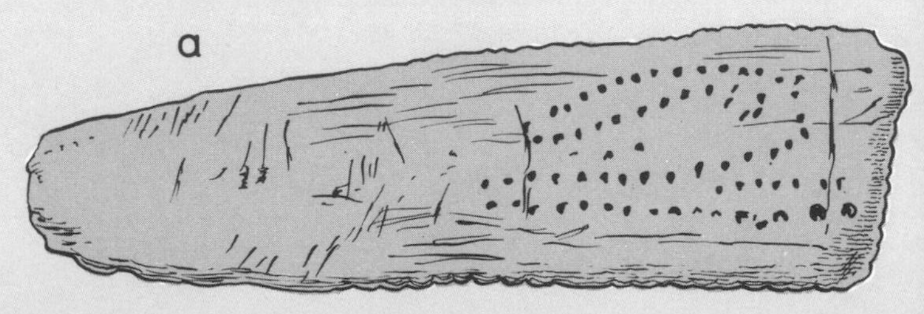

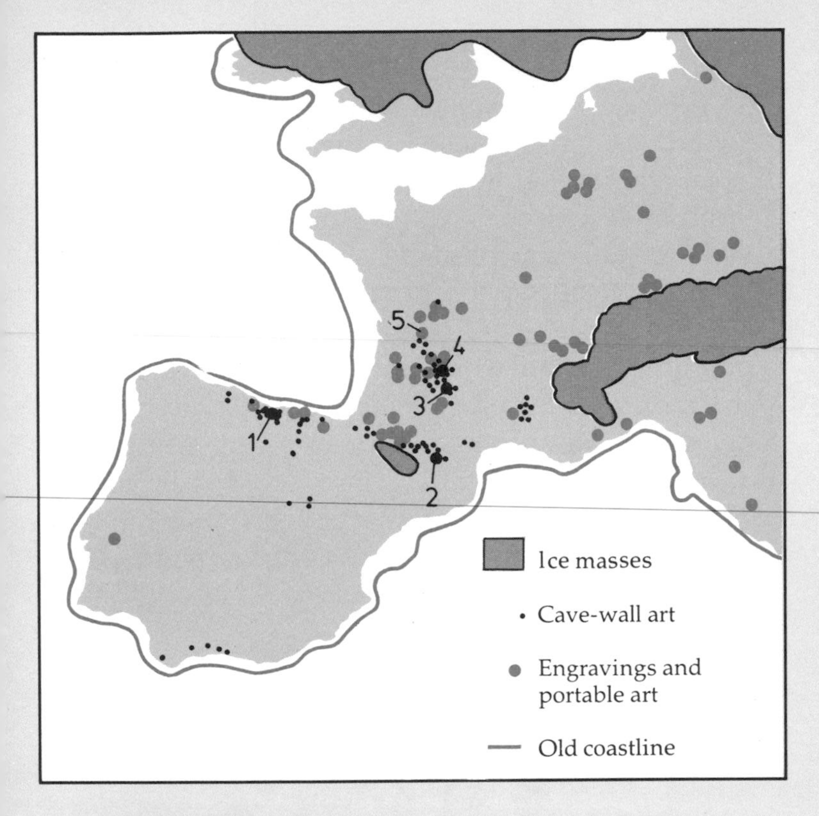

West European art sites (left)
This map of Late Paleolithic Western Europe includes ice masses and the coastline then and now. Symbols show cave-wall art clustered in northern Spain and South-west France, with engravings, figurines, and other items farther north and east. Selected sources of major art objects include:
1 Altamira
2 Niaux
3 Pech-Merle
4 Lascaux
5 Montgaudier

Cro-Magnon masterpieces (below)
Three items from France:
1 Bison licking its back, a reindeer-antler carving
2 Painted horse outline with dots and hand prints added, on a wall in Pech-Merle cave
3 Seals engraved on an antler baton, from Montgaudier. Such perforated batons perhaps served to straighten spears.

Cro-Magnon burials

Burials as far apart as France and Russia hold clues to life among Cro-Magnon peoples.

Studied skeletons suggest that two-thirds of Cro-Magnons reached the age of 20 against under half of the Neandertals before them; and one in 10 reached 40 compared with one in 20 Neandertals. Although based on scanty samples, these figures might imply that life expectancy was rising.

More reliably, Cro-Magnon burials speak of symbolic ritual, and evolving wealth and status. Many mourners had sprinkled the dead with red ocher, presumably symbolizing blood and life, and so hinting at belief in afterlife. Some corpses had gone richly ornamented to their graves – early signs that hunter-gatherer societies had begun producing rich, respected individuals.

Refined grave goods (above)
Two Late Paleolithic necklaces buried with their owners hint at accumulating personal possessions.
1 A necklace of perforated bone beads and pendants, buried with a child in Siberia
2 Part of a three-row necklace with two rows of perforated fish vertebrae, one row of perforated shells, and perforated deer's teeth linking all rows. It was buried with a young man in South-east France.

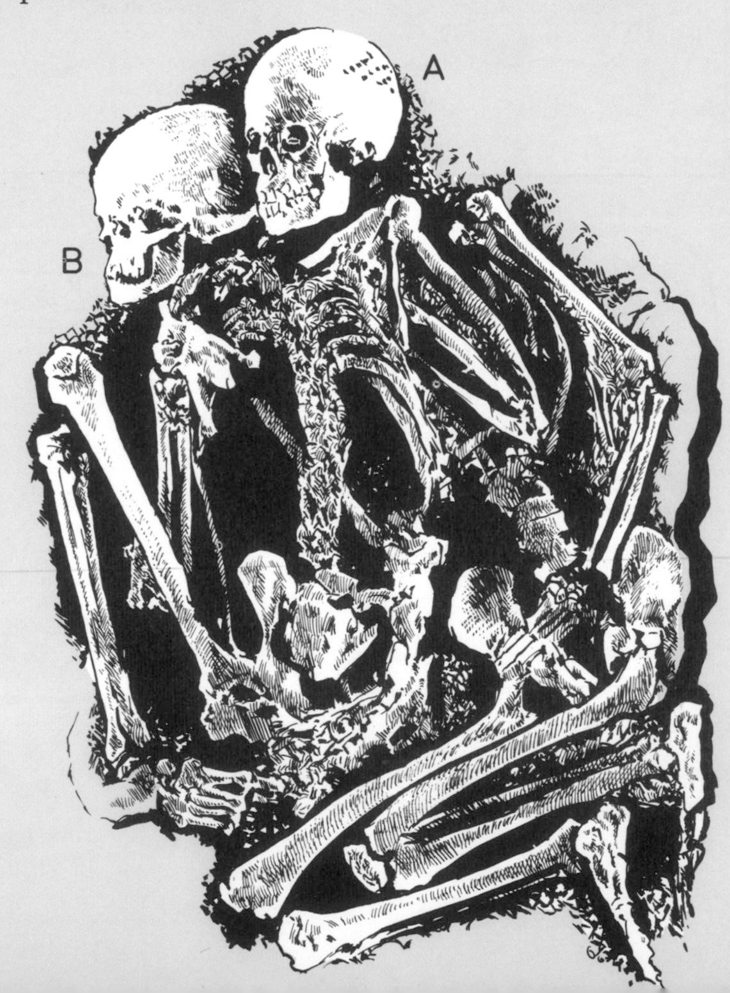

Grimaldi "Negroes" (right)
These skeletons were found in a single, shallow burial inside the Grotte des Enfants, a cave in Italy. Distortion after death produced projecting jaws that misled experts into identifying them as immediate ancestors of modern Negroes.
A Youth 5ft 1½in (1.56m) tall.
B Old woman 5ft 3in (1.6m) tall.

Here are just a few burial examples.

The Cro-Magnon rock shelter near Les Eyzies in South-west France contained five adults and an infant, colored with red ocher, and buried with Aurignacian flint tools and pierced sea shells.

The Grimaldi fossils from Riviera Italy comprise 16 individuals buried in several caves up to 30,000 years ago. The Grotte du Cavillon contained a man with a stag's teeth crown and a bonnet embroidered with hundreds of whelk shells. The Grotte des Enfants included two children with whelk-shell ornaments. The Barma Grande held individuals with necklaces of animal vertebrae.

Předmost in Czechoslovakia revealed 29 deliberately buried skeletons, with clay figurines, and tools of horn and bone.

Perhaps the most striking finds of all came from a 23,000-year-old hunters' burial ground at Sungir east of Moscow. Here lay one old man in elaborately beaded fur clothes. Nearby, two boys wore beaded furs, ivory rings and bracelets; beside them lay long mammoth-tusk spears and two of the strange, scepter-like carved rods called *bâtons de commandement*.

A princely burial (above)
Two skeletons reveal remains of boys aged 12 to 13 buried head to head 23,000 years ago at Sungir in Russia. The wealth of grave goods buried with them foreshadowed royal burials of much later times.

Ornamented corpse (left)
More than 1000 beads and other ornaments embellished the fur clothes of an old man whose buried skeleton was excavated at Sungir in 1955. The furs had rotted many centuries ago.

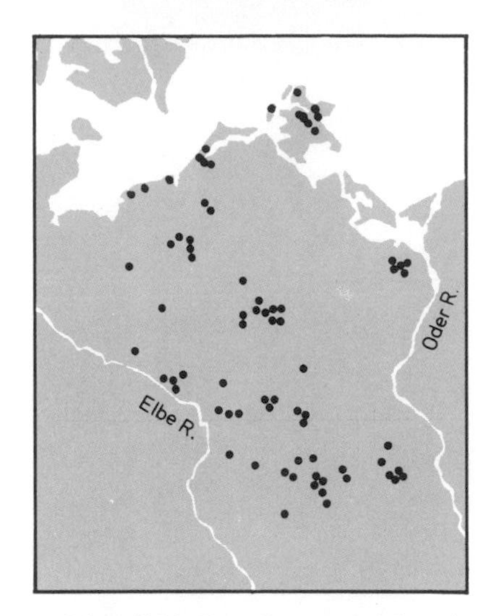

Mesolithic East Germany (above) Dots show contemporary sites, most evidently small. An area of 18,000sq mi (47,000sq km) held perhaps only 1500 people.

The hunters' prey (below) This shows how many of 165 European sites held each of the most widely hunted mammals.
1 Red deer
2 Wild boar
3 Wild ox
4 Roe deer
5 Fox
6 Badger
7 Wild cat
8 Beaver

The Mesolithic period

By 10,000 years ago the cold Pleistocene epoch had given way to the Holocene or "Recent" epoch – the mild phase we live in now. As Europe's climate warmed up, expanding forests invaded huge tracts of former tundra and rising sea drowned low shores and river valleys. Climatic change and hunting pressure destroyed the great grazing herds that Cro-Magnons had depended on for food. But, between them, land and water now teemed with forest mammals, fish, and waterfowl.

Tools and weapons depicted on the next two pages enabled northern Europeans to exploit this food supply. Their distinctive hunter-gatherer societies produced the cultural Mesolithic Period or "Middle Stone Age," so named because it followed the herd-hunting Late Old Stone Age and preceded the entry of New Stone Age farming into northern Europe. Lasting only from about 10,000 to 5000 years ago, the Mesolithic occupied a mere eyeblink of prehistoric time.

Bones found at Mesolithic sites reveal that hunters killed and ate such prey as red and roe deer, wild boar, wild cattle, beaver, fox, ducks, geese, and pike. Huge heaps of shellfish show that these were widely eaten on Atlantic and North Sea coasts. But Mesolithic peoples also gathered roots and fruits including nuts. Groups of people evidently moved from place to place as different foods reached peaks of seasonal abundance.

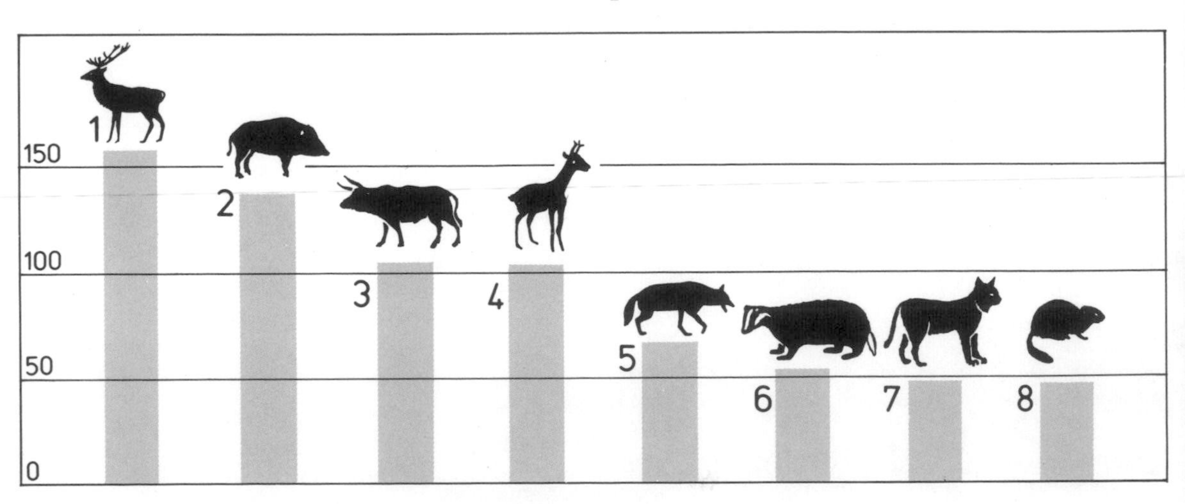

Archeologists believe that Mesolithic people lived in smaller groups than their likely ancestors, the Cro-Magnon big-game hunters. But year-round food supply was now more stable, which helps to explain why camp sites and evidently population multiplied. It also seems that life expectancy increased. At Vlasac on the River Danube in Yugoslavia, several dozen adult skeletons show that, once past childhood, females at this favored spot attained an average age of 35, males 55. Few known Paleolithic individuals lived as long as this.

The yearly cycle
Five seasons formed a year's food cycle in North-west Europe 8000 years ago. Deer and ox were eaten all the year, but varied in abundance.

a Red deer
b Small game
c Wild boar
d Wild ox
e Fish
f Plants

Mesolithic toolkits

New stone-edged tools and weapons helped Mesolithic peoples exploit the forest and sea that encroached on parts of North-west Europe after northern ice sheets melted.

Key hunting aids included bows and arrows, probably a Late Paleolithic invention. Lighter to carry than a sturdy spear, a bow and arrows were also more accurate, afforded greater range, and required less muscle power. A skilled archer could shoot an ibex 35 yards (32m) away, and if his first arrow missed, had time to fire another.

Mesolithic arrows were mostly barbed or tipped with small, sharp flints called microliths. Many were no more than one centimeter long, and trimmed in standard forms: as crescents, rods, triangles, or trapezes. Resin-cemented into shafts of antler, bone, or wood, microliths

Early archery (above)
A bowman grips three arrows while he fires a fourth. Pictures of warring archers figured on cave walls in Spain by Epipaleolithic times, southern Europe's post-glacial equivalent of the Mesolithic period farther north.

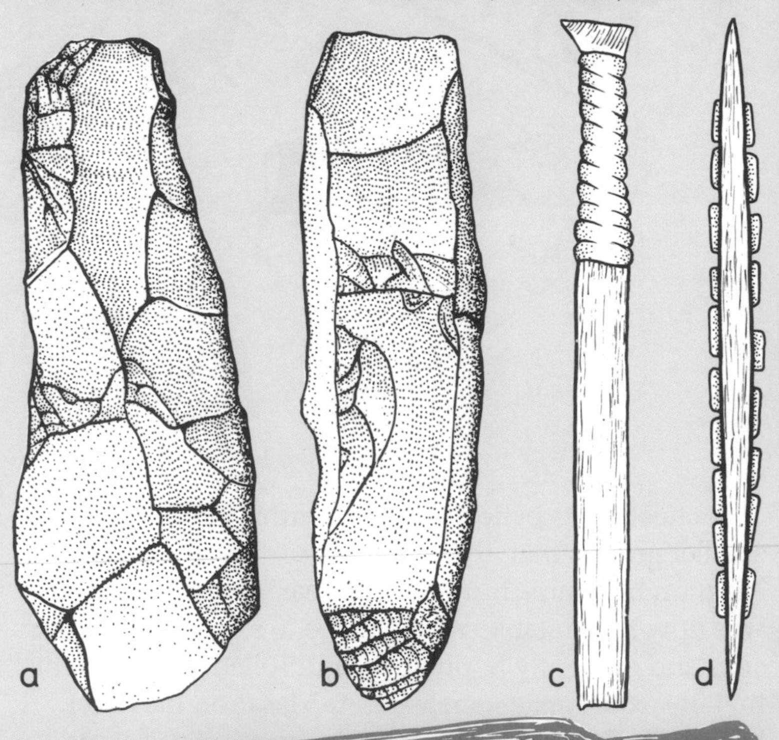

Flint implements (right)
These are not to scale:
a Axe
b Adze
c Arrow with microlith head
d Bone fish spear with microlith barbs

Dugout canoe (below)
Flint blades socketed in handles of antler, bone, or wood sculpted a solid tree trunk into this open boat, discovered in the Netherlands.

could simply be replaced if lost or blunted. Moreover, they were economical to make: a skilled microlith maker could produce 300 feet (91.5m) of cutting edge from just two pounds (987gm) of flint.

New types of large stone tool helped Mesolithic peoples fell and fashion timber. They chopped down trees with stone-headed axes, and shaped trunks and branches with stone-bladed adzes. To sharpen these tools they removed flakes or ground cutting edges on hard stone slabs. Then they hafted the heads in antler sleeves inserted into wooden handles.

Axes and adzes enabled Mesolithic peoples to produce dugout canoes, paddles, skis, and sledges. Between them, these opened up huge water areas for fishing, and made it easier to travel over snow and boggy ground.

Making microliths (below)
1 Flint blade
2 Blade notched on opposite sides
3 Blade snapped in three
4 Central piece retouched as a rhomboidal microlith
5 Blade notched on one side
6 Blade snapped in three
7 Central piece retouched as:
a Trapeze
b Lunate
c Triangle

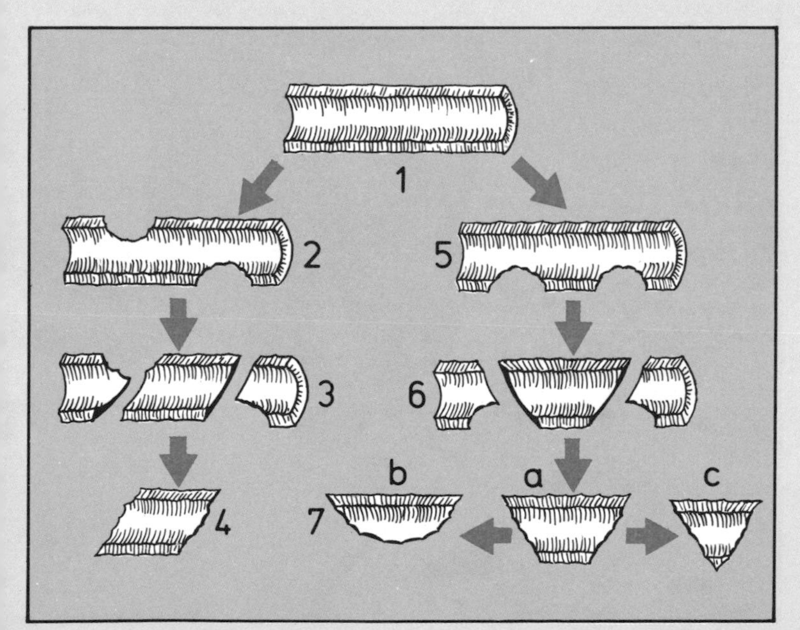

Microlith types (below)
These represent four Mesolithic cultures:
1–3 Azilian, named after the Mas d'Azil cave in the French Pyrenees.
4–6 Maglemosian (or Forest culture), named after Maglemose in the Dutch province Zeeland
7–9 Sauveterrian, named after Sauveterre-la-Lemance in France, but widespread in western Europe
10–12 Tardenoisian, named after Fère-en-Tardenois, in northern France.

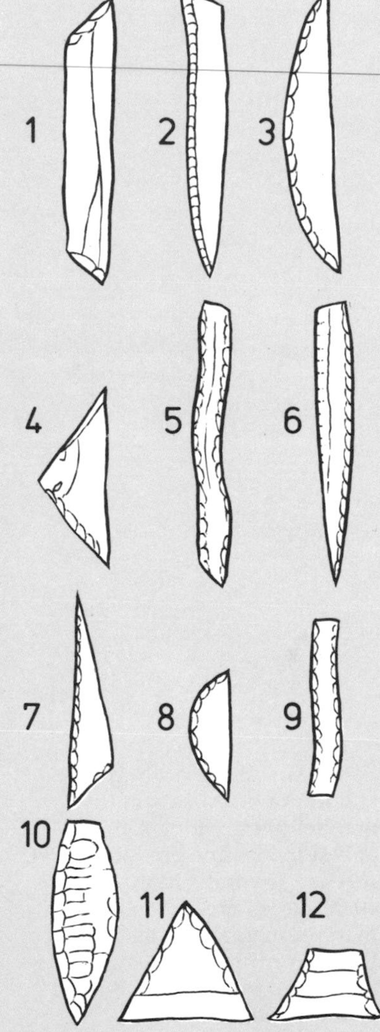

Chapter 9 MODERN MAN WORLDWIDE

Our own subspecies' emergence in Europe coincided with its rapid spread in Africa, Asia, and Australia. Indeed, modern man probably evolved in Africa 100,000 years ago or more from a local archaic form of *Homo sapiens*.

This chapter briefly surveys early finds of modern man around the world and shows how local differences of climate contributed to racial differences observed today.

Some Australian Aborigines still hunt with spears in the manner of our Stone Age ancestors. People probably reached Australia more than 40,000 years ago, when modern man was colonizing all the Old World continents. (Australian News and Information Bureau photograph)

Origins of modern man

By the 1980s it seemed increasingly unlikely that Neandertals had given rise to Europe's Cro-Magnons, or that these had been the world's first fully modern humans. Skeptics claim that any Neandertal-Cro-Magnon evolutionary leap would have been too great to have occurred as fast as fossil evidence implies. Instead, most argue, our subspecies arose from an archaic form of *Homo sapiens* in Africa, about 100,000 years ago.

"Modern" fossil skulls that old indeed suggest that Africa is where our ancestors originated (see pp. 178–179). Other evidence is living all around us in the genes inherited by different racial groups. Scientists once classified these groups by unreliably superficial characters like body form and color. Later, biologists compared more stable features such as blood groups and their proteins. But various DNA sequences can produce genes with the same protein products. The surest way to trace long-term inherited relationships is to compare the DNA sequences themselves. Now biologists have done that, too.

In 1986 British scientists James Wainscoat and Adrian Hill reported using biological agents called enzymes to isolate five DNA fragments from a gene responsible for producing part of the hemoglobin molecule of red blood cells. They recorded the different patterns that these fragments formed in 600 people including Africans, Britons, Indians, Melanesians, and Thais. It emerged that non-Africans shared a limited number of common patterns, while Africans stressed a pattern not found in other groups. The scientists concluded that one small, inbred group of prehistoric Africans had given rise to all the other peoples of the world.

Many subscribers to this theory believe that modern man arose in Africa by 100,000 years ago, entered Asia not very much later, and by 30,000 years ago had reached all continents except Antarctica.

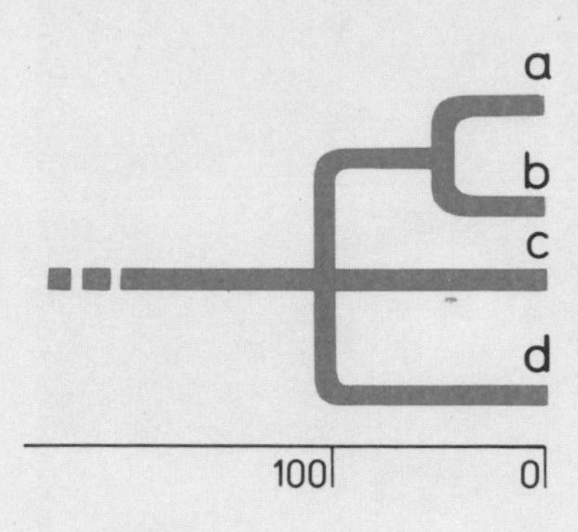

One view of racial origins
Above we show racial origins according to studies of proteins and DNA in chromosomes. Numbers are in thousands of years ago.
a Australian Aborigines
b Orientals
c Blacks
d Whites

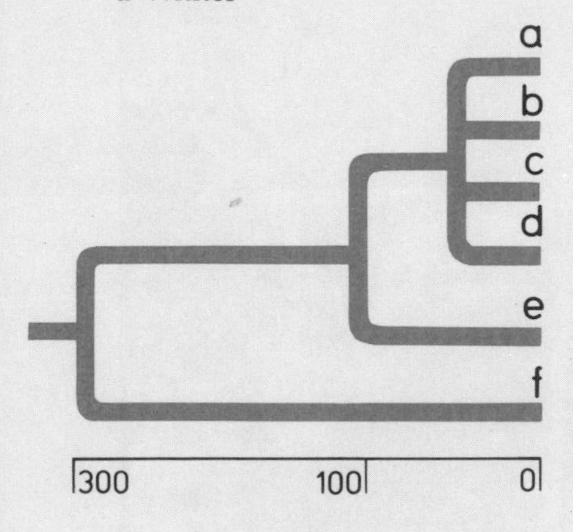

Racial origins: another view
One study of mitochondrial DNA produced this rival theory of racial origins. Again, numbers are in thousands of years ago.
a Most Australian Aborigines
b Most Orientals
c Blacks
d Whites
e Some Orientals
f Some Australian Aborigines

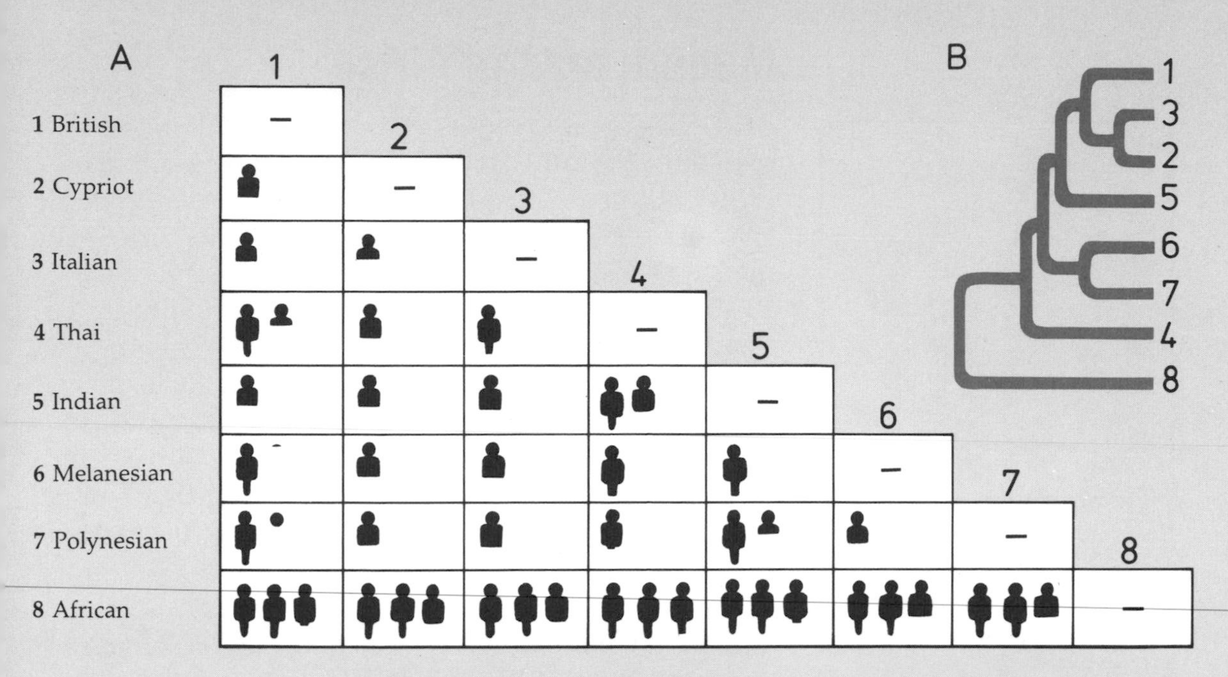

US scientists Rebecca Cann and Mark Stoneking reached geographically similar conclusions by studying mitochondrial genes, passed on by females only. These imply that our subspecies began with a woman who lived in Africa 200,000 years ago – much earlier than fossil evidence suggests.

More genetic research should shed fresh light upon our hazy origins.

Genetic distances (above)
A Symbols represent genetic distances among eight population groups based on the 600 individuals mentioned in our text. Between any two groups, the fewer the symbols the shorter the genetic distance and the closer the presumed relationship.
B This diagram arranges the eight groups as a family tree.

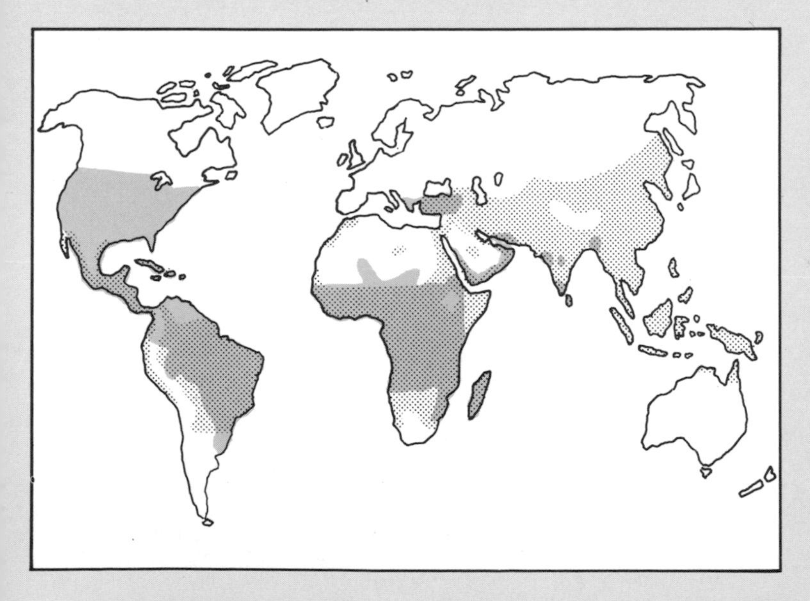

Sickle-cell trait (left)
This map shows recent distributions of sickle-shaped red blood cells and malaria in human populations. Sickle cells originated with a mutation in one individual. The gene producing sickle cells confers resistance to malaria but causes anemia if inherited from both parents.

Sickle cell distribution

Malaria

Modern man in Africa

Africa has yielded older fossil finds of fully modern man than any other continent. Thus from southern Ethiopia comes Omo I, a probably more than 60,000-year-old partial skull with many modern features. The mouth of South Africa's Klasies River produced 100,000-year-old "modern" fragments, and Border Cave, a "modern" 90,000-year-old lower jaw.

Some other ancient bones foreshadow today's Negroes and Bushmen. Supposed ancestral features of both groups figure in a more than 100,000-year-old skull from Florisbad, South Africa. But distinctive, childlike Bushman skulls predate the 17,000-year-old first-known fossil Negroes. The Bushman-Hottentot group seemingly once occupied all southern Africa, and bands of Bushmen hunter-gatherers still live an almost Paleolithic lifestyle in Botswana's Kalahari Desert.

Sites and centers (above)
1 Fish Hoek
2 Klasies River Mouth
3 Florisbad (archaic *sapiens*?)
4 Border Cave
5 Bushmen today
6 Mumbwa
7 Lukenya Hill
8 Elmenteita
9 Ishango
10 Omo
11 Negroes' alleged center of origin
12 Afalou-bou-Rhummel

Hunter-gatherers today (below) Like their Paleolithic forebears, some Bushmen still hunt with bow and arrows and chiefly eat wild plants.

Numbered illustrations depict examples of past and present Africans of modern form.

1 **Omo I:** a thin-boned, partial skull, with bulging upper sides, rounded back, slight brow ridges, well developed chin, and 1400cc brain capacity. Some experts rank it as archaic *sapiens*. Age: over 60,000 years. Place: Omo River, Ethiopia.

2 **Florisbad skull:** a large, low-vaulted skull, with brow ridges, broad face, protruding mouth, and low, rectangular eye sockets. Age: over 100,000 years. Place: Florisbad, South Africa.

3 **Bushman:** a short, slim, yellowish, wiry racial form of modern man with high cheekbones and childlike facial features including a small face; flat, broad nose; and prominent forehead. Time: perhaps dating back 40,000 years. Place: Southern Africa.

4 **Negro:** a dark-skinned modern form of variable size, with distinctive broad nasal opening, outturned lips, and projecting lower face. Time: perhaps dating back 20,000 years. Place: Sub-Saharan Africa (with emigrants worldwide).

Africans past and present
Numbers correspond to items featured in the text.
1 Omo I: Reconstruction based on a partial skull
2 Florisbad skull: Reconstruction showing old-fashioned features that might merit its inclusion in archaic *Homo sapiens*.
3 Modern Bushman
4 Modern Negro

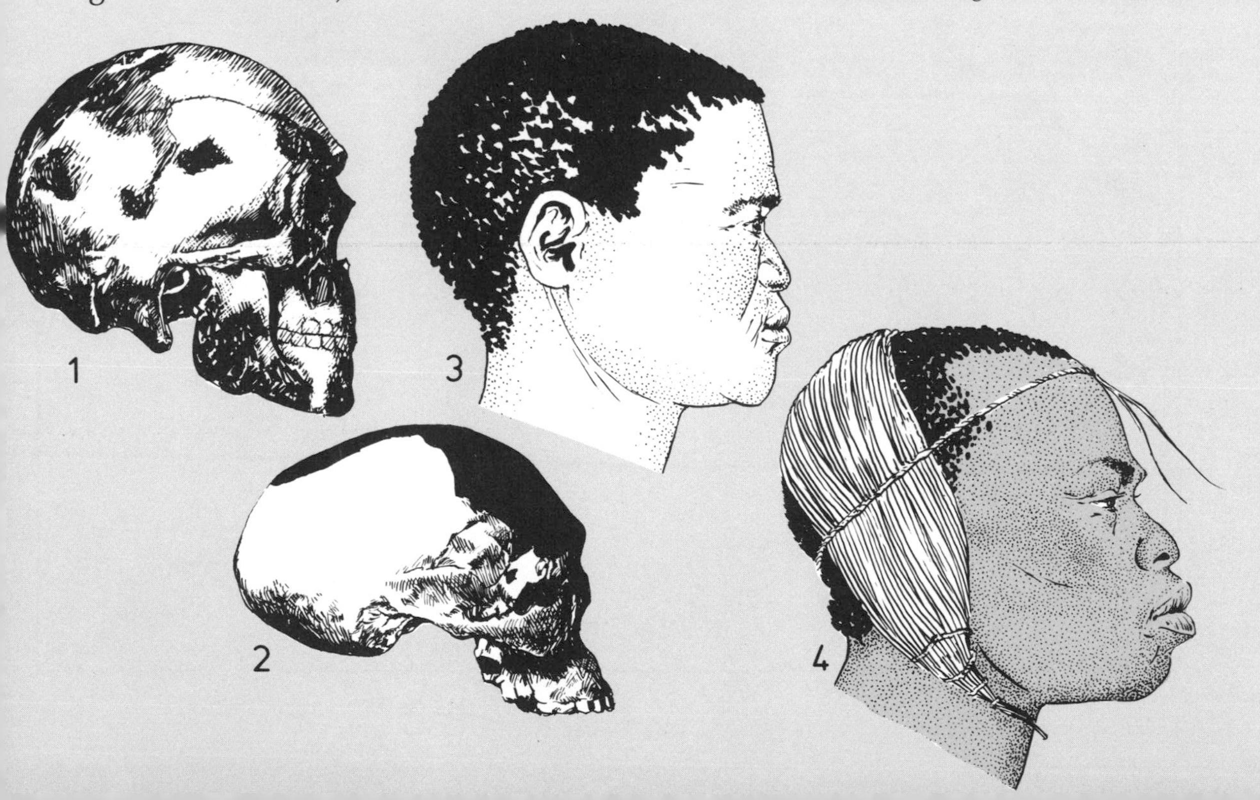

Modern man in Asia

Where they lived
This map shows 15 sites of finds of fully modern man of 100,000–50,000 years ago.
1 Qafzeh
2 Ksar Akil
3 Mahadaha
4 Attirampakkam
5 Batadombalena
6 Novoselovo
7 Malta
8 Ordos (artefacts)
9 Zhoukoudian (Choukoutien)
10 Ziyang
11 Liujiang
12 Minatogaa
13 Tabon Cave
14 Niah Cave
15 Wadjak

Fossil skulls up to 40,000 years old and attributed to fully modern man crop up in Asia as far apart as Israel and Java. All have a chin or other crucial "modern" features. Yet there are West Asian and South-east Asian forms with old-fashioned details found in the Neandertals (see pp. 142–143), and some Far Eastern forms that confusingly recall the much earlier *Homo erectus*. Experts have disagreed about what this implies.

Certain Chinese skulls suggest that by 20,000 years ago there lived ancestors of today's Asian Mongoloids – short, stocky people ranging from white to brown, with straight dark hair, sparse body hair, a fold in the upper eyelids, flat faces, wide cheek bones, and narrow noses.

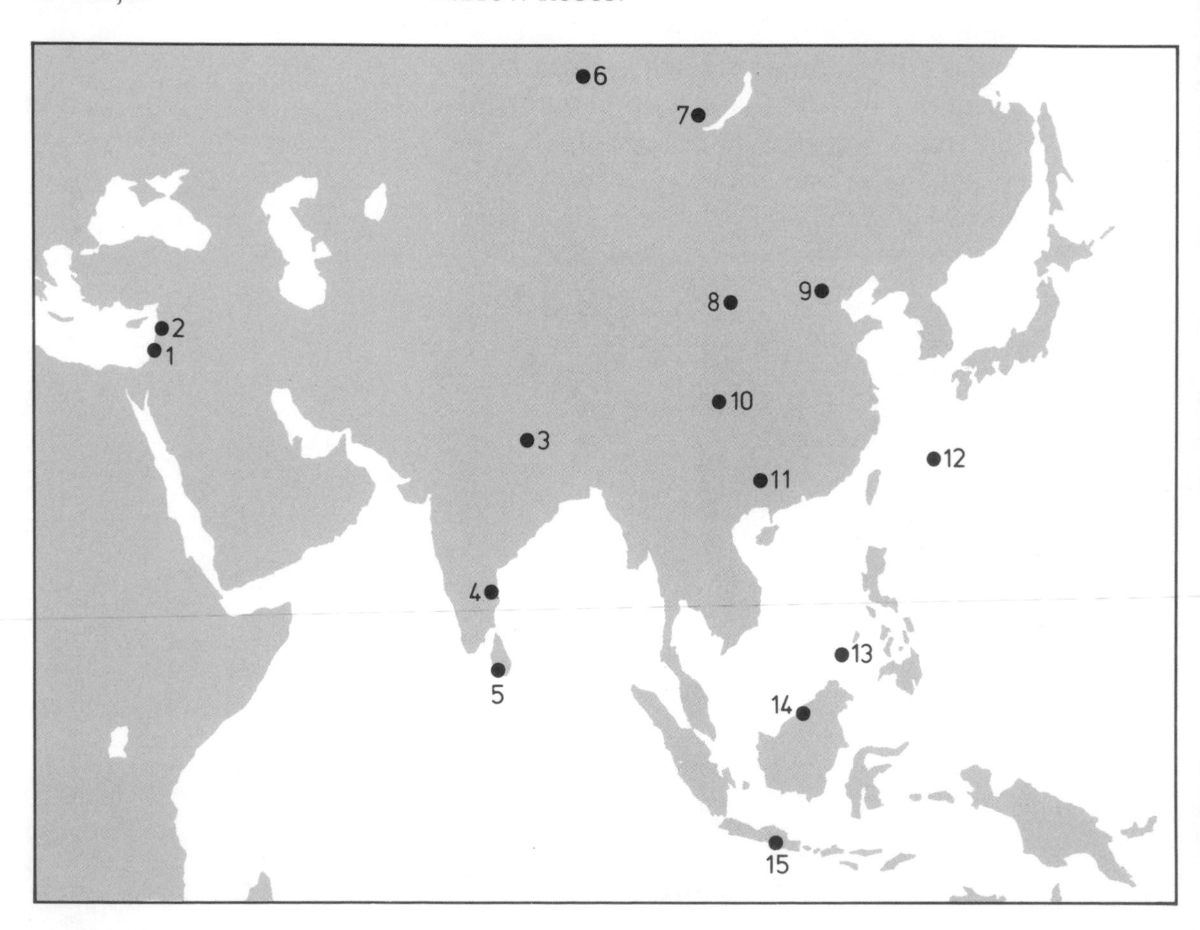

Numbered illustrations depict past and present Asian examples of fully modern man.

1 Qafzeh Despite brow ridges recalling the Neandertals, this skull has been identified with fully modern man. (But see also pp. 142–143.) Age: 33,000 years. Place: Qafzeh Cave, Israel.

2 Niah Man is known from skull fragments, teeth, and a foot bone from a delicately built young youth. Age: 40,000 years. Place: Niah Cave, Borneo.

3 Wadjak Man had a long skull, sloping forehead, brow ridges, jutting mouth, and heavy jaws. Age: uncertain. Place: Wadjak, Java.

4 Liujiang Man represents anatomically an early modern man from China. Age: 20,000 years. Place: Liujiang, Kwangsi Province, China.

5 Asian Mongoloid a modern skull with flat face, narrow nose, and angulated cheekbones – features foreshadowed in some fossil Chinese skulls.

Asians past and present
Numbers correspond to items featured in the text.
1 Qafzeh skull
2 Niah man
3 Wadjak man
4 Liujiang man
5 Asian Mongoloid

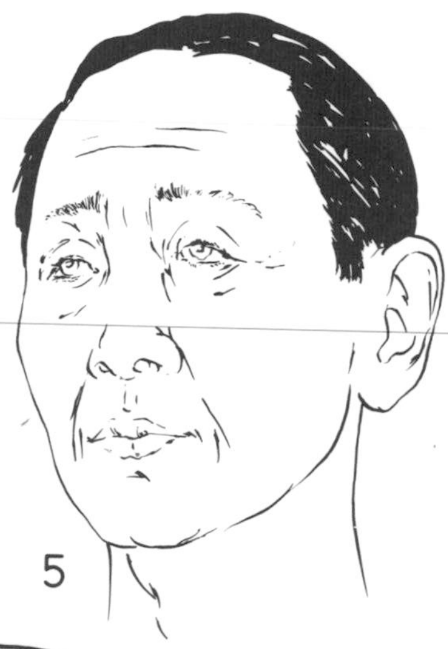

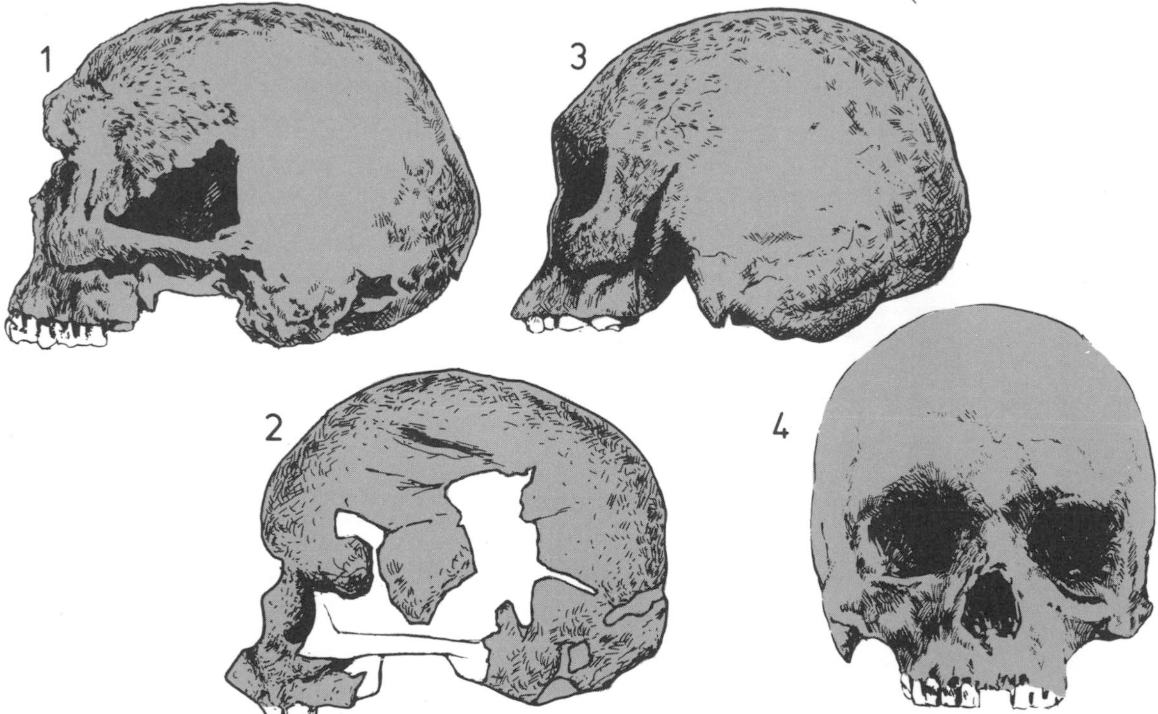

Del Mar Man (above)
A Californian coastal cliff yielded this skull of the "oldest American," since demoted to less than a quarter of its alleged antiquity.

Early artefacts (below)
1 Clovis point of 10,000 BC, fluted for hafting
2 Bone flesher of 25,000 BC, serrated at one end for scraping hides

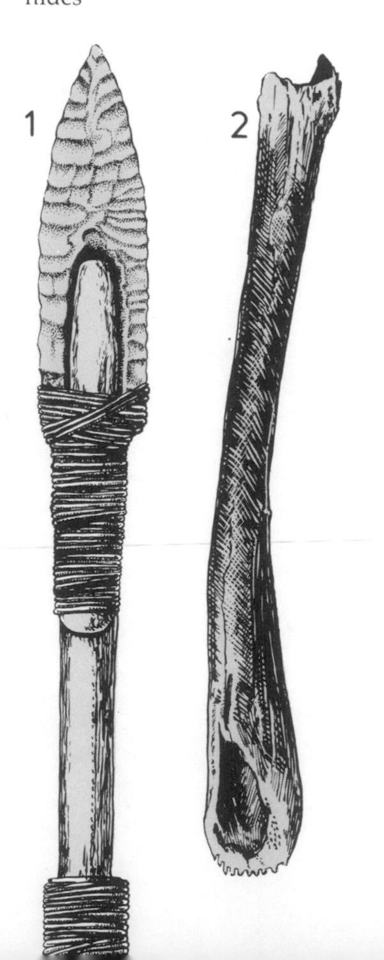

Man enters the Americas

The first humans to reach North America probably arrived some time between 70,000 and 12,000 years ago. At the coldest intervals between those dates, sea shrinkage exposed a broad land bridge, Beringia, now drowned beneath the Bering Strait. Hunter-gatherers probably walked dry shod from Siberia to Alaska. Moving south through one of two ice-free routes, they fanned out across the Great Plains, then penetrated south to Patagonia.

Anthropologists believe these so-called Paleo-Indians had straight black hair, coppery skin, dark eyes, broad cheekbones, and shovel-shaped incisor teeth. Such Mongoloid features occur in American Indians alive today.

But just when their ancestors arrived remains debatable. Late arrivalists have argued that the first firm evidence goes back a mere 12,000 years. About then, distinctive stone spearheads with delicately fluted bases appeared as far apart as Alaska and Mexico. These so-called Clovis points may represent an explosive population growth and spread as Paleo-Indians' reached game-rich lands south of the northern ice sheets.

Early arrivalists claim man entered the Americas more than 40,000 years ago. But redating of Del Mar Man, a Californian skull supposedly 48,000 years old, has reduced its age to 11,000 years.

That leaves the "middle entry" theory: arrival 30,000 years or so ago. Supporting this are finds from Yukon's Old Crow Basin, notably an evidently 27,000-year-old bone "backscratcher" designed for scraping hides. Other well accepted finds aged 20,000 years or more include 22,000-year-old hearths and bones from Mexico.

Paleoanthropologists still argue about when humans first set foot in the Americas, and whether immigrants (before those very late arrivals, Eskimos) arrived in several successive waves. One fact is not in doubt: by 11,000 years ago these two last, empty, habitable continents were peopled end to end.

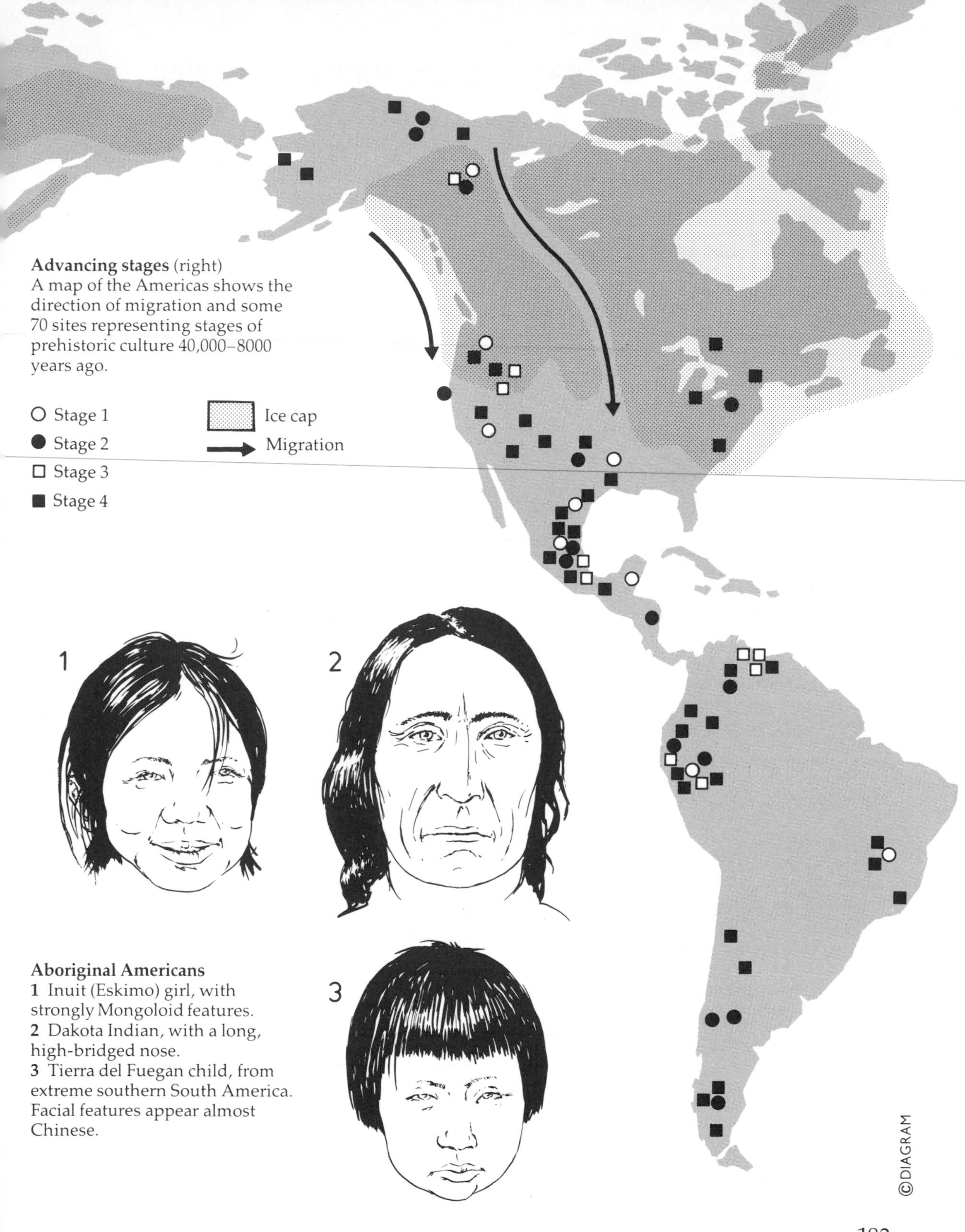

Advancing stages (right)
A map of the Americas shows the direction of migration and some 70 sites representing stages of prehistoric culture 40,000–8000 years ago.

Aboriginal Americans
1 Inuit (Eskimo) girl, with strongly Mongoloid features.
2 Dakota Indian, with a long, high-bridged nose.
3 Tierra del Fuegan child, from extreme southern South America. Facial features appear almost Chinese.

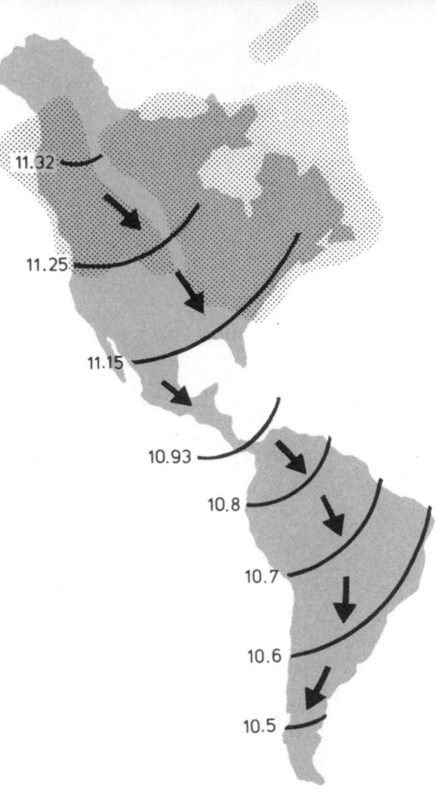

Waves of extinction (above)
The map above shows waves of extinction spreading south through the Americas with man. Bands show one theorist's notion of the rate of spread. Numbers represent thousands of years ago.

Man and animals in America

Between about 12,000 and 5,000 years ago, a great wave of extinctions swept away many of the larger animals of North and South America. Victims included seven prehistoric kinds of elephant (four mammoth species and three types of mastodon), three kinds of camel, the long-horned bison, the horse, certain pronghorns and musk oxen, giant ground sloths, those gigantic "armadillos" the glyptodonts, and sabertooth cats.

Extinctions peaked between about 12,000 and 10,000 years ago, as Paleo-Indians increased their hold upon both continents. So some people see man as the exterminator. This overkill theory holds that stone-tipped spears launched by spear-throwers accomplished the annihilation. One expert claims that 100 humans advancing south from Canada 10 miles (16km) a year could have given rise to 300,000 people in 300 years by when man would have reached the Gulf of Mexico and could have killed 100 million big mammals on the way.

But extinctions also coincided with dramatic climatic changes, as northern ice sheets melted. Fluctuating temperature and rainfall drastically affected vegetation, robbing certain herbivores of food, and so indirectly starving the carnivores that preyed upon them.

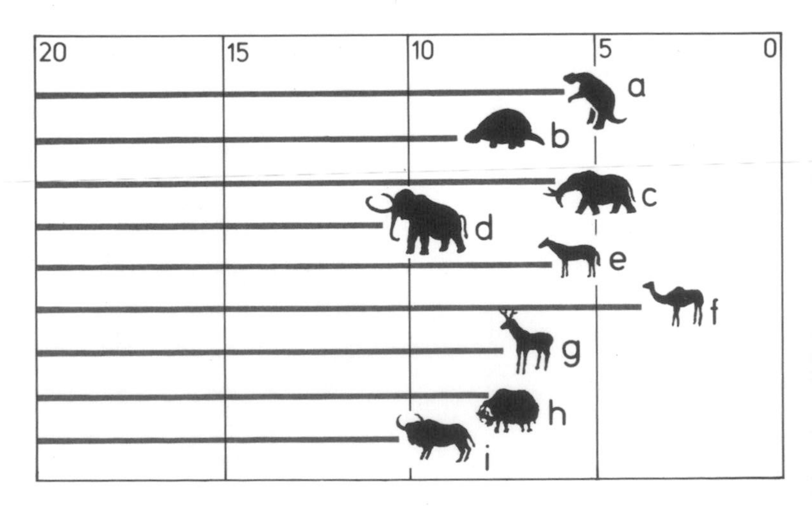

When they went (right)
These large mammals all died out in North America in relatively recent times. Some experts group their deaths later than the map above suggests. Here, numbers show one expert's estimates of how many thousand years ago each group disappeared.

a Ground sloths
b Glyptodonts
c Mastodons
d Mammoths
e Horses
f Prehistoric camels
g Prehistoric pronghorns
h Prehistoric musk-oxen
i Long-horned bison

A third extinction theory blames man and climate jointly. This theory holds that climatic change affecting vegetation reduced the feeding areas available for certain animals. Concentrated in these zones, and weakened by a lack of food, whole herds would have fallen prey to hunters.

As mammoths disappeared and human numbers multiplied, the successive mammoth-hunting Clovis and Folsom cultures gave way about 10,000 years ago to the Plano culture, combining bison hunting with foraging for wild-plant foods. North America's Paleo-Indian hunting cultural stage was superseded by the more mixed economies of the so-called Archaic stage which lasted locally until about 2300 years ago.

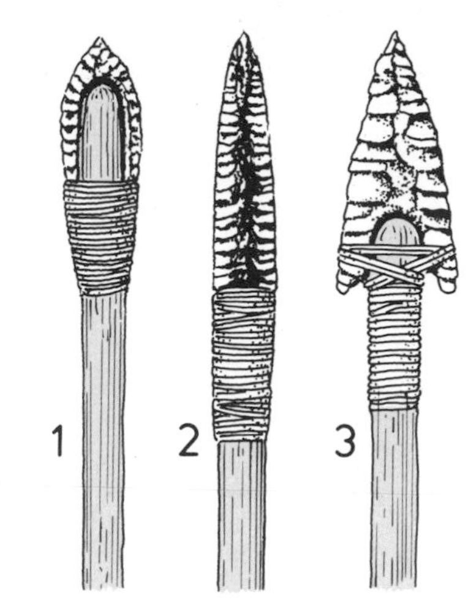

Hafted spear points (above)
1 Folsom point: fluted
2 Eden point: slender
3 Eva point: barbed

Points and prey (left)
Here are implements and prey of six North American prehistoric hunting cultures. Differences between dates here and in the text reflect different expert views.
A Plano, about 7500–5000 BC
B Plainview, 8000–5500 BC
C Folsom, 9000–7000 BC
D Clovis, 13,000(?)–9000 BC
E Sandia, 23,000(?)–10,000 BC
F Pre-projectile, 36,000 (?)–20,000(?) BC
a Modern bison
b Pronghorn
c Long-horned bison
d Mammoth
e Prehistoric camel
f Horse
g Sabertooth cat
h Dire wolf

The first Australians

Dated hearths and fossils prove that modern man lived in Australia at least 40,000 years ago. Some experts think humans had moved in by 120,000 years ago, when New South Wales saw a sharp rise in brush fires and the spread of fire-resistant plants of modern kinds. But Australia has yielded no human skeletons or tools that old.

Most likely, people first arrived 45,000 to 55,000 years ago when ocean level stood 160ft (50m) lower than today and many islands coalesced. South-east Asian migrants on canoes or rafts had to cross a mere 40 miles (65km) of sea to get to Meganesia – the landmass that embraced Australia.

Early Australians were by no means uniform. About 30,000 years ago, Lake Mungo in the South-east was home to people slightly built on modern lines, with some skeletal resemblance to today's Chinese. But about 13,000 years ago Kow Swamp, not

Four Australians
1 Robust skull, about 9500 BC
2 Gracile skull, 28,000 BC
3 Robust skull, elongated by being squeezed in infancy
4 Modern Australian Aborigine, an Australoid, with strong brow ridges, projecting jaws, large teeth, slim build, dark skin and wavy hair.

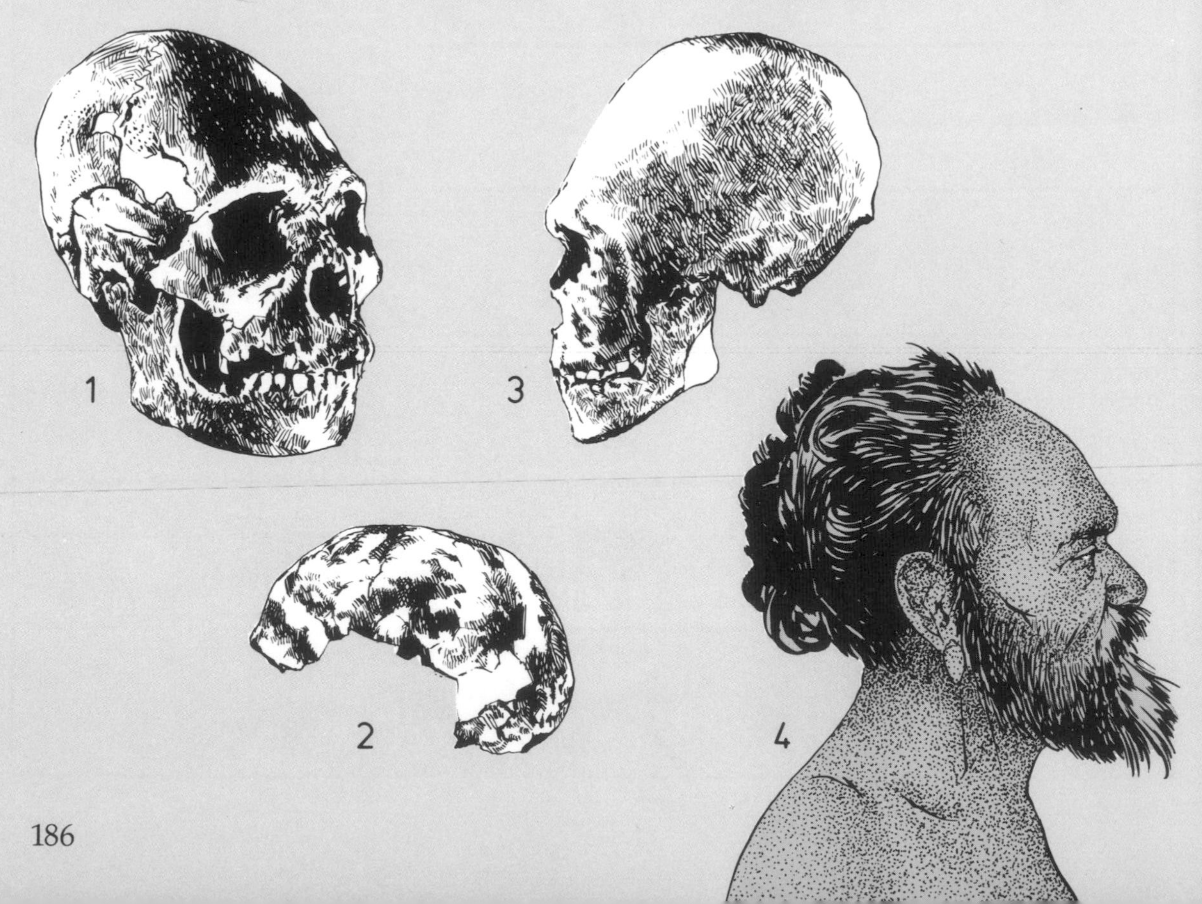

far away, had more strongly built individuals with thick skulls, jutting jaws, ridged brows, and sloping foreheads – features found in Solo man – archaic *Homo sapiens* from Java.

Some experts therefore think Australia received two waves of immigration. Others argue that varied prehistoric forms were just extremes among ancestors of today's Aborigines, whose ridged brows suggest retention of an ancient trait.

Until Europeans landed two centuries ago, all Aborigines still lived a Late Old Stone Age way of life. Small seminomadic bands of near-naked individuals roamed fixed territories. Men armed with spears or boomerangs hunted game. Women gathered plant and insect foods. Myth and ritual played a major role in life, which featured dancing, painting, and verbally transmitted songs and legends. Such practices survive as insights into prehistoric life worldwide.

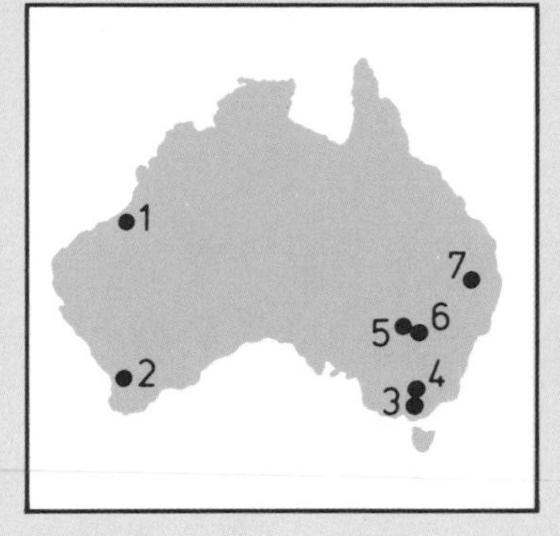

Where they lived (above)
Dots mark selected prehistoric sites in Australia.
1 Cossack
2 Devils Lair
3 Keilor
4 Kow Swamp/Cohuna
5 Lake Mungo
6 Willandra Lakes
7 Talgai

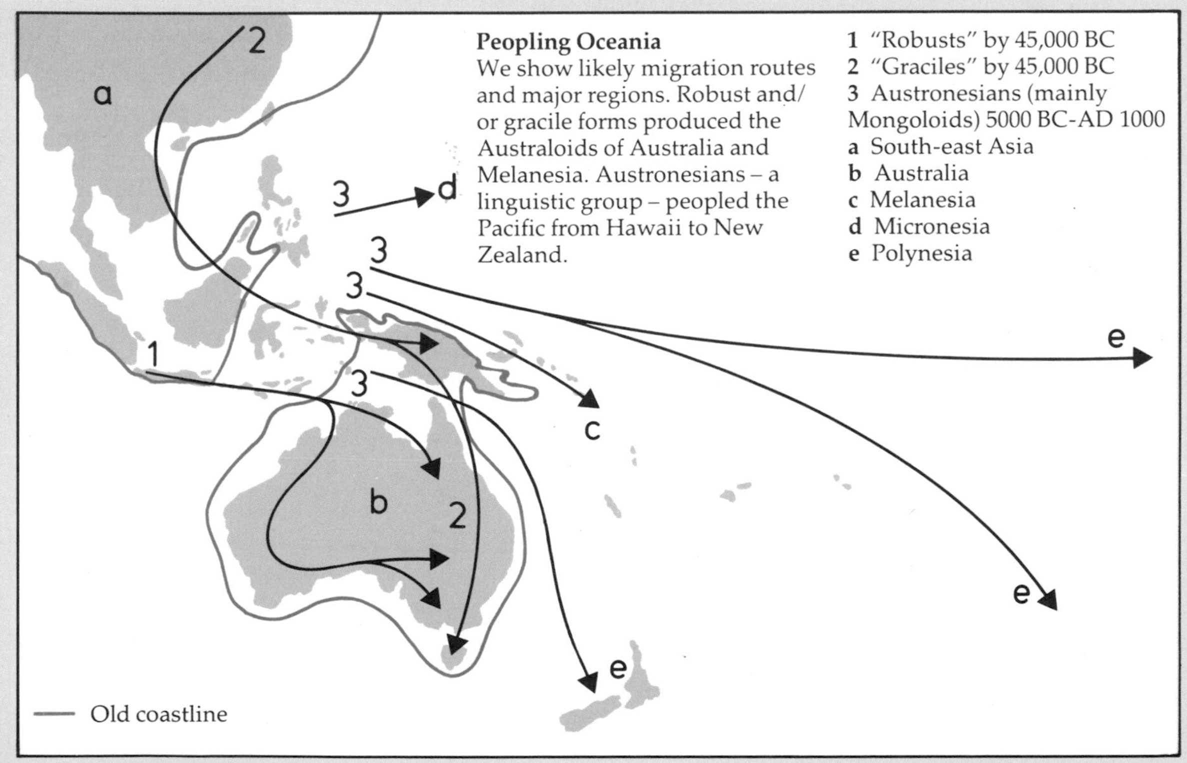

Peopling Oceania
We show likely migration routes and major regions. Robust and/or gracile forms produced the Australoids of Australia and Melanesia. Austronesians – a linguistic group – peopled the Pacific from Hawaii to New Zealand.

1 "Robusts" by 45,000 BC
2 "Graciles" by 45,000 BC
3 Austronesians (mainly Mongoloids) 5000 BC-AD 1000
a South-east Asia
b Australia
c Melanesia
d Micronesia
e Polynesia

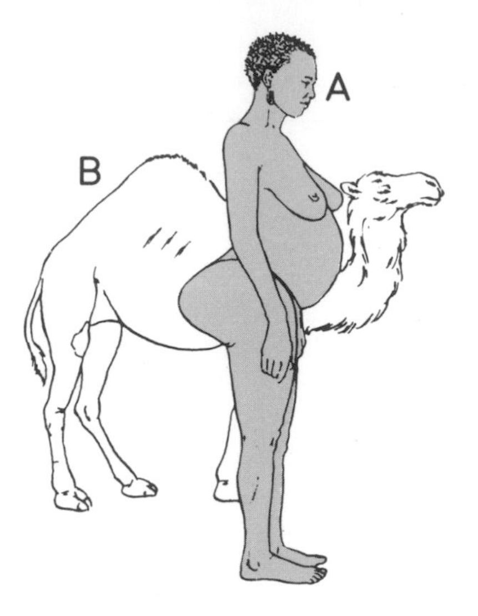

Physique and body fat (above)
Fatty humps or buttocks store food against emergency.
A Female Bushman
B Arabian camel

Skin color (below)
Aboriginals' skin tones tally with the strength of sunshine except where recent migration has given natural selection too little time to work.

Climate, color, and physique

As our subspecies spread around the world, groups of people found themselves in different climatic zones. Natural selection produced physical adaptations to differing conditions. The result: the world's Blacks, Whites, and Mongoloids. Each label by no means indicates a race (a biological breeding group). Negroes and Melanesians are Blacks as unrelated to each other as they are to Whites or Mongoloids. And of course all humans still belong to one subspecies.

With that in mind these two pages examine briefly how major differences between Blacks, Whites, and certain Mongoloids reflect the way their bodies have evolved to cope with hot or cold climates.

Take skin color first. Among Whites, strong tropical sunshine makes pale skins peel and blister, and intense ultraviolet radiation tends to cause skin cancers. Among Blacks, skin darkened by the brown pigment melanin is protected from these risks. Yet in cloudy lands outside the tropics, pale skin is an advantage for it absorbs enough of the weak ultraviolet radiation to make Vitamin D, a substance promoting healthy bone growth. In cool, cloudy lands, young Blacks can suffer rickets.

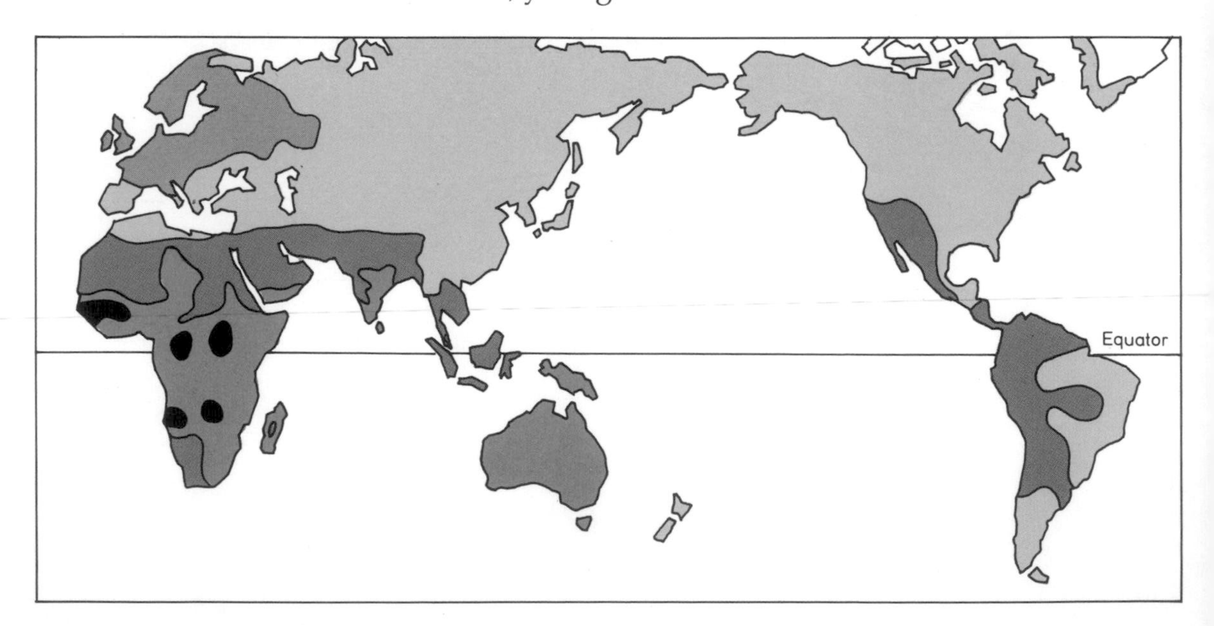

Sunlight and skin cancer (right)
Diagrams contrast the incidence of skin cancers in Europeans and Cape-Coloreds.
1 Rodent epithelioma
2 Malignant lip
a Male European
b Female European
c Male Cape-Colored
d Female Cape-Colored

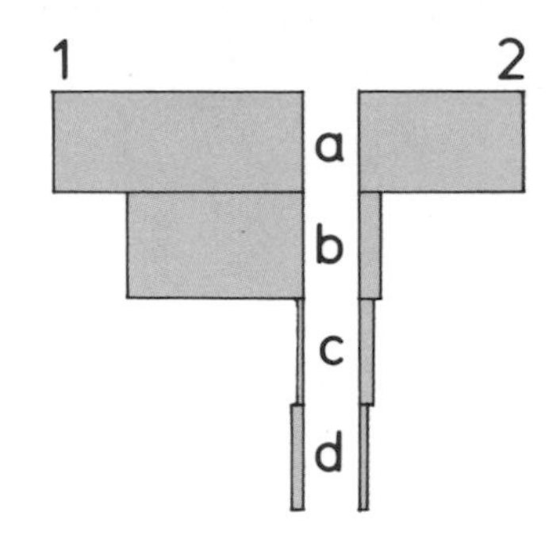

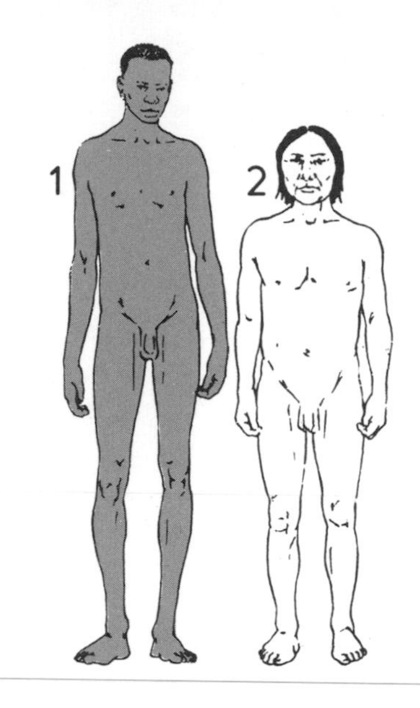

Physiques compared (above)
1 (Equatorial) Nilotic Negro
2 (Polar) Inuit (Eskimo)

There are equally good reasons for differences in body shape and size. In relation to body size, a tall, slim, long-limbed Nilotic Negro has a large surface area, ideal for keeping cool in the tropics by radiating surplus body heat. In similar conditions a (Mongoloid) Eskimo's short, squat body with its small surface area would suffer overheating. But this same body form, with its insulating fat beneath the skin, conserves essential body heat in Arctic winters where a lanky body could suffer fatal chilling.

Of course this oversimplifies the world picture. For instance, by no means all Mongoloids are built like Eskimos. But biological adaptations like those described undoubtedly helped Paleolithic modern man to populate all habitable continents.

Hair form (below)
Hair form seems related to solar heat. The crinkly hair of some tropical peoples protects their brains against heat radiation from the Sun.

Key for Skin color map (left)

- Very pale
- Fairly pale
- Medium
- Fairly dark
- Very dark

Key for Hair form map (right)

- Crinkly or woolly
- Straight
- Wavy-straight

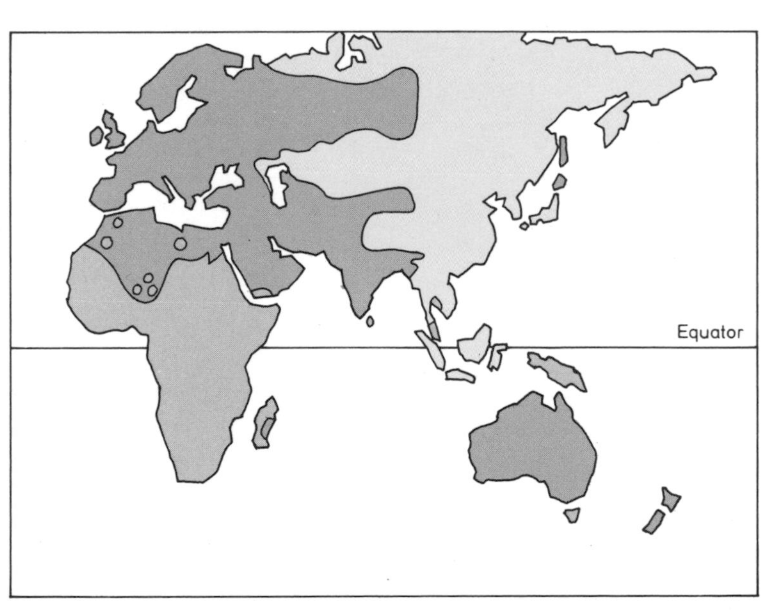

Chapter 10 SINCE THE ICE AGE

Humans have altered physically little since the Ice Age ended about 10,000 years ago. But cultural evolution has transformed humankind. This chapter charts major prehistoric innovations, from food production and the growth of metallurgy to the rise of cities and the birth of writing. We see how cultural change has modified physique and vastly multiplied our species' numbers, and we end with a (daunting) glimpse into the future.

A man plowing with a team of oxen figures in this Swedish Bronze Age rock-engraving. Farming superseded hunting in most of Europe by about 5000 years ago. (Illustration from *The Testimony of the Spade* by Geoffrey Bibby)

A warmer world

Sweeping cultural advances that would bring humans dominance on Earth marked the Holocene or "Recent" geological epoch, which began about 10,000 years ago. This human progress coincided with climatic warming after the last long wave of Ice Age cold that closed the Pleistocene. (In fact the Holocene is probably a brief mild interval with further cold to come.)

Global change was under way soon after northern ice sheets began melting 15,000 years ago. Meltwater pouring into oceans slowly raised their level 430 feet (130m). Sea severed land bridges that had joined Asia and North America, Indonesia and Malaya, the British Isles and Europe. One-twentieth of Earth's land surface drowned. Yet, as ice sheets dripped

Emergent lands (below) Maps show changes to Scandinavia since 8300 BC.
1 8300 BC: Ice weighs down much of Scandinavia
2 7000 BC: Ice has melted but water covers depressed land
3 5000 BC: Much land has bounced back up again.

Ice
Dry land
Water
Modern coastline

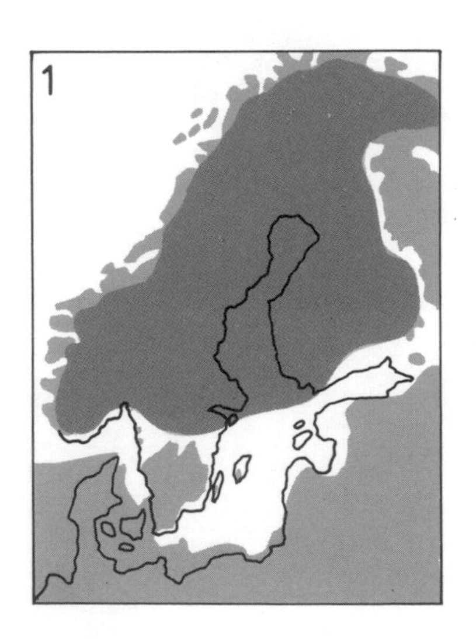

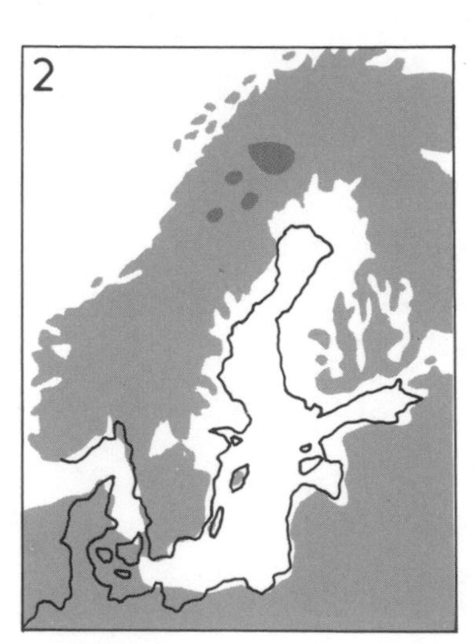

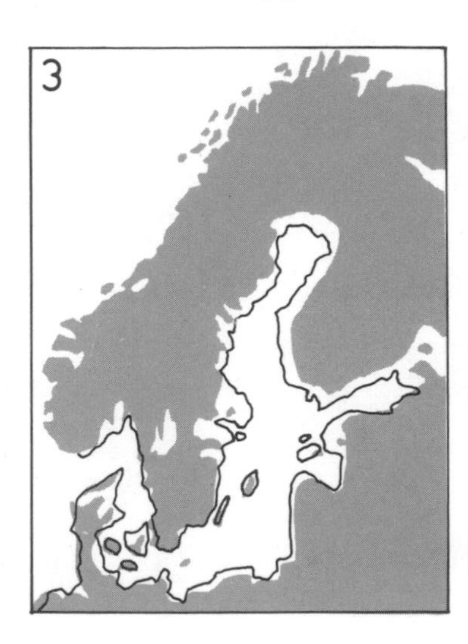

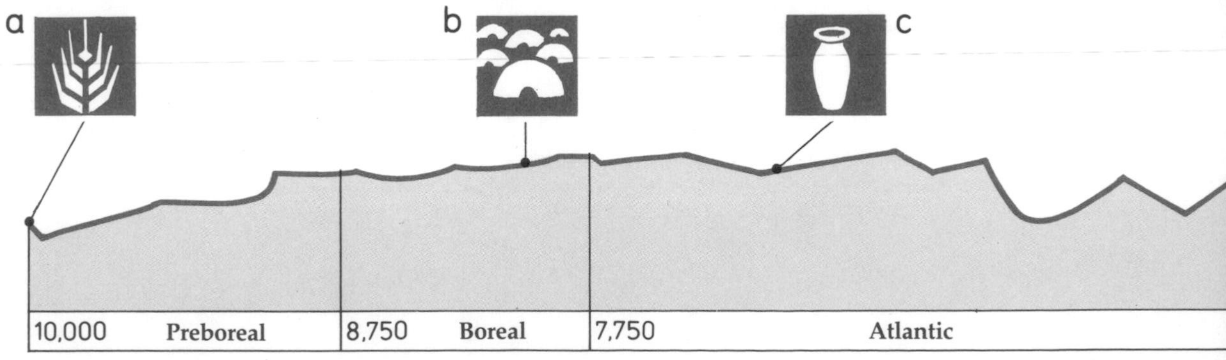

away, the lands they had pressed down bobbed up. In parts of Canada and Scandinavia old shorelines now stand 980 feet (300m) above the present level of the sea.

European studies reveal several climatic stages in the Holocene – each affecting lands and living things. In the Preboreal and Boreal stages (10,000–7750 years ago) ice sheets and mountain glaciers shrank, and the midlatitudes and tropics received hugely increased rainfall. As lakes and rivers rose, moist grassland invaded the Sahara and some other arid regions of the world. Wild Eurasian cattle, goats and sheep moved north, locally replacing elk and reindeer. Now early farming villages took root in South-west Asia.

The Atlantic Stage (7750–5100 years ago) was warmer than today in northern lands. Vegetation spread worldwide, except up to the highest mountain tops, and forest ousted much northern tundra. Indian hunters peopled once glaciated parts of North America, and farming spread across Eurasia.

The Subboreal stage (5100–2200 years ago) brought cooler, drier winters to the hearts of midlatitude continents. Steppe and prairie plants and animals multiplied. Farming started in Central America, and city life emerged in several continents. In places written records of events appeared: humankind began emerging from its prehistoric past.

About 2200 years ago began the climatically fluctuating Sub-Atlantic stage we live in now.

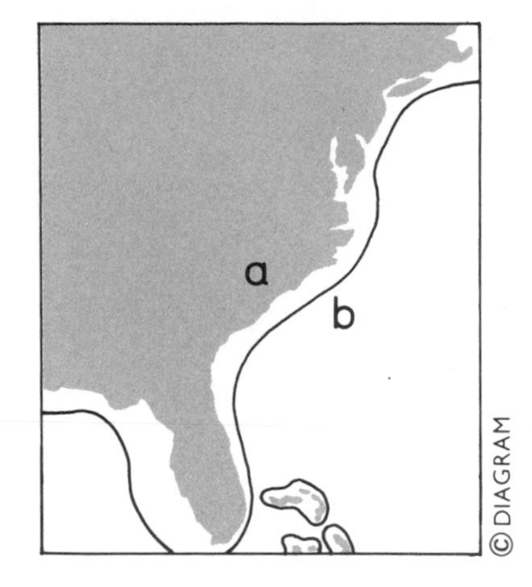

Drowned lands (above)
South-eastern North America shrank when melting ice raised the level of the sea.
a Shoreline now
b Shoreline in glacial times

Cultural landmarks (below)
Symbols of cultural innovations appear above a time scale showing temperature fluctuations in the last 10,000 years – since major warming closed the Pleistocene.
a Farming
b Towns
c Turned pottery (also metal smelting, and weaving)
d The wheel (also writing and sailing boats
e Nation states
f Classical civilizations
g Age of Discovery
h Industrial Revolution

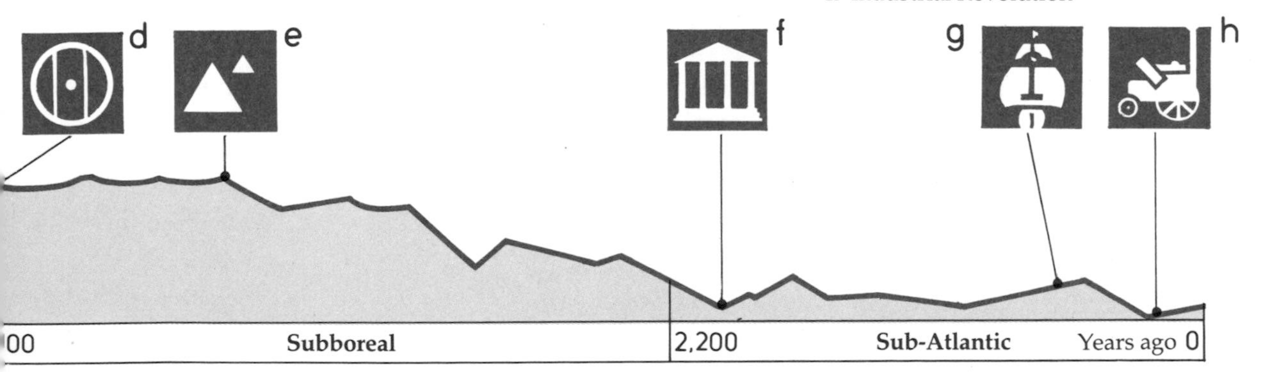

The rise of farming

By 10,000 years ago, the discovery of farming had begun transforming humankind. A given area of fertile land feeds far more food-producers than hunter-gatherers; so where farming spread, populations multiplied. Then, too, tending crops encouraged a trend toward a settled, village way of life with craftsmen who produced and bartered such articles as ground stone tools and pottery.

Such Neolithic (New Stone) Age farming and/or stock raising cultures followed Paleolithic or Mesolithic lifestyles but predated the metal-using Bronze Age and Iron Age cultures.

Different Neolithic cultures sprouted, independently it seems, at different times and places. Farming first became widespread around 10,000 years ago in and near the so-called Fertile Crescent that runs from Egypt through South-west Asia to the Persian Gulf. Here early farmers grew wheat, barley, lentils and peas. Beginning about 7000 years ago the Chinese produced such crops as millet, rice, soybeans, taro, and yams. By 5000 years ago Mesoamerica (southern Mexico, Guatemala, and Honduras) was a third great farming nucleus, with maize, beans, squash, and cotton.

Experts debate just how and why Stone Age peoples took up sedentary food production. Many hunter-gatherers already had enough to eat, and leisure time in plenty. Early farmers worked longer hours for a less varied diet at the risk of famine if their crops should fail.

In places, though, population growth outstripping food supply perhaps forced hunters to seek extra food from plants. Another powerful influence must have been the spread of grasses with big edible seeds.

Cradles of cultivation (above)
This world map shows three major regions where early farming emerged.
a Mesoamerica (maize, beans, squash, peppers, etc)
b Fertile Crescent (wheat, barley, peas, lentils, etc)
c China (rice, millet, soybeans, tea, etc)

Wild harvest (left)
In a good season one person could cut 6lb (almost 3kg) of wild wheat in an hour, and a family could harvest a year's supply in three weeks.

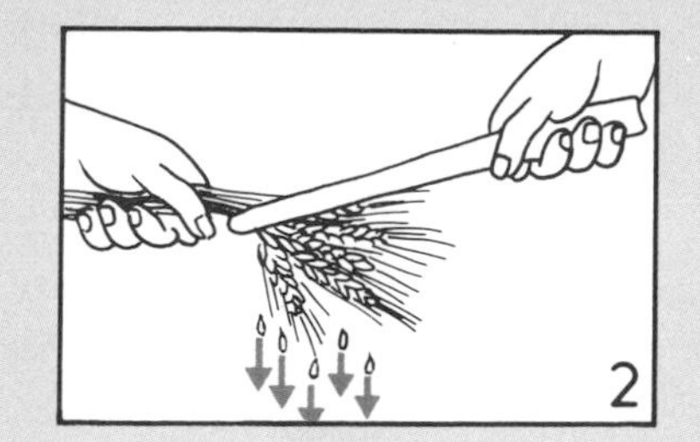

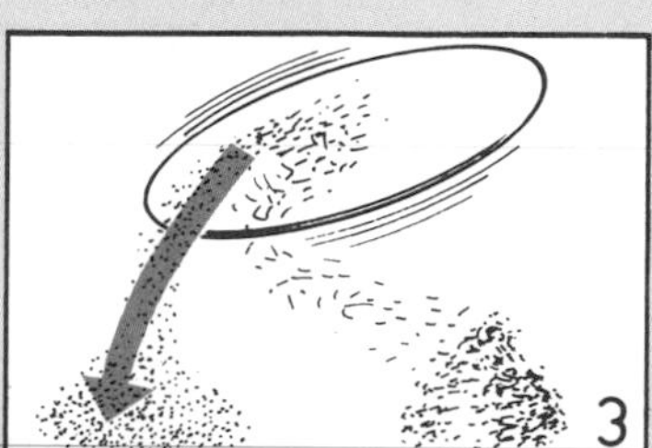

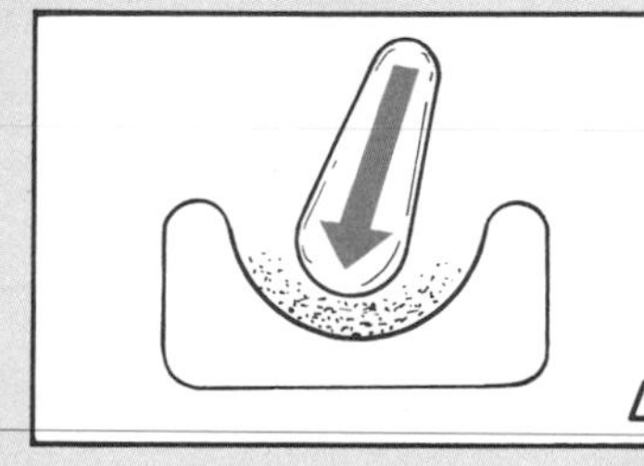

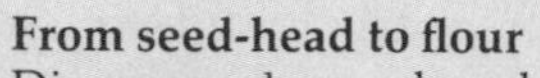

From seed-head to flour
Diagrams, above, show how early farmers processed ripe wheat.
1 Cutting spikes (ears) of wheat with a sickle. Only tough spikes stayed intact, and were easily harvested.
2 Threshing: beating spikes to break them up into spikelets
3 Winnowing: tossing spikelets to let the wind blow away unwanted pieces of straw
4 Pounding spikelets to free grains of wheat from protective husks. More winnowing separated grain from husks. The grain could be boiled to make porridge, or ground into flour and baked to make bread.

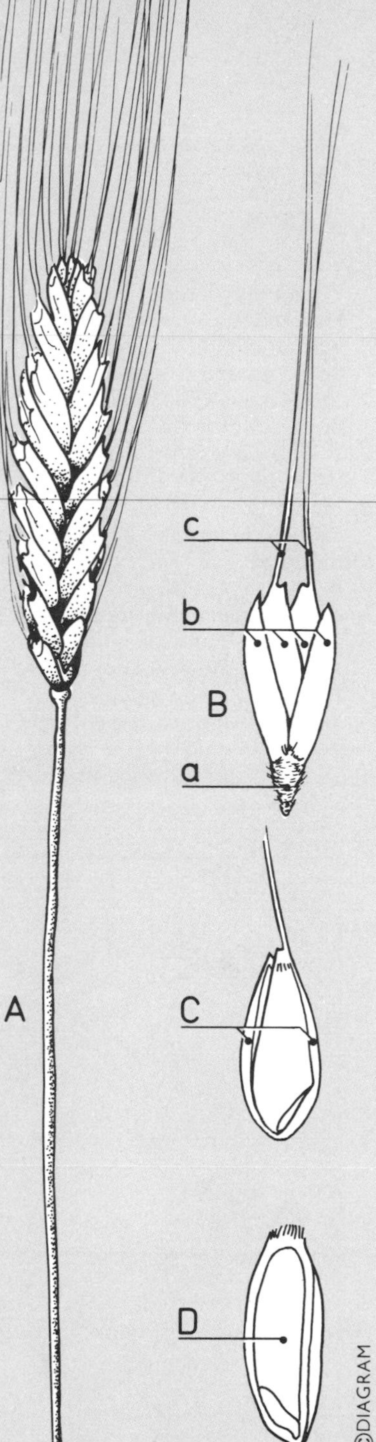

An ear of wheat (right)
Four illustrations show structures involved in processes mentioned above.
A Wheat spike
B Spikelet, with:
a Internode
b Glumes (four husks)
c Awns
C Floret: two inner husks, containing:
D Grain – the edible kernel rich in protein and carbohydrate

By 13,000 years ago harvesting wild cereals had become a mainstay of at least one group in Israel. Natufians used flint-bladed sickles to cut the ripe seed-heads of wild relatives of modern wheat and barley. In time such systematic gatherers began actually sowing seeds, selecting the strains of wheat or barley that proved easiest to harvest. Selective breeding of this kind was to transform small wild seeds, fruits, roots, and leaves into the big domesticated kinds we eat today.

Living larders

Much as food gathering led to farming, so hunting led to the domestication of animals. Again South-west Asia formed a major nucleus. By 13,000 years ago, dogs descended from tamed wolves were probably assisting South-west Asian hunters to drive herds and flocks of wild hoofed mammals into ravines for slaughter. By 8500 years ago, some hunters had turned pastoralists: keeping sheep, goats, or cattle as "living larders and walking wardrobes." Each group conveniently fed on plants indigestible to man: goats browsing on trees and shrubs, sheep grazing hillside grasses, and cattle thriving on lush valley pastures. Meanwhile pigs domesticated from wild boar emerged as useful village scavengers.

Selective breeding of all these produced docile creatures with shorter horns or tusks than their wild ancestors, and higher yields of meat, milk, or wool. Evidence of this comes indirectly from old barnyard bones.

By 5000 years ago, tame cattle, camels, donkeys, horses, and (in South America) llamas were shifting loads to transform travel overland.

Where they lived
Four maps give distributions of wild mammals ancestral to domesticated species. All but the aurochs still largely occupy these zones today.
1 *Ovis*, sheep. Only (**a**) and (**b**) made much contribution to modern breeds.
a *Ovis orientalis*, the West Asian mouflon
b *Ovis vignei*, the urial
c *Ovis ammon*, the argali
2 *Capra aegagrus*, a wild goat known as the bezoar or pasang
3 *Sus scrofa*, the wild boar
4 *Bos primigenius*, the aurochs, an extinct wild ox. The map shows its distribution in the Pleistocene.

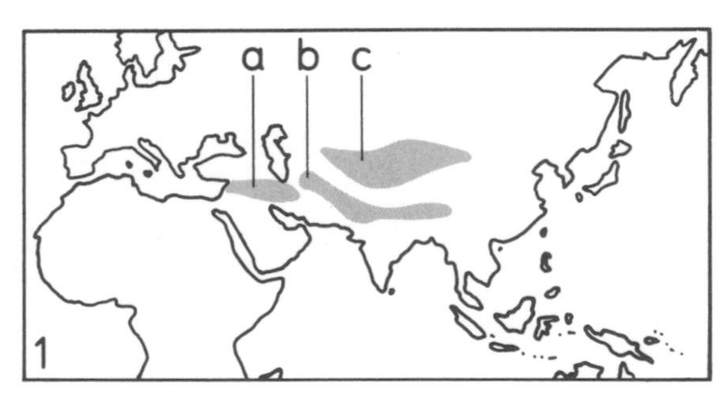

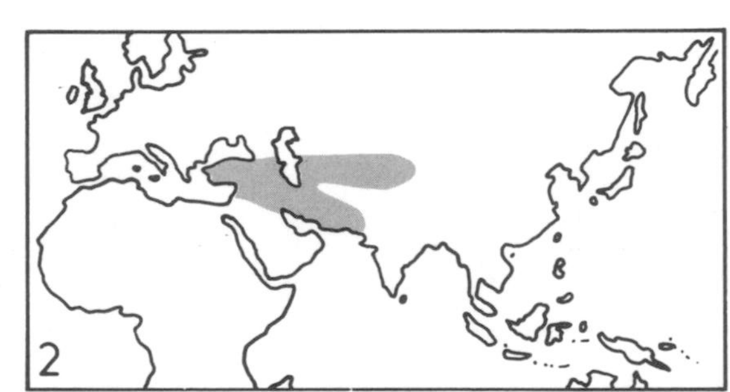

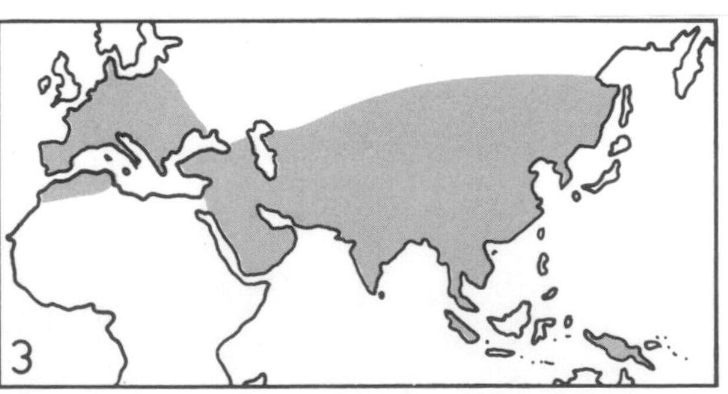

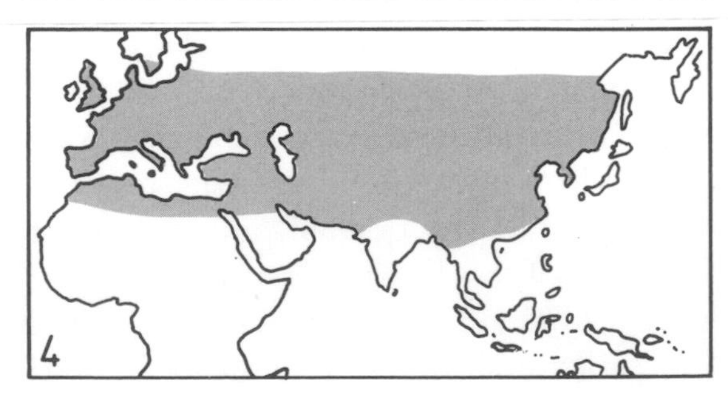

Here are brief details of the wild ancestors of four food animals now found worldwide.

1 **Ovis orientalis,** a wild sheep, has longer horns, hair, and limbs than domesticated sheep. Domestication date: about 11,000 years ago. Place: probably Turkey, Iraq, or Iran.

2 **Capra aegagrus,** a wild goat, has horns that curve straight back without the spiral twist seen in domesticated goats. Domestication date: about 10,000 years ago. Place: probably Iran.

3 **Bos primigenius,** the aurochs, was larger, longer, longer-horned and fiercer than early domesticated cattle. Domestication date: about 8500 years ago. Place: probably Turkey.

4 **Sus scrofa,** the wild boar, has a longer snout and denser coat of bristles than domesticated pigs. Domestication date: about 9000 years ago. Place: perhaps Turkey.

An ancient harness? (above) Rope-like features on this carved horse's head from Southwest France hint that Paleolithic hunters rode tamed horses 14,000 years ago.

Wild ancestors (below) Numbers correspond to items featured in the text.
1 *Ovis orientalis*
2 *Capra aegagrus*
3 *Bos primigenius*
4 *Sus scrofa*

Neolithic toolkits

Archeologists dismiss an old belief in a standard Neolithic toolkit, with ground stone axes, pottery, and stones for grinding grain. We now know, for example, that farming predated pottery in Greece, but followed pottery in Spain and Scandinavia. Yet with early farming came a spread of artefacts for clearing land, preparing soil for planting, harvesting crops, storing surplus grain, baking, weaving, and sheltering the living, dead, and images of gods.

Neolithic villagers made most articles themselves, but many bartered with outsiders for imported goods like salt, hematite (for use in rouge and colored pottery), and obsidian – a glassy stone for making tools and ornaments. Certain mines supplied places several hundred miles away.

Neolithic trade
This map reveals the Mediterranean trade in obsidian in Neolithic times. West and east formed two main trading areas, with most exports moving from islands to other islands or the European mainland. Major export centers included:
1 Sardinia
2 Pontine Islands
3 Lipari Islands
4 Pantelleria
5 Melos
6 Giali
7 Çiftlic
8 Acigöl
9 Lake Van

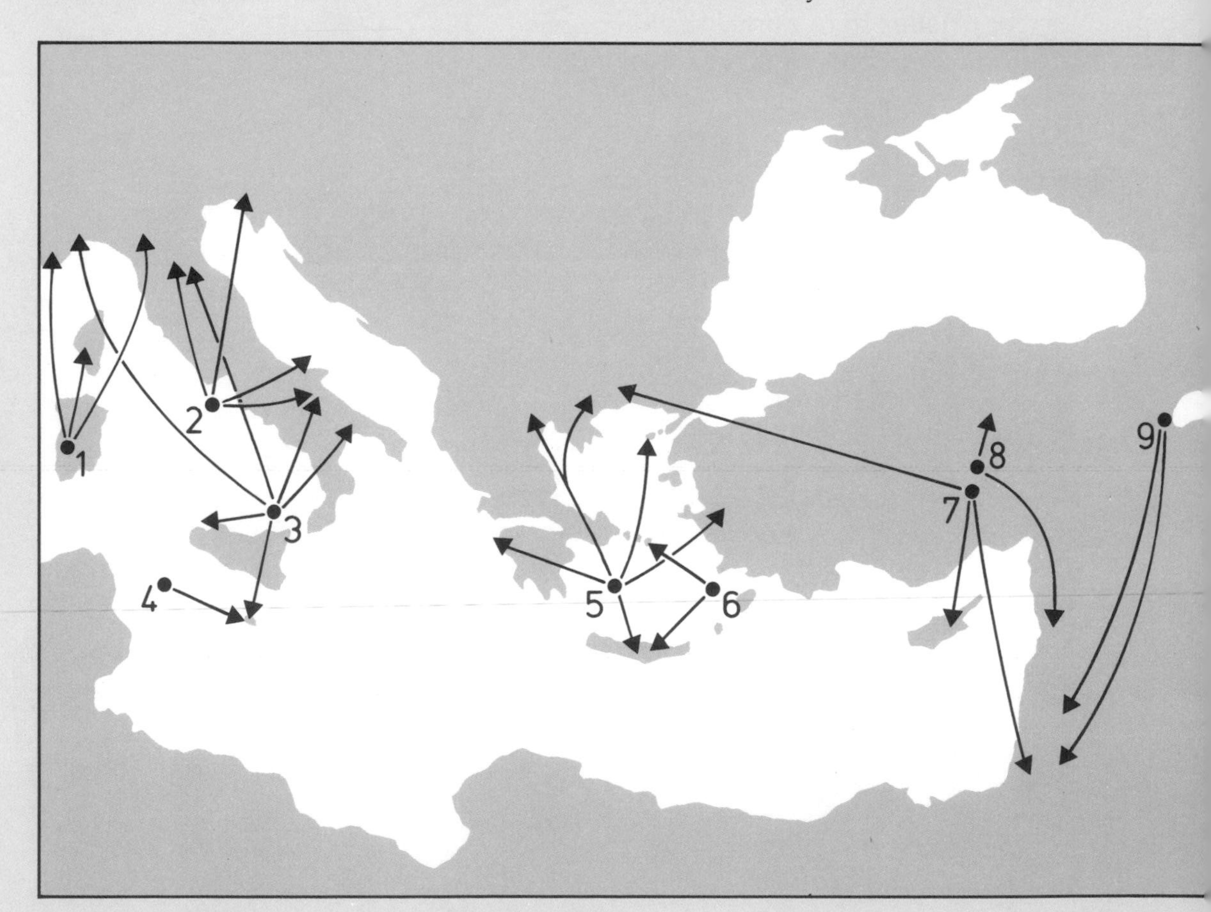

Here are six examples of artefacts of types produced by early farming cultures.

1 **Axe with polished stone head** used for forest clearance. This specimen comes from Denmark.

2 **Wooden hoe** used to clear land for planting. This specimen comes from Egypt.

3 **Flint-bladed sickle** used to harvest wheat or barley. This specimen comes from Egypt.

4 **Clay pot** for food or water storage or cooking. This specimen from Kenya bears an incised basketwork design. Fired, durable, pottery perhaps arose with the chance burning of a clay-lined basket, more than 9000 years ago.

5 **Quern stone and rubbing stone** for grinding grain. Such stones were widespread.

6 **Ground loom,** depicted on an Egyptian pottery dish. Spinning and weaving date from at least 7000 years ago.

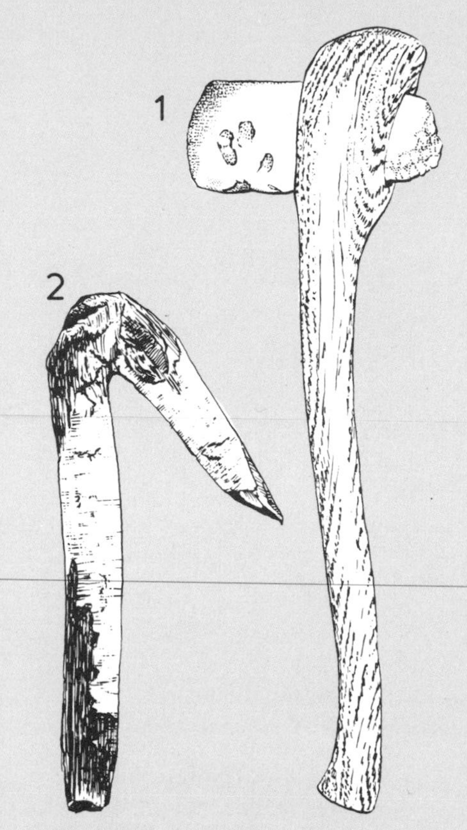

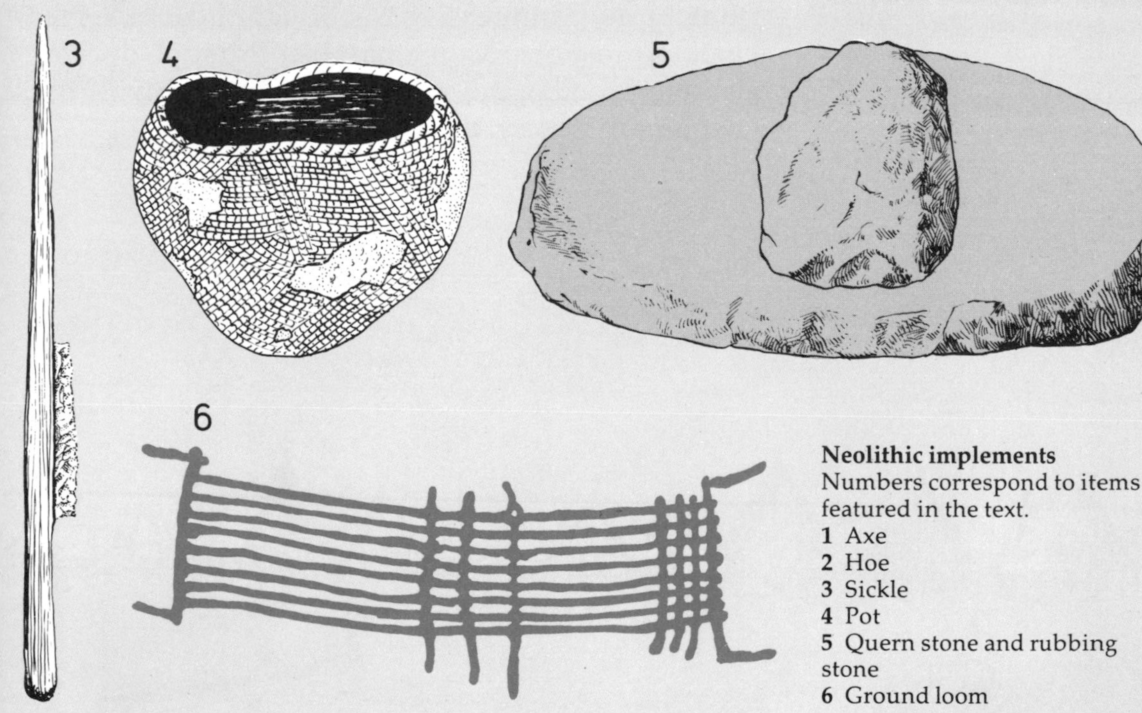

Neolithic implements
Numbers correspond to items featured in the text.
1 Axe
2 Hoe
3 Sickle
4 Pot
5 Quern stone and rubbing stone
6 Ground loom

Homes, tombs, and temples

Old World Neolithic cultures never matched the splendid wall art of the Late Old Stone Age, although they did yield figurines and highly decorated pottery. Perhaps their creators' greatest triumphs lay in building in more ways, and more splendidly, than anyone before.

Craftsmen erected villages and towns of locally available materials. Thus Jericho in Palestine had mud-brick homes with plaster floors and walls, protected by an outer wall of stones. Mud walled houses served in dry South-east Europe and South-west Asia. But timber figured heavily in rainy western Europe. Here, homes had wooden posts and ridge poles supporting sloping roofs that shed rain from eaves jutting over woven wattle walls waterproofed by clay or dung. In Switzerland, wooden stilts protected lakeside dwellings against the risk of flooding. Stone walls roofed with driftwood, skins, and turf formed Skara Brae, a Neolithic village in the treeless, windswept Orkney Islands.

But the most impressive Neolithic structures served religious functions. More than 8000 years ago dozens of shrines stood in the early town of Çatal Hüyük, in what is now Turkey. By 6000 years ago, rock-cut

A Neolithic village
Excavated posts and post-holes enabled artists to reconstruct this Neolithic lakeside village, found in South-west Germany.
A A plan view shows how huts were grouped and aligned.
B Seen from the lake, the huts appeared like this. Wooden frameworks produced quite sturdy structures.

An early shrine (right)
Bull images and sculptures figure strongly in this shrine discovered during James Mellaart's excavations at the early Turkish town of Çatal Hüyük. Perhaps Neolithic men worshiped bulls as symbols of virility and strength.

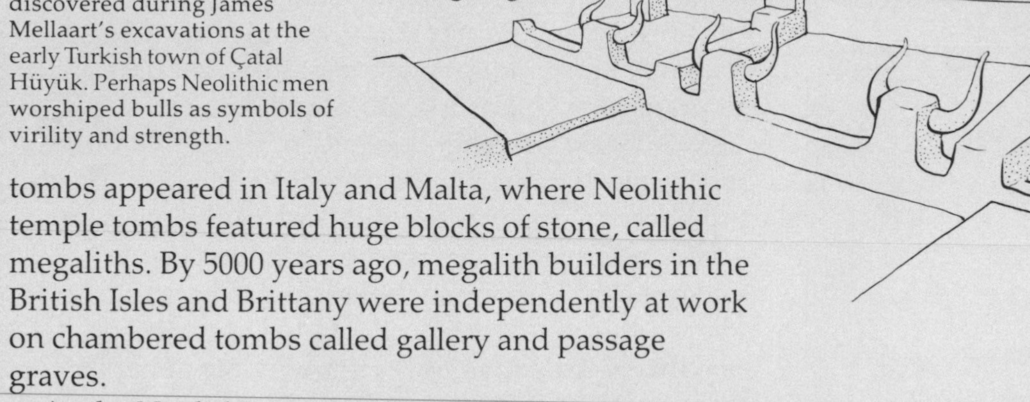

tombs appeared in Italy and Malta, where Neolithic temple tombs featured huge blocks of stone, called megaliths. By 5000 years ago, megalith builders in the British Isles and Brittany were independently at work on chambered tombs called gallery and passage graves.

As the Neolithic merged into the pre-bronze Copper Age, large communities devoted millions of man-hours to erecting rows and rings of standing stones respectively at sites like Carnac in Brittany and Stonehenge in southern England. Such stone alignments show some knowledge of astronomy.

By 2700 BC in southern England, communal burial in long barrows (mounds) was giving way to single burials in round barrows. Gold and other rich grave goods hint at the high status of some occupants, perhaps paramount chiefs powerful enough to organize construction of the local monuments. Society was growing stratified.

Megalith builders (below)
Bars show when people raised different megalithic building types at various localities in Europe.
1 Temple complexes
2 Passage graves
3 Gallery graves
4 Stone circles
5 Stone alignments

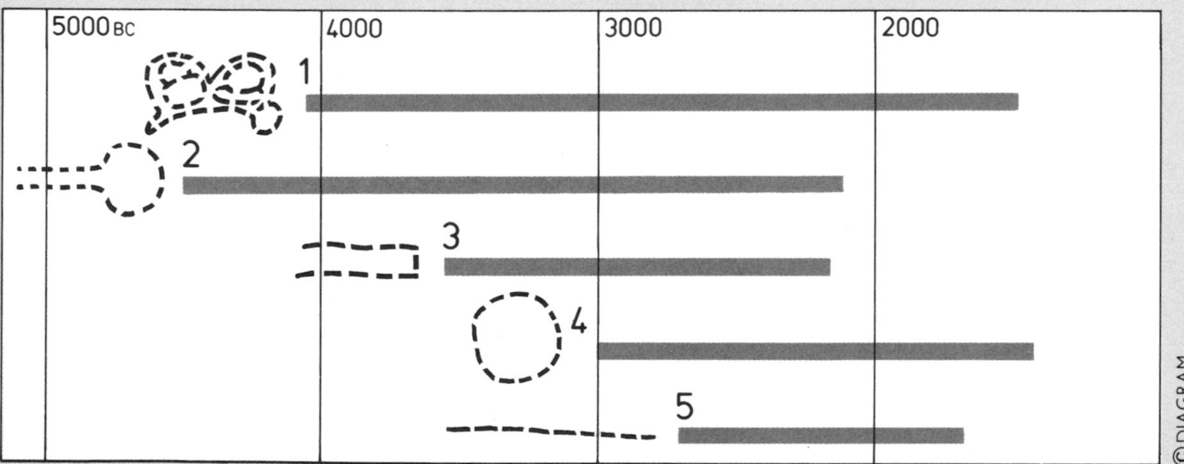

From stone to metal

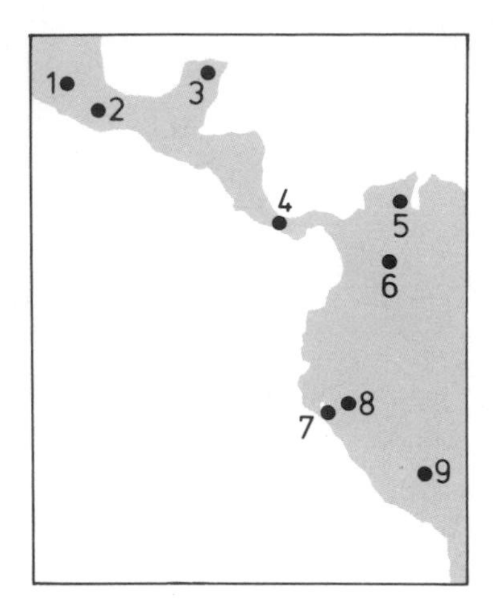

Realms of gold (above)
Pre-Columbian metallurgy stressed gold ornaments. Here are selected gold-working centers and peoples.
1 Aztec
2 Mixtec
3 Maya
4 Veraguas
5 Tairona
6 Muisca
7 Chimú
8 Mochica
9 Inca

For more than two million years, chipped stone provided the hardest, sharpest tools and weapons. But in the last few thousand years, metal has transformed technology. Like farming, metallurgy evidently arose independently at different times in different parts of Europe, Asia, and America.

About 9000 years ago, people in South-east Turkey began cold-hammering pure copper into pins, beads, and borers. Copper is tougher and less brittle than stone, and can be more readily reshaped or resharpened. But pure copper is much scarcer than its ore – the metal chemically joined to various impurities. True metallurgy started only about 6600 years ago when South-east European and South-west Asian smiths began extracting copper from its ore by smelting.

Furnaces for smelting probably evolved from potters' kilns heated to 1470°F (800°C) – hot enough for melting copper. Metalsmiths learned to pour molten copper into stone molds. By 5000 years ago, they were mixing one part tin to nine parts copper to produce the alloy bronze, easier to cast than copper and yielding harder tools and weapons.

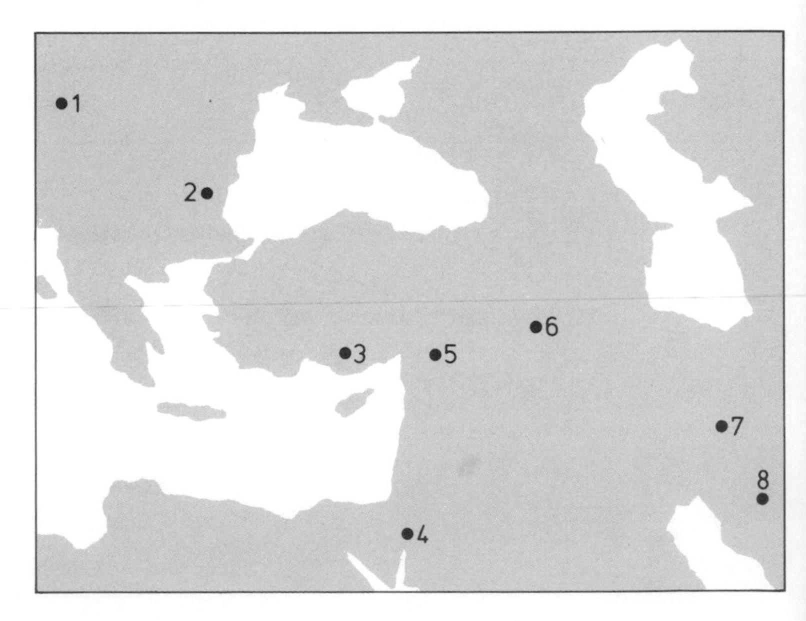

Early metalworking (right)
This map shows selected early sites with evidence of metalworking. Most lay in ore-rich mountain ranges. Despite some very early South-west Asian dates for hammered copper, South-east Europe perhaps discovered smelting independently.
1 Tiszapolgar, 4500 BC
2 Gulmenitsa, 4400 BC
3 Çatal Hüyük, 6000 BC
4 Timna, 4000 BC
5 Cayönü Tepesi, 7200 BC
6 Shanidar, 9500 BC
7 Tepe Sialk, 5000 BC
8 Tal-I-Iblis, 4000 BC

By about 1500 BC the Hittites in Turkey had begun producing iron – far more plentiful than copper, but more difficult to smelt. Hittite furnaces did not reach 2880°F (1535°C) – the melting point of iron. So the product was a "bloom" – a solid mass of iron and slag. Smiths hammered this red hot to separate the slag, added charcoal to the iron to harden it, beat the hot iron into shape as tools or weapons, and made these harder still by quenching them in water.

Forging stayed the only way of shaping iron until high-temperature Chinese furnaces began producing cast-iron objects about 2500 years ago.

By then, Iron Age cultures had succeeded Bronze Age cultures in much of Europe, Asia, and North Africa. In all three, mass-produced iron tools and weapons were promoting farming, war, and trade, and deforesting the land. Elsewhere metals had made less impact. Sub-Saharan Africa was passing straight from stone to iron, and Pre-Columbian America never got beyond a Bronze Age level.

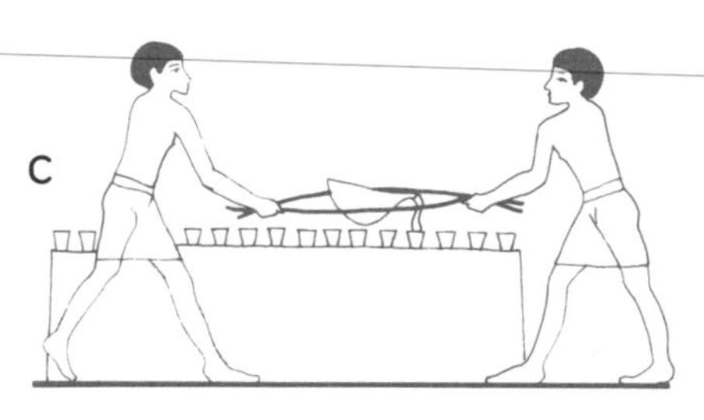

Bronze casting (above)
Scenes from ancient Egypt show:
a Tending a charcoal fire
b Lifting a crucible of molten bronze from the fire
c Pouring molten bronze into a mold to make a big bronze door

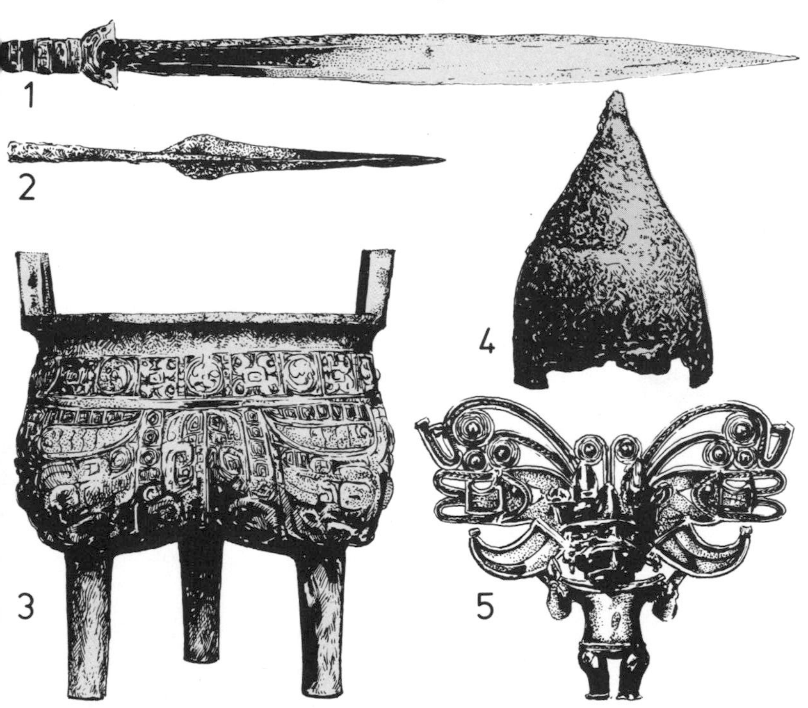

Metal masterpieces (left)
Five products, shown here not to scale, represent the fine craftsmanship of ancient metalworkers.
1 Two-edged sword from Bronze Age Europe
2 Broad-bladed Austrian dagger of the sixth century BC
3 Bronze tripod cooking vessel from China's Shang Dynasty (1600–1100 BC)
4 Conical Assyrian iron helmet of the eighth century BC
5 Tairona cast-gold figure of a god – actual height 5¼in (13cm) – from AD 13th century Colombia

The rise of cities

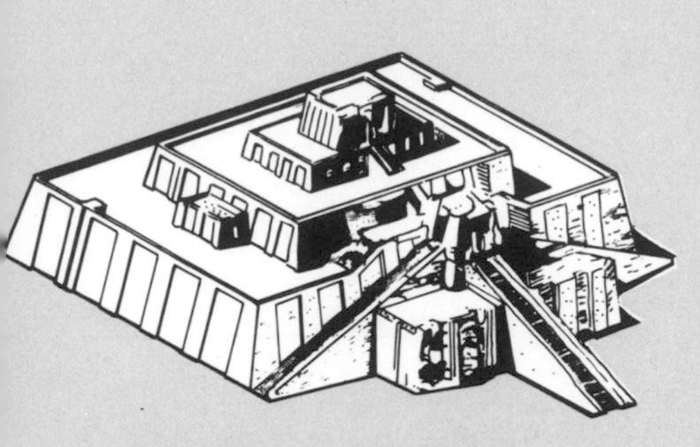

The Ur ziggurat (above)
This stepped temple pyramid dominated the Sumerian city of Ur 4000 years ago. Organized mass labor must have made and raised millions of bricks to build such man-made "mountains."

By 5000 years ago food production and metallurgy had helped some towns expand into the world's first cities. Civilization – our city-oriented way of life – first emerged with the Bronze Age Sumerians of Mesopotamia – the land between the Tigris and Euphrates rivers in what is now Iraq. Later, early cities sprang up in Egypt's Nile Valley, Pakistan's Indus Valley, by China's Yellow River, and in favored regions of Peru and Mexico.

In all these places, irrigation of rich soils helped farmers to produce enough spare food to feed big urban concentrations of full-time non-food-producing specialists. Potters, carpenters, metalworkers, jewelers, scribes, merchants, and others exchanged manufactured goods or services for food produced by local farmers, and for raw materials like stone, wood, bitumen, and metals imported from much greater distances.

With such economic change, societies evolved from tribal bands via chiefdoms into states with layered social classes. Kings or priests controlled most wealth and power, and ruled through bureaucrats who evidently organized irrigation, taxation, and communal food storage.

Cradles of civilization (below)
A map shows where city life began in four great Old World river valleys.
1 Nile Valley
2 Mesopotamia
3 Indus Valley
4 Yellow River Valley

New, imposing building types reflect this social structure. Dominating early Mesopotamian cities were ziggurats: huge stepped brick pyramids crowned by temples. Temple administrators with knowledge of writing, mathematics, and astronomy seemingly used calendars to tell farmers when to sow, measured fields, and gathered and stored taxes in the form of grain. By 3000 BC the building of large palaces suggests a shift of power from priests to kings who ruled small city-states. High walls defended their cities' goods from raiding nomads and rival city-states.

In time, wars between rival city-states fused some into the first small nations. By 2279 BC the ambitious Akkadian ruler Sargon had conquered lands from Turkey to the Persian Gulf to forge what might be called the world's first empire.

Early empire builder (above)
Sargon of Akkad is probably the subject of this fine bronze head, sculpted more than 4000 years ago. Under Sargon, Akkadians imposed Semitic-speaking rulers on their Sumerian neighbors to the south.

City planning (left)
An excavated grid of streets and buildings in Mohenjo-daro suggests that town planners laid out Indus Valley cities more than 4000 years ago.

Bronze Age toolkits

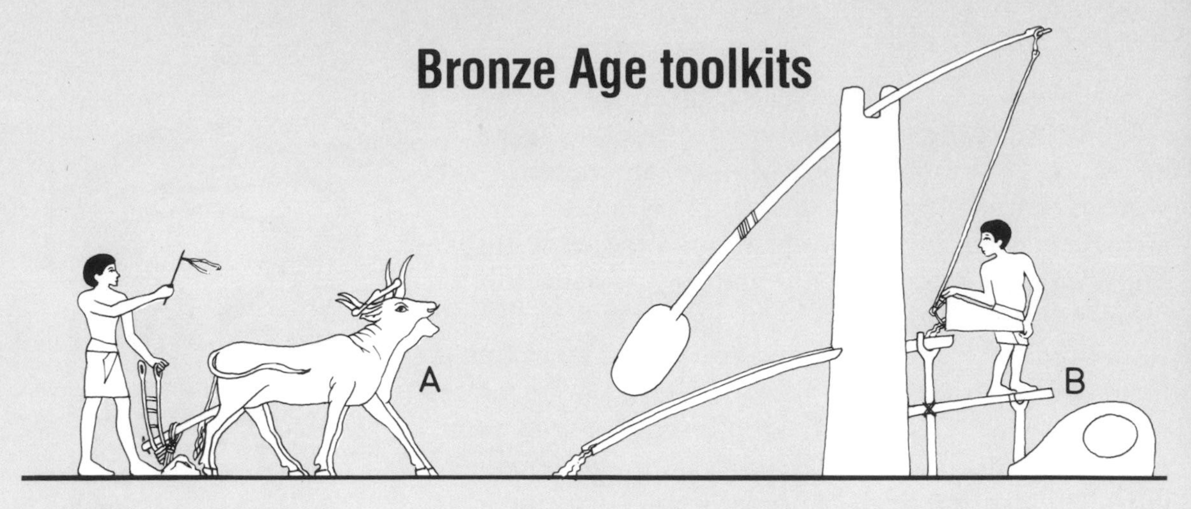

Tending the soil (above)
We show two farming innovations as pictured by artists in Bronze Age Egypt.
A Plowing with a team of oxen
B Irrigating fields with water by *shaduf* – a device with a vessel dipped in water that is raised by counterbalanced pole, then poured via trough onto the land.

The transformation of some Old World Neolithic towns into the world's first, Bronze Age, cities owed much to several inventions and discoveries.

Bronze itself provided points for spears and arrows, but remained too scarce and costly to replace stone farm implements.

The enlarged food surplus that fed Sumerian city populations owed more to oxen, plows, and irrigation. An ox harnessed to a wooden plow tilled far more land each day than a farmer with his hoe. Cattle were plowing fields by 6000 years ago. Ox power quite likely also helped construct irrigation ditches and canals, and raise water to the fields. Warm, fertile irrigated soil yielded several heavy crops each year.

Another major innovation was the wheel. At Ubaid in Mesopotamia craftsmen were turning pottery on revolving disks 7000 years ago. By 5800 years ago Sumerians evolved a fast-revolving spinning wheel. By 5500 years ago, wheeled carts began transforming land transportation.

The wheel (below)
A Early three-piece wheels were clamped with wooden struts and metal-rimmed (at first with copper nails).
B Four-wheeled battle chariots like this were used in Sumer 5000 years ago.

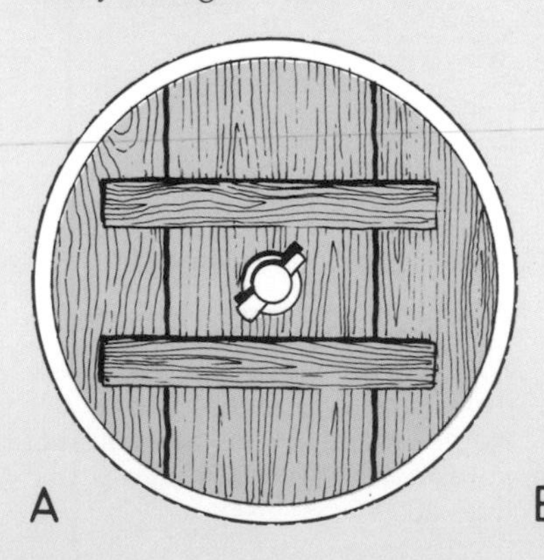

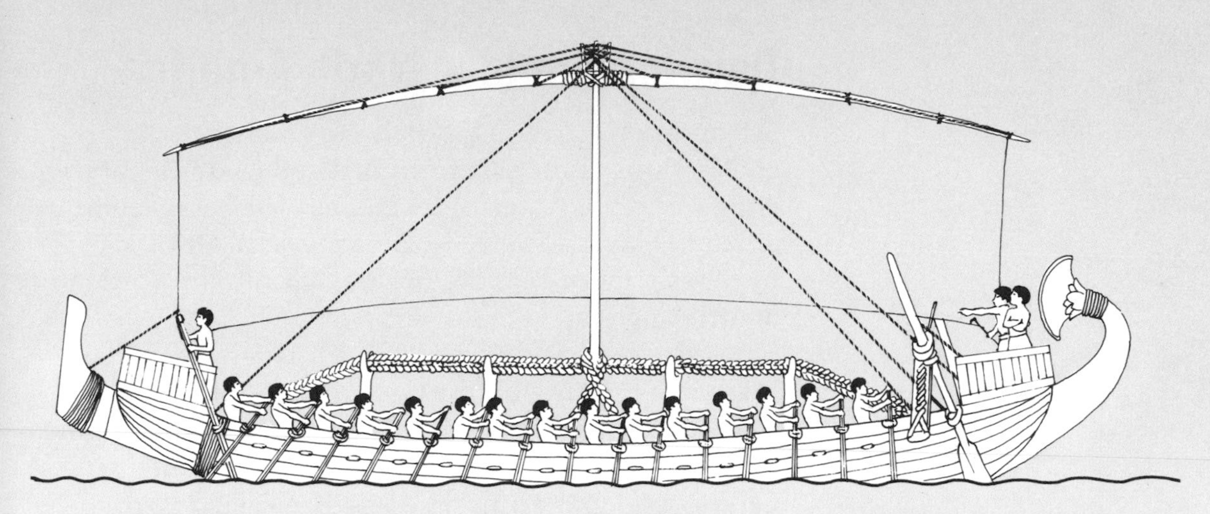

By 5000 years ago water travel, too, was being revolutionized. Now, sailing ships were shifting heavy loads up rivers. By putting wind and animals to work, people had begun to tap far greater energy supplies than human muscle power.

Meanwhile writing revolutionized communication. By 5000 years ago, Sumerians used reed pens to draw picture signs called pictograms on soft clay tablets that became durable when dried and hardened. Later, scribes reduced the 2000 pictograms to fewer phonograms – word/sound symbols based on pictograms. Abbreviation then converted these into a written script that reproduced the spoken language. Scribes used their Sumerian cuneiform (wedge-shaped) script to record law codes, calendars, taxes, and significant events.

Written records meant that new discoveries in science, mathematics, and astronomy could be set down accurately for posterity. Writing thus boosted cultural evolution. And where written records of events began, prehistory ended.

Sea travel (above)
Big wooden merchant ships like this could be propelled by sail, or oars, or both. By 3500 years ago such seaworthy Egyptian craft had sailed south down the Red Sea to the land of Punt (Somalia).

Cuneiform writing (below)
1 A reed pen inscribes wedge-shaped characters in a soft clay tablet.
2 Horizontal strips show (left to right) two cuneiform characters evolving from pictogram to phonogram:
A Fish
B Grain

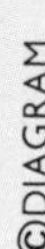

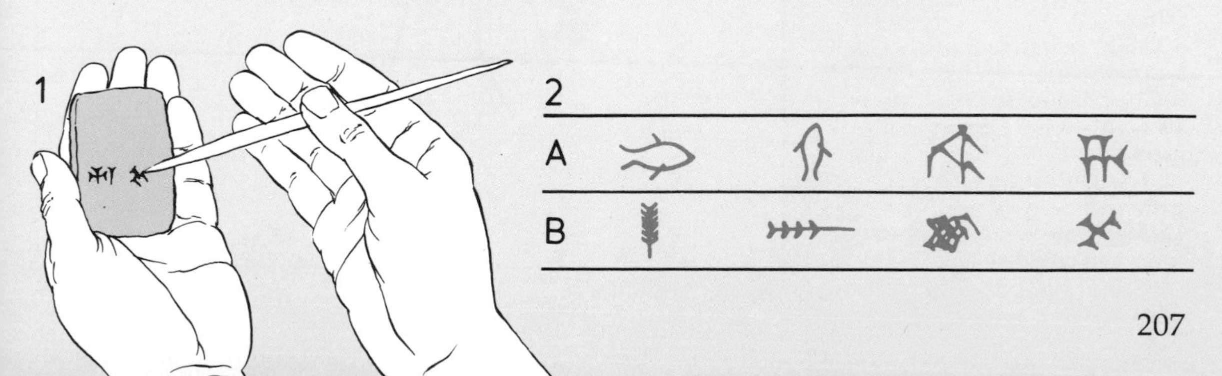

Developments in North America

Sea or desert barriers isolated the emergent civilizations of Eurasia from Australia, sub-Saharan Africa, and the Americas. Yet these lands saw cultural advances, too – especially the Americas. Here, four pages summarize trends in Pre-Columbian North and South America.

North of Mexico, by 5000 BC Indians were still big-game hunting on the Great Plains. But sea and river fishing featured strongly in the North-west. The Eastern Woodlands economy stressed hunting, fishing, and gathering. The South-west's desert tradition involved seed gathering and small-game trapping. These patterns persisted, but by the early centuries AD agriculture, pottery, plant-fiber textiles, burial mounds, and ceremonial centers spread north from Mexico to influence the South-west's Anasazi and Hohokam cultures and the Ohio basin's Adena and Hopewell cultures. After AD 700 the Hopewell gave way to the Mississippi culture, with sizable maize-based communities in the Eastern Woodlands zone.

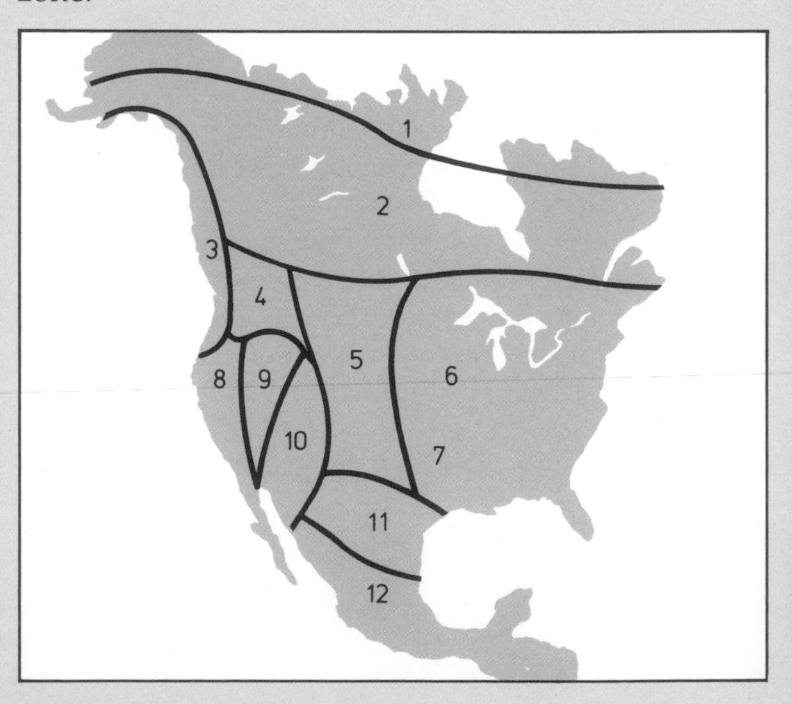

Where they lived
This map shows cultural areas established in North America by AD 1500.
1 Arctic: (Eskimo) hunters
2 Sub-Arctic: hunters
3 North-west coast: fishers and hunters
4 Interior plateau: fishers and hunter-gatherers
5 Great Plains: bison hunters, and farmers
6 Eastern Woodlands: hunter-fisher-gatherers, and farmers
7 Ohio and Mississippi Valleys: farmers
8 California: hunters, fishers, and gatherers
9 Great Basin: desert gatherers
10 South-west: farmers and gatherers
11 South-west deserts: gatherers
12 Mesoamerica: farmers and town and city dwellers

These decorative objects represent six local cultural traditions.

1 **Storage jar** of Arizona's Hohokam culture (about 300 BC-AD 1400), noted for irrigation canals.

2 **Textile** of the Ohio-centered Hopewell culture (about 200 BC-AD 700), renowned for carved stone pipes, copperwork, and ceremonial earth mounds.

3 **Travois** Paired, dog-hauled poles like these moved loads for the tepee-dwelling Great Plains bison hunters.

4 **Spoon handle** This stylized Tlingit "beaver" represents the elaborately carved woodwork produced by North-west Coast Indians.

5 **Birchbark box** Eastern Woodlands Indians used bark for boxes, roofing, and canoes.

6 **Ivory head** The Eskimos of Arctic North America were carving walrus ivory 2000 years ago.

Pre-Columbian artefacts
Numbers of items below correspond to items featured in the text.

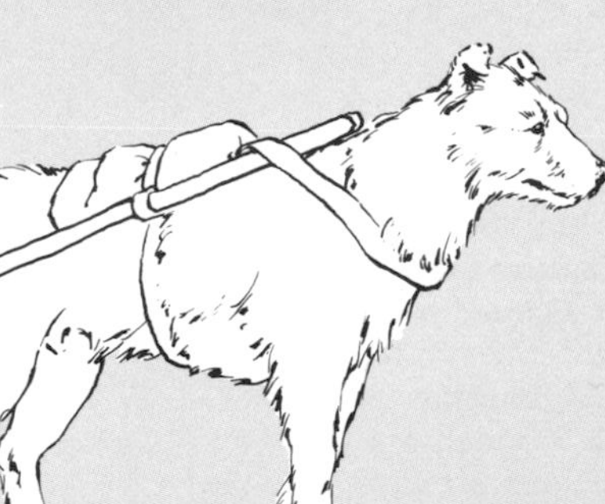

Mesoamerica and South America

North of Mexico, farming villages and towns remained the largest settlements until the Europeans came. But farther south Indians built cities, states, and empires based on agriculture, trade, and conquest. This happened in two regions: Mesoamerica (Mexico through Honduras) and in and near the Central Andes mountains. Independently of early Old World centers, these areas developed social classes with monarchs, priests, soldiers, artisans, and peasants. Independently, they made fine pottery, sculpture, and painting, discovered bronze technology, and raised pyramids and platform mounds of brick or stone. Mesoamericans even devised pictographic writing. Yet American Indians never mastered iron technology or used the wheel, except for toys.

Centers of civilization (above)
Five major areas of Pre-Columbian civilization figure in this map.
a Classic and Post-Classic Mexican
b Olmec
c Classic Maya
d Post-Classic Maya
e Inca Empire at its height

1

Pre-Columbian works of art
Numbers correspond to items featured in the text.
1 Carved head
2 Maize god
3 Obsidian knife
4 Stone puma
5 Portrait vase
6 Gold mask
7 Gold llama

Our numbered illustrations represent major stages of cultural development after 2000 BC.

1 **Carved head** This monolith 8ft (2.4m) high was produced by the Olmec (1300–400 BC) of Gulf Coast Mexico, likely founders of Mesoamerican culture.

2 **Maize god** from a Mayan religious book. Mayas built great ceremonial lowland centers in Mesoamerica's Classic Epoch (about AD 300–900).

3 **Obsidian knife** used to kill sacrificial victims by the Aztecs. Their Mexican empire climaxed Mesoamerica's Post Classic Epoch (AD 900–1520).

4 **Stone puma** of the Andes-centered Chavín (900–200 BC), the first highly developed culture in Peru.

5 **Portrait vase** of the Mochica, a rich north coast culture of Peru's Early Intermediate Period (200 BC–AD 600).

6 **Gold mask** made by the Chimú of Peru. Their coastal state (about AD 1300–1465) had fine metalwork irrigation canals, and a huge capital: Chan Chan.

7 **Gold llama** made by craftsmen of the Andes-based Inca Empire which briefly dominated 2000 miles (3200km) of western South America before Spanish conquerors arrived in 1532.

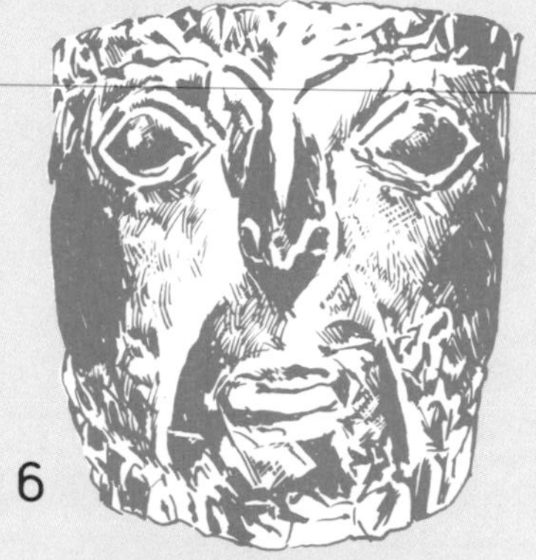

Food and physique

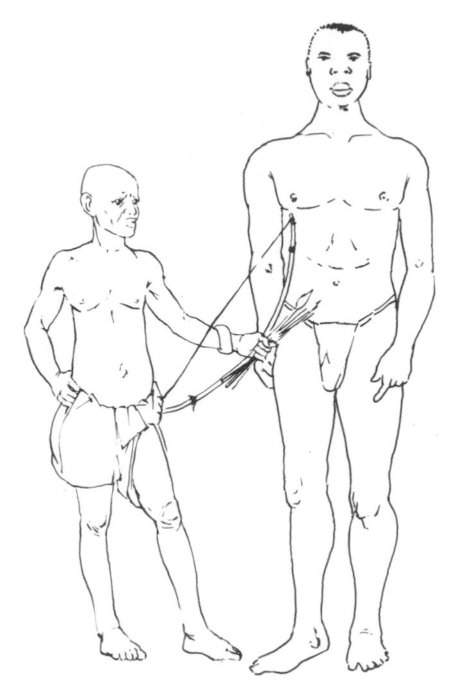

Negrillo and Negro
A tiny Negrillo is shown here beside a Negro of normal stature. Natural selection dwarfed certain groups of hunter-gatherers, probably by weeding out individuals too big to survive food shortages.

The previous chapter ended by showing how climate helped local racial groups evolve. In and since the Ice Age, food and its preparation have also affected body form and size.

At least some populations with unreliable or inadequate supplies of food are shorter or more slightly built than groups who always get enough to eat. Genetically tiny peoples include the Central African Negrillos, and the Negritos found as scattered, mainly island, groups in South-east Asia and New Guinea. Collectively called pygmies, all were once supposed to be descendants of a single group. Modern thinking sees them rather as hunter-foragers derived from local stocks of normal height, but dwarfed by long dependence on uncertain food supplies: tiny people can survive on less food than tall people.

From Egypt east to China, overpopulation has produced huge populations of normal height but very slender build. Arguably long dependence on inadequate subsistence crops has fashioned individuals big enough to produce food for their

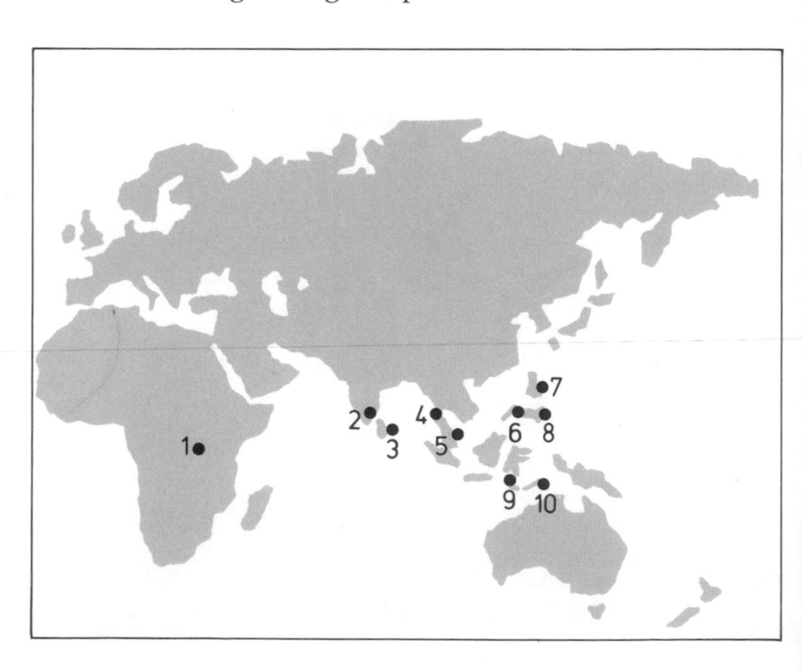

Where they live
This map shows the distribution of the world's pygmies.
1 Congo (Zaire) Basin
2 Kerala State, India
3 Sri Lanka
4 Andaman Islands
5 Malay Peninsula
6 Palawan
7 Luzon
8 Mindanao
9 Sumbawa
10 Timor

families but not so big that they must consume more food than they can grow, or so small that they cannot do the necessary work.

But improved diets increase stature in groups where this is not genetically curbed. One 1960s study showed that children of Italian immigrants to the US tended to outgrow their parents.

Cooking and tool use have transformed the human face. Eating soft, cooked foods, and using tools for tasks once done by teeth have reduced most facial structures, whose major role is the support of chewing apparatus. Teeth have "shrunk," especially in populations with a long history of using cooking pots. The smallest teeth occur in Europe, the Middle East, and China – the largest among Australian Aborigines. Since Bronze Age times, lower jaws have tended to retreat until top teeth overlap bottom teeth and edge-to-edge bite has given way to overbite.

Since Late Paleolithic times, increasing use of tools and machines has generally reduced the need for powerful muscles and strong bones. Accordingly skeletons have tended to become less robust.

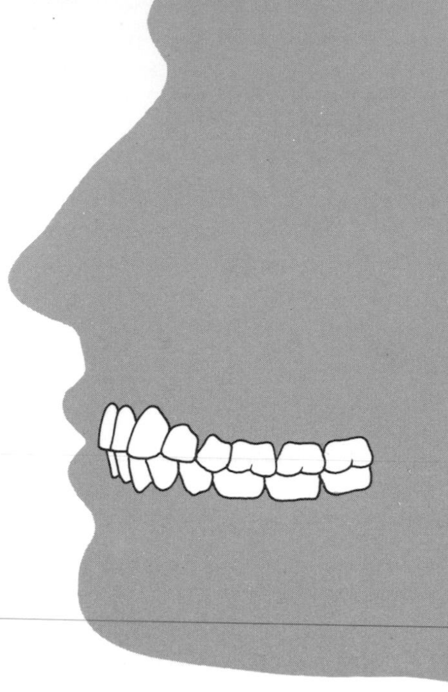

Overbite
A view inside a modern human's mouth shows upper teeth overlapping lower teeth. Jaw reduction, responsible for this phenomenon, can cause overcrowding and the failure of some people's wisdom teeth (back molars) to erupt at all.

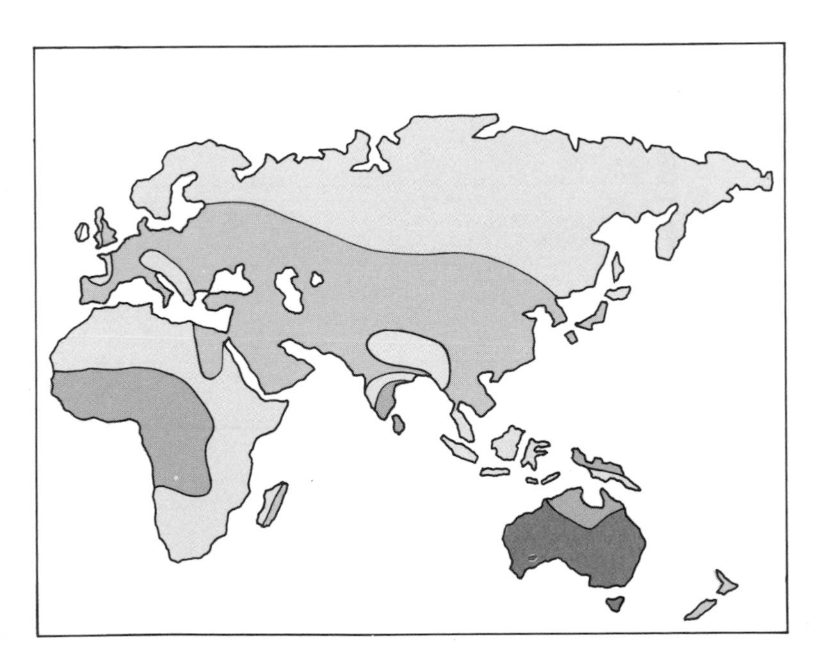

Tooth size
A map of four continents shows the distribution of tooth size before European influence affected population characters. The largest teeth occurred among some peoples used to eating tough foods raw.

Small teeth
Medium teeth
Large teeth
Very large teeth

Population growth

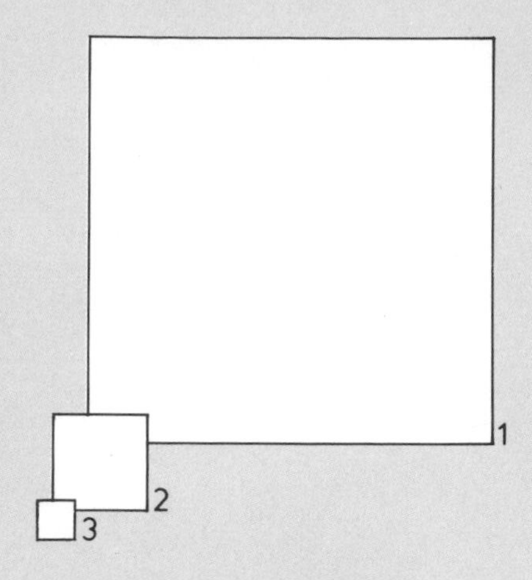

Food, land, and people
Three squares show amounts of land needed to feed three individuals obtaining food in different ways.
1 Hunter-gatherer: 3.86sq mi (10sq km)
2 Dry farmer: 0.19sq mi (0.5sq km)
3 Irrigation farmer: 0.04sq mi (0.1sq km)

Three million years ago the world held probably a few score thousand hominids ancestral to ourselves. By AD 1980, 4500 million people inhabited the Earth. Yet numbers had risen slowly through most of prehistoric time. Growth came in stages as our ancestors evolved, colonized new lands, and stepped up their food supply.

One calculation suggests that two million years ago a million herbivorous australopithecines occupied a few habitats in just one continent.

By 500,000 thousand years ago the hunter-gatherer *Homo erectus* had peopled a variety of landscapes in Africa and much of Eurasia. But a given tract of land feeds even fewer hunter-gatherers than herbivores. So some people think the world held only 1.7 million *Homo erectus*.

By 10,000 years ago *Homo sapiens sapiens* had colonized all continents and climates except Antarctica. Yet most people were still nomadic hunter-gatherers unable to control their food supply. The world held 10 million at most.

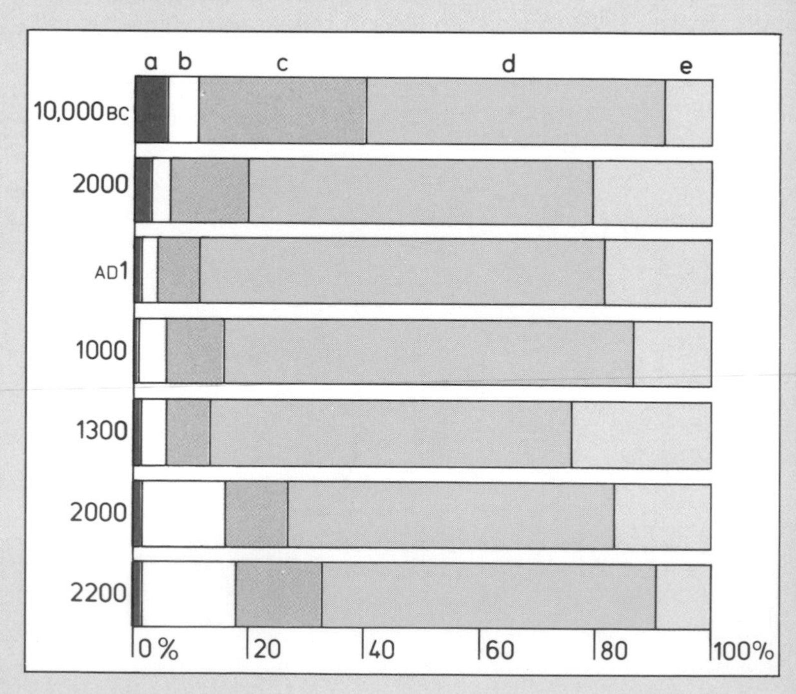

Continental populations
Subdivided bars show estimated continental shares of the world's population at seven dates beginning 10,000BC.
a Oceania (including Australia)
b Americas (two continents)
c Africa
d Asia
e Europe

Population began rocketing only about 7000 years or so ago, as cultivated crops enlarged the food supply, and people lived who would have starved to death before. New tools and know-how outlined in this chapter further boosted food supply and human numbers. By 500 BC (some say 2500 BC) these passed the 100 million mark. Later, a global interchange of food plants and animals, industrial technology, and scientific medicine sent the population soaring faster still. By AD 2000, 6000 million individuals could be rubbing shoulders on this overcrowded planet.

World population (below)
Two bar diagrams show world population growth as estimated for the following periods:
A 15,000 BC to AD 1
B AD 1700 to AD 2000
(Vertical figures relate to time, horizontal population figures are given in millions.)

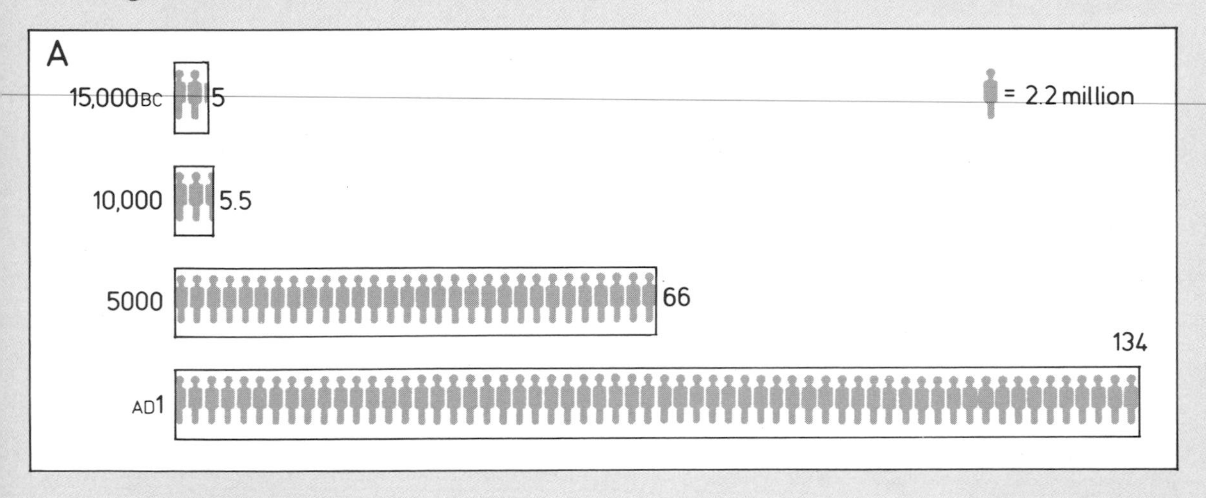

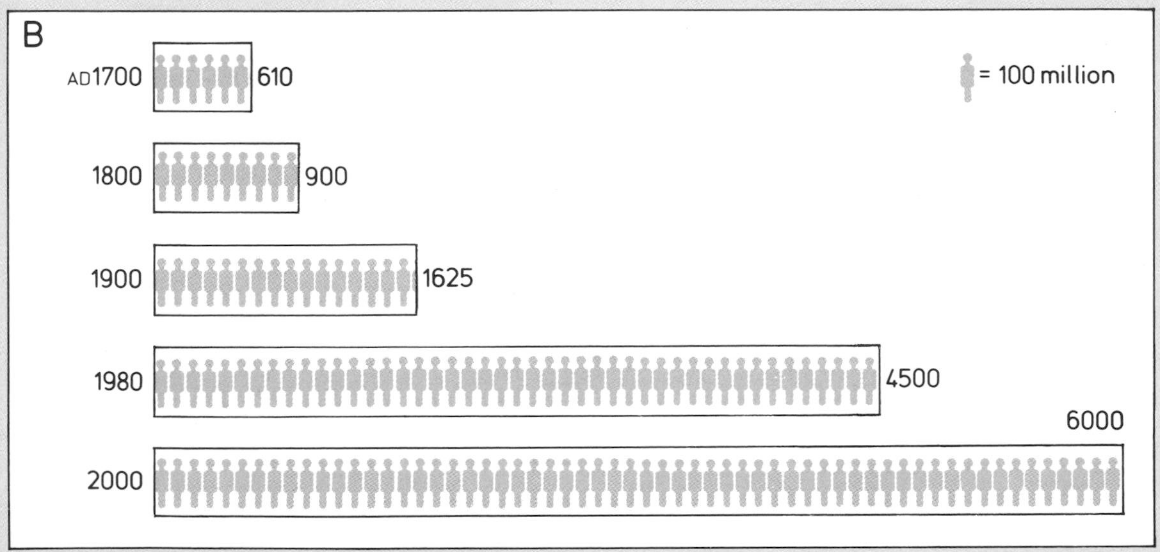

The future of mankind

From a vulnerable group of primates evolution produced our teeming species with its unique ability to multiply, migrate, and mold our planet.

Will humankind continue to evolve? The present answer must be "no." Cultural evolution has buffered us against biological pressures that weeded out the feeble, slow, or stupid. Now, power tools, computers, clothes, spectacles, and modern medicine devalue the old inherited advantages of powerful physique, intelligence, pigmentation, visual acuity, and resistance to diseases like malaria. Societies hold high percentages of physically weak or ill-proportioned people, and people with poor eyesight, or skin color and disease resistance unrelated to the climates where they live. Some individuals who would have died in infancy a century ago survive to breed, handing on genetic faults to future generations.

Migration, too, has helped halt human evolution. No group lives isolated long enough to evolve into a new species as happened in the Pleistocene. And racial differences will decline with increased

Vanishing hunters
A world map shows how hunter-gatherer societies have dwindled. Disappearance of their lifestyle marks an end to selective pressures that helped our species evolve. Tinted areas were still occupied by hunter-gatherers 200 years ago. Dots show surviving hunter-gatherers in recent decades:
1 Warau Indians
2 Shiriana Indians
3 Bororo Indians
4 Negrillos
5 Hadza
6 Bushmen
7 Birhor
8 Sakai
9 Semang
10 Punan
11 Tasaday
12 Australian Aborigines

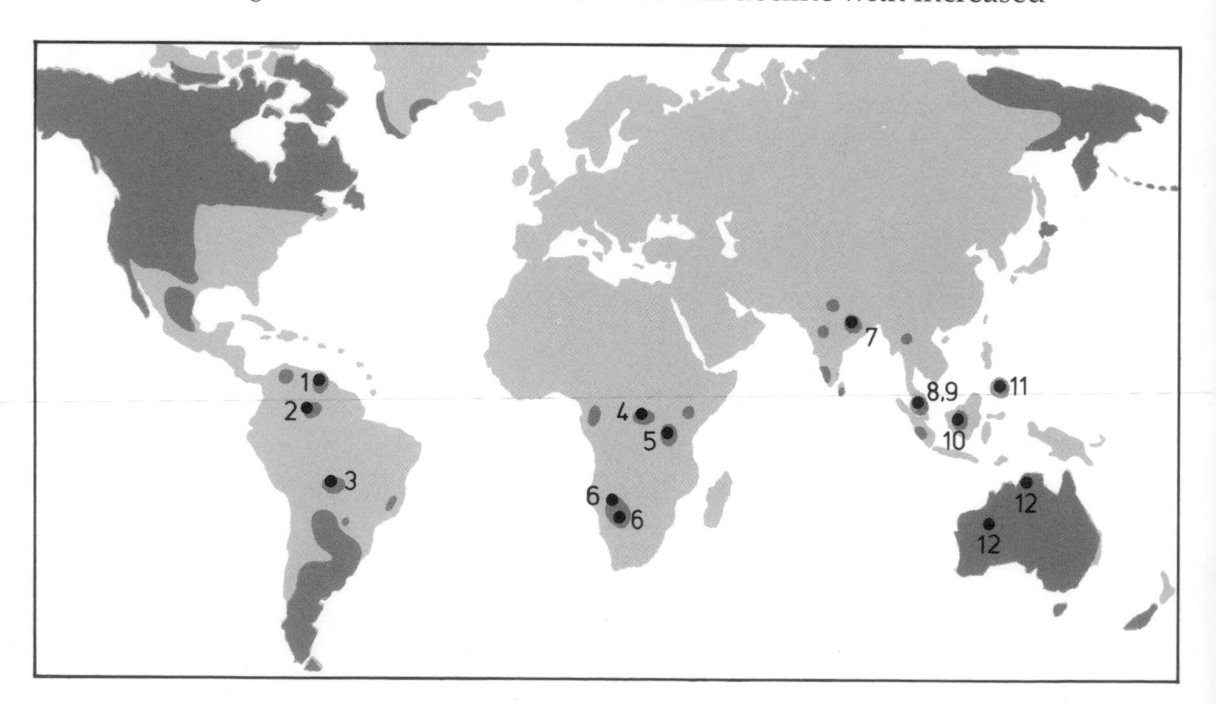

interbreeding of peoples from Europe, Africa, the Americas, India, and China.

In fact extinction seems likelier than further evolution. Soaring populations depend on rising food and energy production. But both encourage overexploitation of our planet. Overusing soil brings soil erosion, reducing Earth's capacity for yielding food; and fossil fuel depletion threatens energy supplies. Both problems could worsen with climatic change. Overcrowded, underfed, and underfueled, *Homo sapiens* might fizzle out in famine, war, and pestilence.

Most species last under three million years. Our own – the most intelligent – will need all its ingenuity to manage half as long as that.

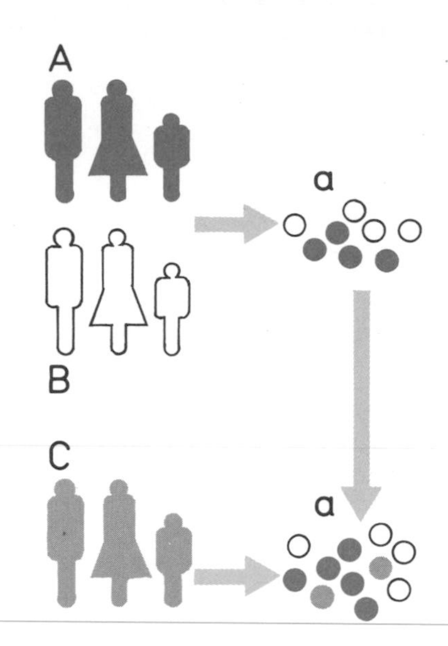

Racial mixing (above)
Interbreeding between racial groups (**A, B, C**) increases the variety of genes (**a**) in a population. Genes inherited by individuals dictate their characteristics. Thus race crossing can produce blond Negroes or red-haired Chinese.

Migrations (below)
This world map shows major recent migrations from Europe (**A**), Africa (**B**), and Asia (**C**). Migrants interbreeding with each other and with native peoples produce racial mixing.

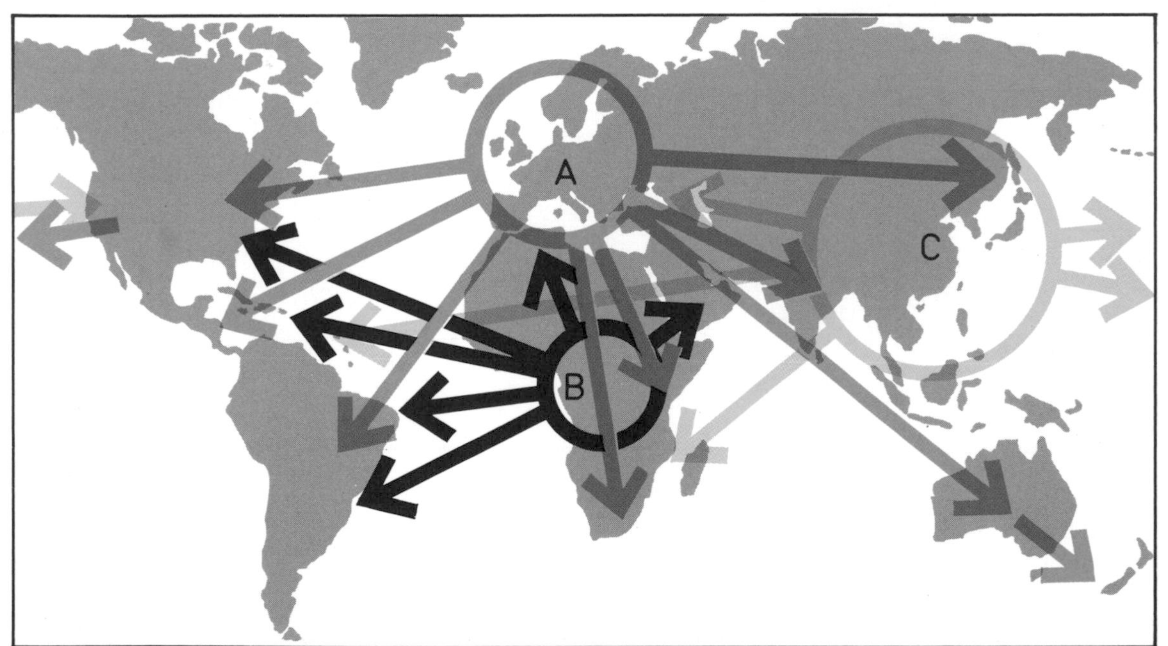

Chapter 11

DISCOVERING MAN'S PAST

These pages summarize the evidence for early man and the tools and methods that help experts interpret it. The book ends with brief biographies of some of the many individuals who have added important pieces to the puzzle of the past, and a worldwide museum guide to selected collections with fossil primates or prehistoric artefacts. Not all these are open regularly to the public, so check access before a visit.

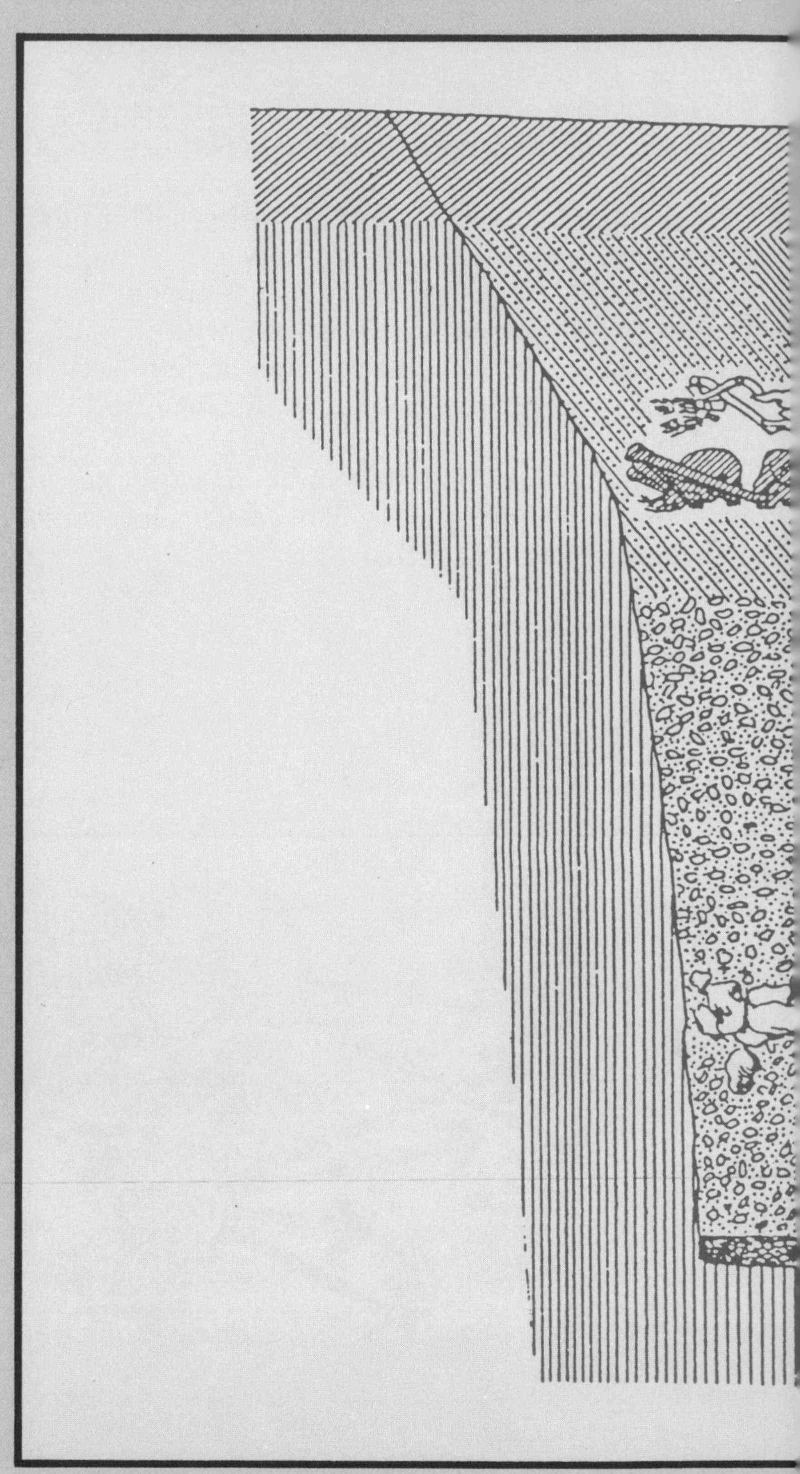

This section through a group of prehistoric burials appeared in a book by the 19th-century British general Augustus Pitt-Rivers – a pioneer of scientific archeology.

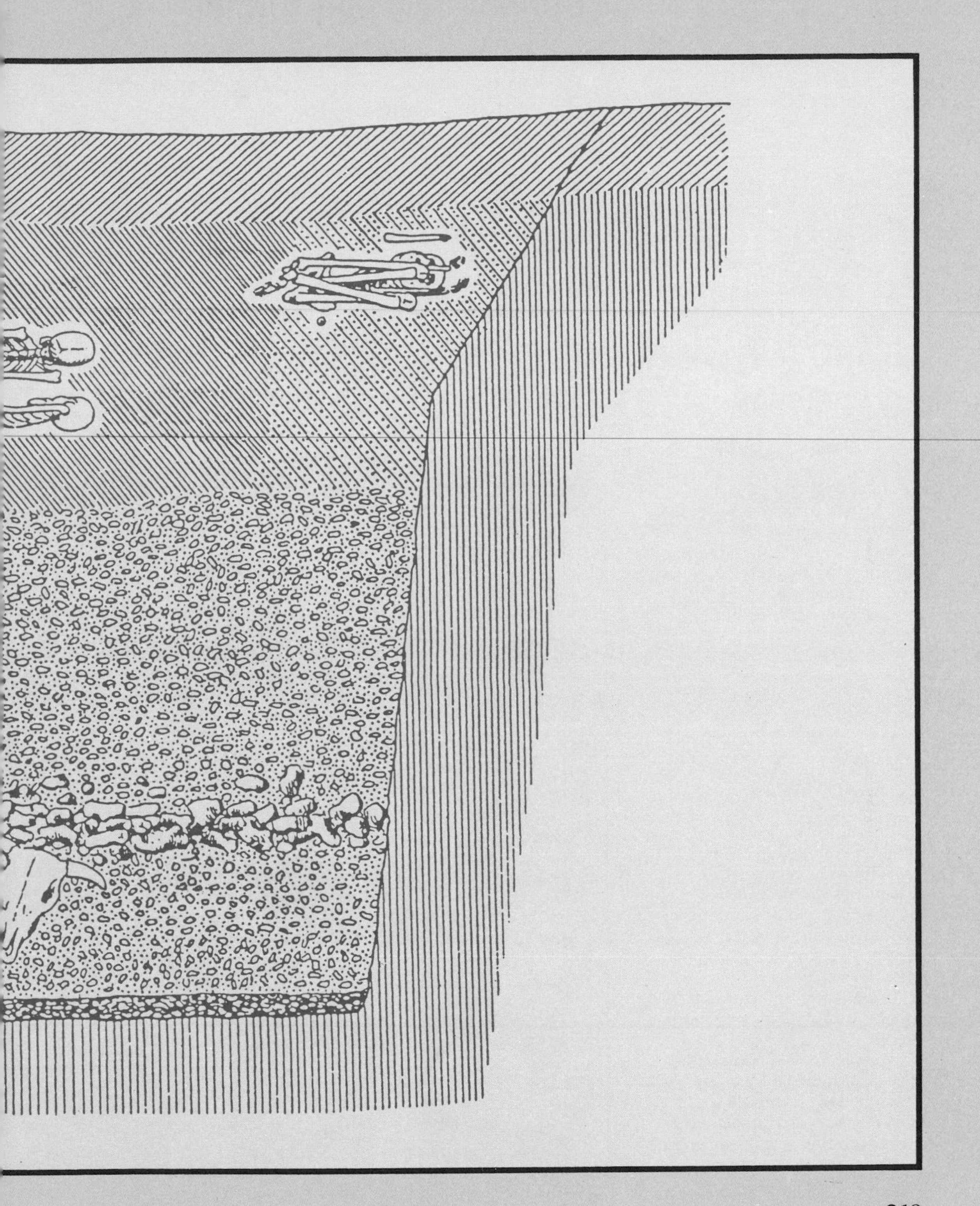

Fossils – how they formed

Much of what we know of very early man derives from fossils – remains of living things preserved in rocks. Only teeth and bones – the body's hardest parts – are usually fossilized. Soft parts just rot away. Indeed, scavengers, decay bacteria, and weathering utterly destroy most corpses.

The new scientific study of taphonomy reveals that the best-preserved remains of early humans and the animals that shared their world were fossilized where burial had quickly followed death. This happened, for example, where a rapid rise in lake level deposited a film of lakeside sediment, or where volcanic ash rained down and blanketed the land. Swift burial protects a corpse from the destructive work of

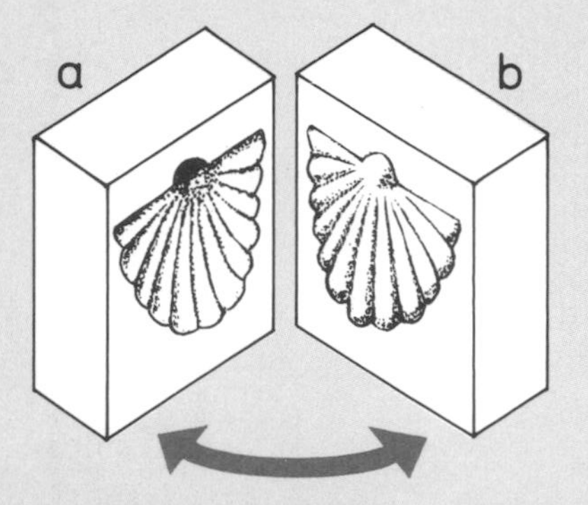

Mold and cast (above)
Split open, some rocks reveal a fossil mold or cast.
a Mold: a hollow in the form of an organism that has dissolved and disappeared
b Cast: a mold filled by minerals

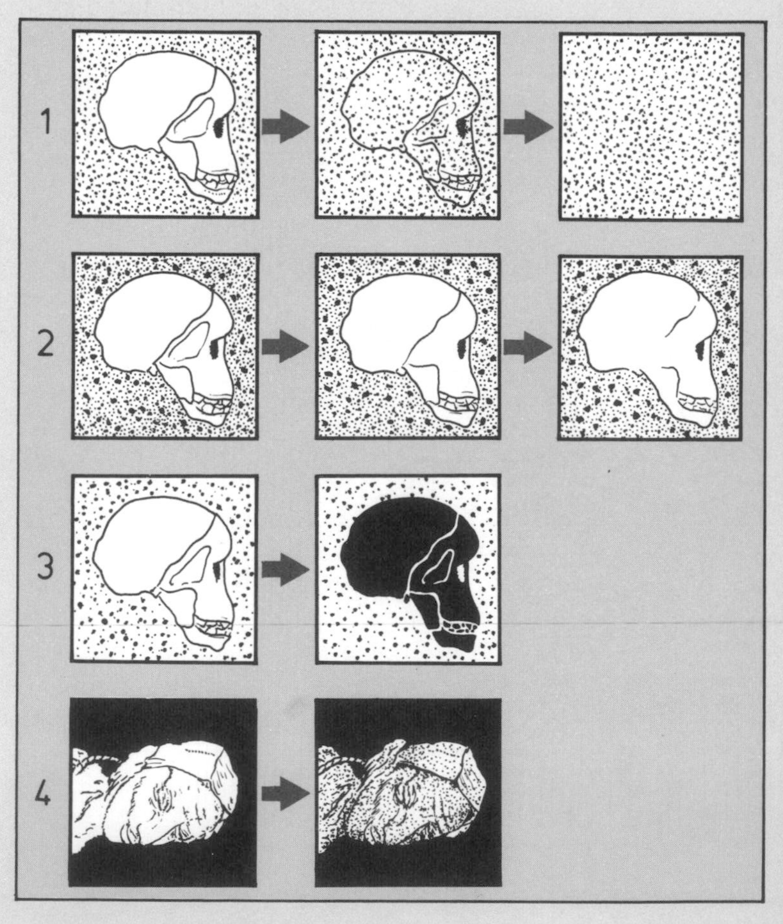

Soils and fossils (right)
Diagrams show how four soil conditions affect a buried organic object (**1–3** an australopith skull, **4** a modern human head).
1 Wet acid soil containing air dissolves all bone.
2 Wet alkaline soil leaves bone intact. This skull may become a fossil.
3 Dry alkaline soil removes some matter from bone, and leaves a light subfossil.
4 Wet airless acid soil such as peat preserves soft tissue as well as bone.

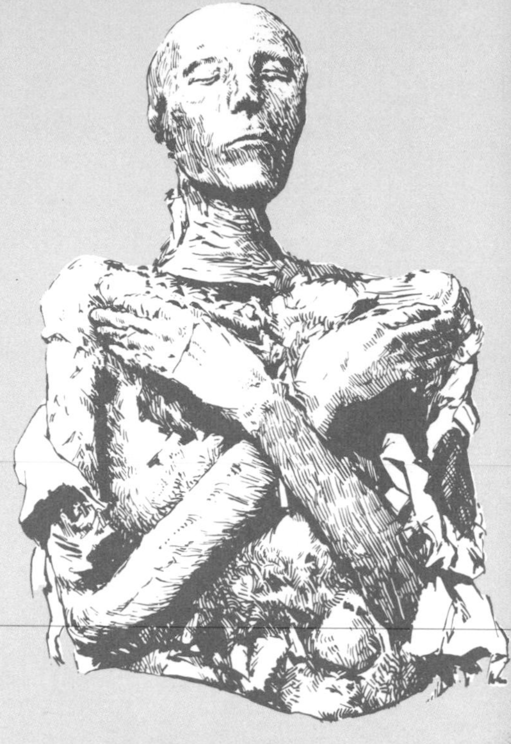

scavengers, frost, heat, wind, and acid water which between them crack, flake, and dissolve bones. Burial may also slow decay.

In time, bone protein vanishes, perhaps to be replaced by dissolved minerals containing such substances as calcium or silica. These make bone harder, heavier, and stronger – better able to resist the weight of underwater layers of mud or sand piling up above. Intense pressure and natural cements slowly compact these sediments into layered limestone, mudstone, or sandstone rocks – massive tombs in which the fossils lie embedded.

Fossil bones and teeth are the sole immediate remains of truly early man. Yet human skin and even flesh have survived 2000 years or more in special circumstances. Withered mummies turn up at ancient burials in Egypt and Peru, where dry, decay-free desert air has preserved their skin and hair. Even more remarkable are the lifelike corpses of garotted victims thrown in bogs as far apart as Scandinavia and Ireland about 2000 years ago. Wet, airless, acid soil kept their skins flexible and soft, yet held decay at bay. Almost equally astonishing are 2400-year-old corpses from the Pazyryk Valley in Soviet Central Asia. Ice forming here in chieftains' tombs has preserved human skin tattooed with a menagerie of animals.

Egyptian mummy (above)
Protected by dry desert air, Pharaoh Seti I's 3000-year-old mummy had almost lifelike features when the French Egyptologist Gaston Maspero unwound its bandages in 1886.

Preserved by ice and peat (below)
A A tattooed horse frisks on 2400-year-old human skin deep frozen in an ice-filled tomb in Central Asia.
B Tollund Man was garotted 2000 years ago and thrown into a Danish bog where acid tanned his skin like leather.

A

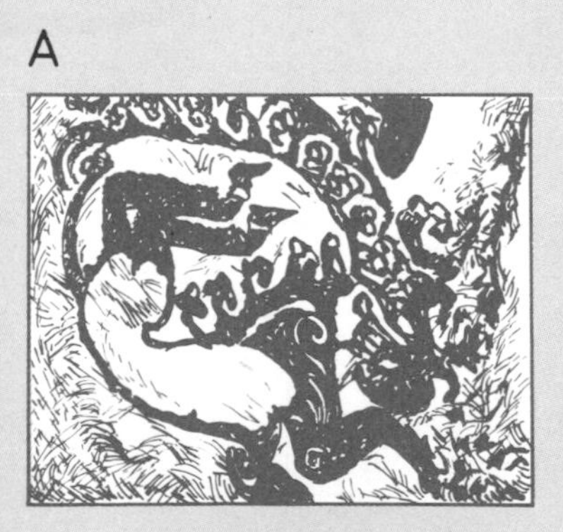

B

Finding fossils

Paleontologists normally find fossils only when these show up on or near the surface. Most fossils get exposed by changes to the Earth's crust. Uplift of fossil-bearing rocks, or a drop in blocks of land nearby starts erosion gnawing into overlying sediments. Between them, heat and cold, wind and rain break up surface rock, then blow or wash it away, revealing the deeper, usually older rocks beneath, together with the fossils they contain. For example, uplift of north Pakistan's Siwalik Hills has revealed 8 to 14 million-year-old fossil pongine apes. Earth movements in Africa's great Rift Valley have lowered Lake Turkana, and drained "Lake Olduvai," exposing old lake-shore sediments containing fossil bones of australopithecines, *Homo habilis*, and *Homo erectus*, together with the bones of animals these killed or scavenged, and the chipped stone flakes they used as tools.

Paleontologists in search of early man scan exposed sediments that formed on continents within the last five million years or so. Besides ancient lake shores, likely sites include the floors of caves, dry gullies, and vast undulating semideserts in Ethiopia and Kenya. Also, sometimes man-made quarries, mines, and cuttings yield unexpected bounty.

A fossil's fate (above)
1 A dead australopithecine lies on a lake bed. Flesh decays, leaving bones.
2 Mud or sand buries bones, protecting from decay.
3 Layers of sediment cover the bones, now reinforced by minerals and fossilized.
4 Weather lays bare the fossil by eroding its covering of sediments, now hardened into rock and raised above water by uplift of the Earth's crust.

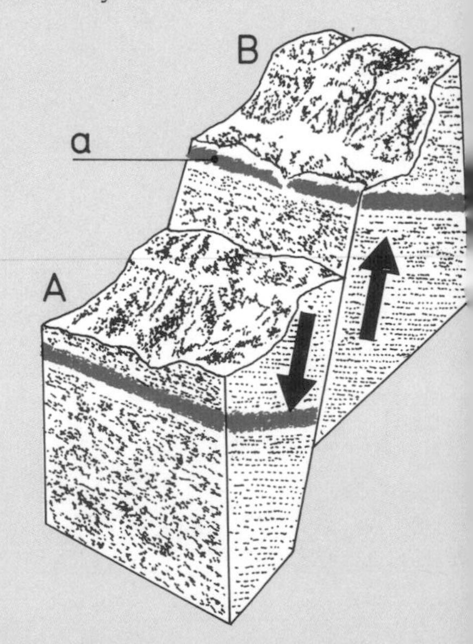

How uplift happens (right)
This diagram shows how slippage along a fault – a line of weakness in the Earth's crust – can reveal buried sediments containing fossils.
A Downfaulted block of land
B Upthrust block of land
a Fossil-bearing rock layer exposed by upthrust.

Hunting fossil hominids is often slow, demanding work. The hunters may have to scout vast slopes and heaps of rock for small objects of unusual shape or color: fossil human bone is sometimes gray or white but often black, and it usually turns up as isolated bits. It takes years to become an expert spotter, and scouring empty wastes of likely land means patient teamwork. In northern Kenya six men scanning 300 square miles (800 sq km) a year on foot took 11 years to find 200 hominid remains. Only nine were skulls; the rest were mainly single teeth or scraps of bone. Whole human skeletons are unknown before those of the Neandertals, the first people to bury their dead.

Hominid hunters (above)
A team of fossil hunters scans weathered stones littering a dry landscape in East Africa. Even a small surface find might help to solve the mystery of human origins.

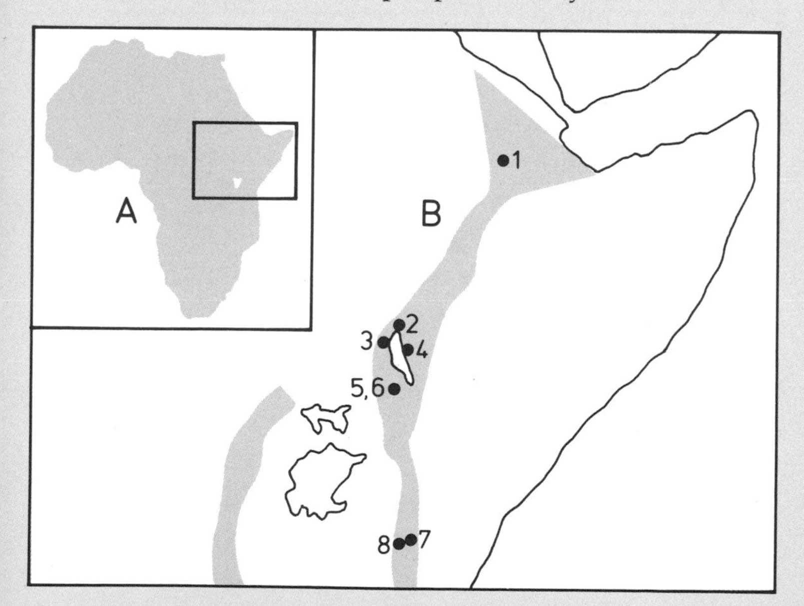

Rift sites in Africa (left)
A Africa
B African Rift Valley system where huge earth movements have bared early fossil hominines at many sites. Here we list eight key localities with ages of fossil finds in millions of years.
1 Hadar area of the Afar Triangle: 3.1–2.6
2 Omo River basin: 2.9–1.0
3 Nariokotome: 1.6
4 East Lake Turkana: 2.4/1.8–1.3
5 Lothagam: 5.5
6 Kanapoi: 4
7 Olduvai Gorge: 1.85-recent
8 Laetoli: 3.5 and less

Excavated burial (above)
The archeologist above brushes soil from a double burial but leaves enough to keep the skeletons intact. After a photographic record has been made the bones may be removed and studied.

Underwater excavation (above)
An archeologist trained in scuba diving uses a grid frame for plotting the positions of old wine jars lying on the sea bed. The Roman ship conveying them foundered off southern Turkey over 1600 years ago.

Scientific excavation

Fossil bones or stone implements found scattered on the ground yield only isolated clues to early man. But chance surface finds may lead to bones, tools, and pollen from a single time and place, still buried undisturbed in ancient sediments. Careful excavation of such sites can tease out information about our early ancestors, their ways of life, surroundings, and antiquity.

Excavation typically starts with mapping of the site to be explored. Paleoanthropologists or archeologists mark it with a grid of meter squares. They gather surface finds. Then they dig out the squares in blocks or trenches, working down through sediments a layer at a time. The team might use picks and shovels to clear overburden. Then work slows as diggers turn to finer tools like trowels and chisels. Dental probes and brushes remove rock matrix stuck to fossils, and sieves reveal small teeth and bits of bone.

Workers patiently number each fossil, artefact, and stone, plot its position on a map, and record the level where it lay. They treat fragile items with a coating of preservative.

The scale of work can be enormous. In 46 months one team recorded 32,000 fossils and 37,000 artefacts from 43 levels at 13 sites in Tanzania's Olduvai Gorge. The spoil removed would fill a trench 1 mile (1.6km) long, 10 feet (3m) wide, and 10 feet (3m) deep.

Some excavations demand unusual methods. For instance South African caves contain frail australopithecine bones set in a natural "concrete" called breccia that can only be drilled out in chunks; weeks of treatment follow to remove the bones intact. Submerged sites involve work under water: relay teams of archeologists trained as divers use special tools for shifting mud or silt, plotting finds, and raising them.

Excavation is just a first step. Later will come the complex work of interpreting and publishing the new discoveries.

Cave excavation
Careful vertical and horizontal measurements accompany the excavation of South African limestone fissures. Here lie scores of australopith remains – fallen fragments from the meals of carnivores.

Reconstructing fossils

Reconstructing fossil skulls and limbs, and restoring lost soft parts of the body helps us visualize the prehistoric ancestors of modern man.

First, museum preparators clean the bones. Drills, vibrating tools, or sandblasting remove adhering sandy, gritty rock. Repeated washing and treatments with acetic acid free fossils stuck in breccia. Wet bones can be dried out by hanging them in *n*-Butyl alcohol. Laboratory experts will harden fragile porous bones with man-made plastics dissolved in solvents or emulsions in water.

Skilled anatomists then join bits of bone like pieces of a jigsaw puzzle to reconstruct a skull or skeleton, perhaps replacing missing bits with plastic imitations. The shapes and sizes of certain bones show how a fossil creature held its head and limbs. But making and mounting one reconstruction can take weeks or months of careful work.

Developing a fossil (above)
Dental picks, powered vibrating tools, scalpels, and magnifying lenses are among the tools and instruments that may be used to help a preparator free a fossil from its matrix.

Acid treatment (above)
Dipping a fossil jaw in acetic acid helps dissolve adhering breccia – a rock that forms in limestone fissures such as Swartkrans in South Africa.

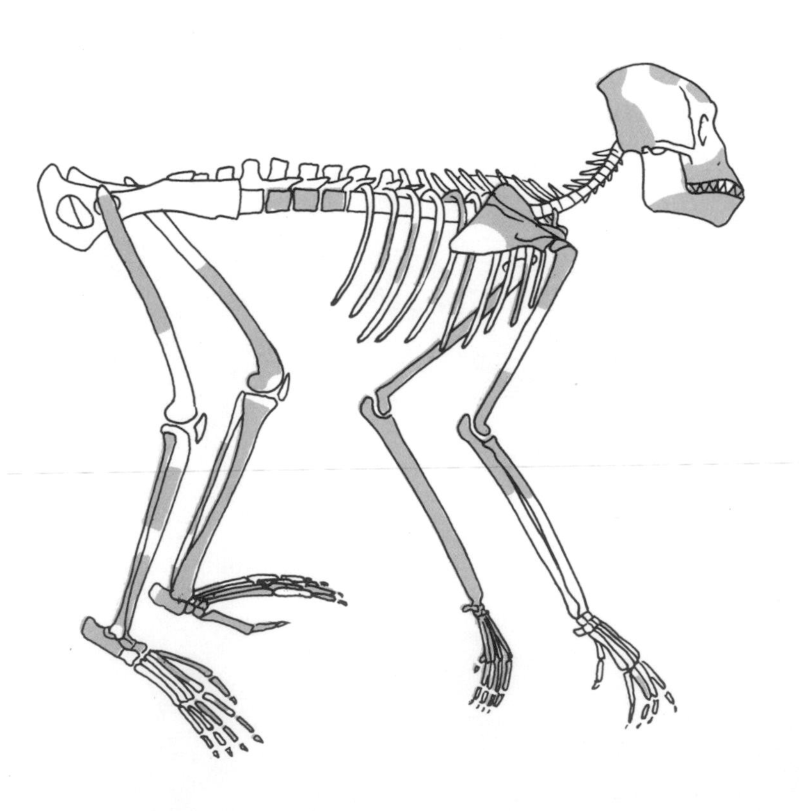

Rebuilding *Proconsul* (right)
In this reconstruction of the fossil hominoid *Proconsul*, tinted areas represent known pieces of bone. The rest has been inferred.

From reconstructed skulls and skeletons, artists and sculptors expert in anatomy can restore whole heads and bodies. The strength and size of limb bones indicates the size of muscles used to operate them. Bumps, grooves, and other features found on bones betray muscles' attachment points and shapes. Thus bony bumps and ridges on the skull show where large muscles held up a forward-jutting head, or operated powerful jaws. All this helps artists to flesh out a skull or skeleton: first, adding muscles and body fat, then skin and hair.

Yet restoring fossil bodies involves more uncertainty than articulating a dismembered skeleton, for not all soft parts leave their marks on bones. Two sculptors working from one hominid skull may produce two very different heads – one more human looking than the other. Similarly, only intelligent guessing can reproduce skin, eye, and hair color, and the amount of body hair.

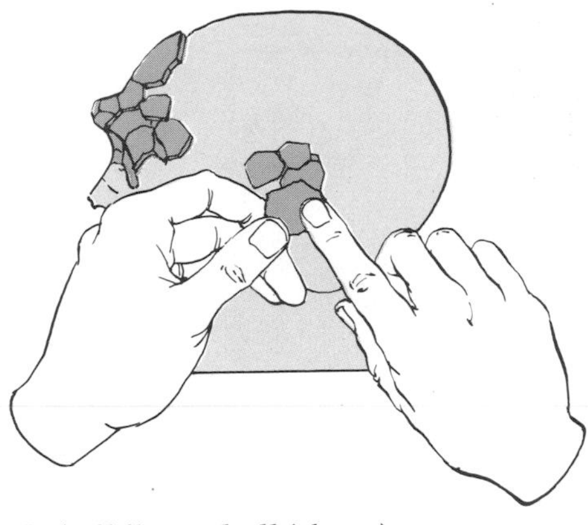

Rebuilding a skull (above)
Piecing together skull bones is like solving a three-dimensional jigsaw puzzle. Clues are the bones' sizes, shapes, and curves. The anatomist arranges bones around a soft clay model head, reshaping this to match the curves of the assembled skull.

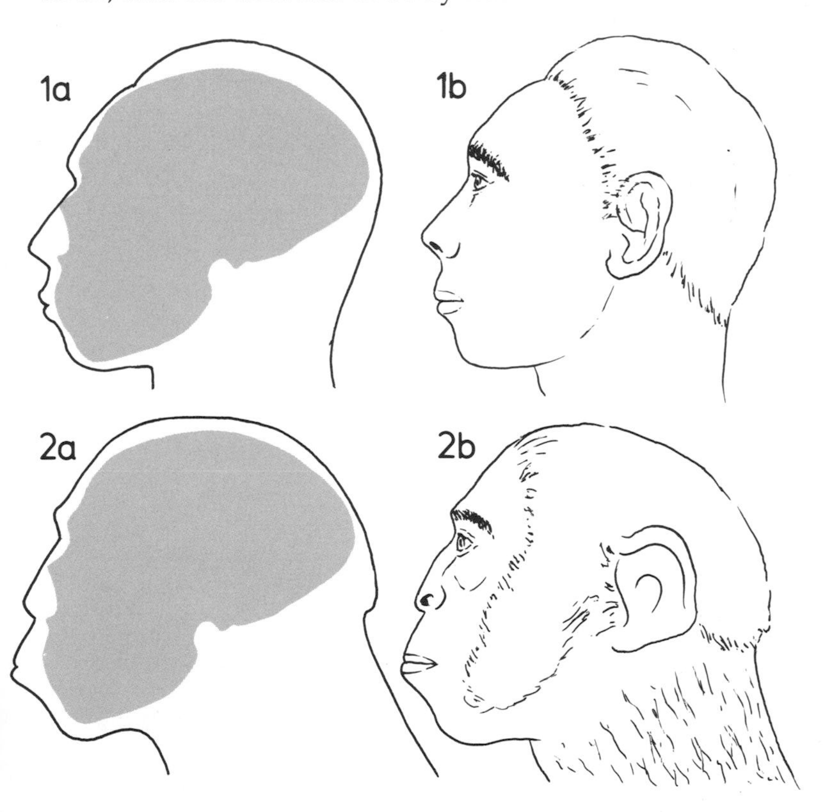

Restoration problems (left)
The same fossil human skull can give rise to several ideas of what its owner looked like, for head shape owes much to soft tissue that is not preserved.
1a Neandertal skull with a modern human profile added
1b The head restored to resemble modern man
2a The same skull with an ape-like profile added
2b The head restored with ape-like features

Facts from bones and artefacts

To skilled interpreters, old bones and artefacts are encyclopedias of ancient life.

Comparing the measurements and structures of scores of bones and teeth has enabled scientists to identify individual bits of skull or limb bone as, say, Neandertal or dryopithecine.

Anatomy also offers clues to physical abilities and modes of life. Long bones indicate body build and height. Shoulder girdles betray which primates swung from trees. The position of the hole that lets the spinal cord in through the skull shows if a creature held its head erect or jutting forward. Thumb and finger bones and joints are guides to power grip, precision grip or – in very early primates – lack of any grip at all. Bones of hips, thighs, and toes distinguish quadrupedal climbers from bipedal walkers, and upright walkers from those whose bodies tilt from side to side.

Skulls are guides to brain capacity and the ability to see and smell. Jaws and teeth – especially tooth structures, sizes, shapes, and wear – are powerful clues to what their owner ate.

Fossil bones can tell us more: from these an expert can deduce some kinds of injury or illness, and if the victim had made a good recovery. Teeth and long bones may pinpoint the age at death.

Further light on prehistoric hominids comes from their enduring artefacts – at first mostly stone, bone,

Measuring bones (above)
Vernier calipers measure the head of a femur (thigh bone). Calipers, calibrated steel rules, and protractors help anatomists assign individual bones to particular species.

Tell-tale tooth wear (below)
Comparing much enlarged views of tooth surfaces in extinct and living species helps experts work out what our ancestors ate.
1 Grass eater's: finely scratched (unknown in hominids)
2 Fruit eater's: smooth with some pitting (seen in orangutan and *Australopithecus*)
3 Carnivore's and omnivore's: deeply pitted and scratched by crunching bones or grit (seen in hyenas and *Homo erectus*)

1
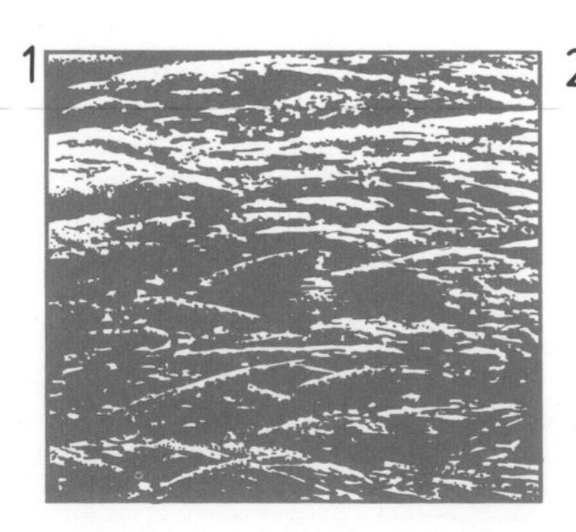

2
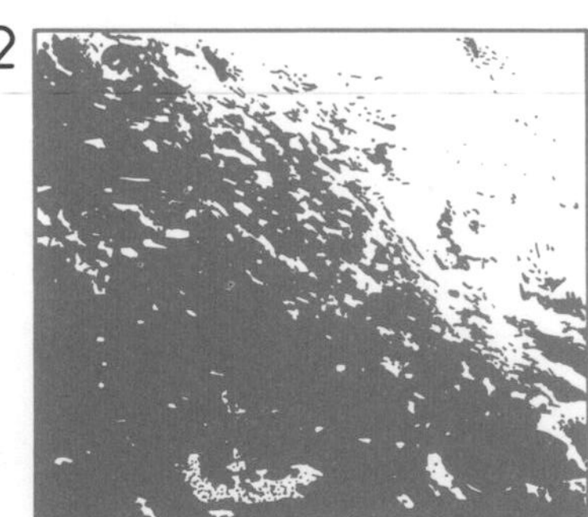

3
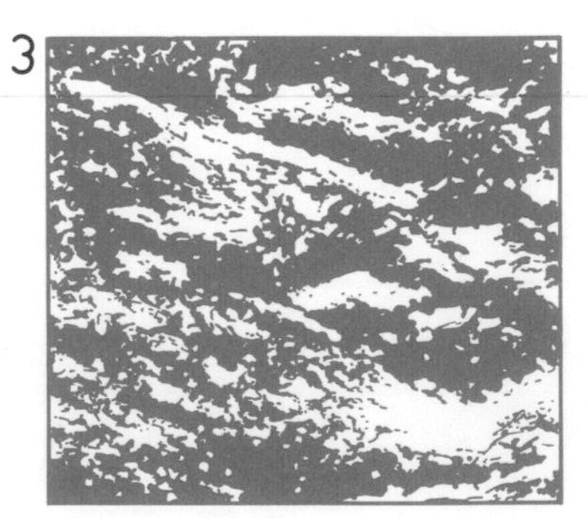

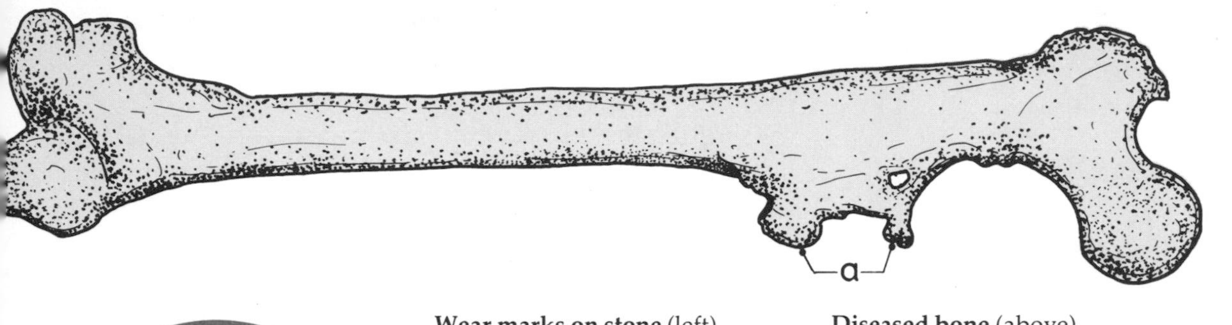

Diseased bone (above)
A bony growth (**a**) deforms this *Homo erectus* femur from Java – a possible result of food contaminated by fluoride compounds from a volcanic eruption.

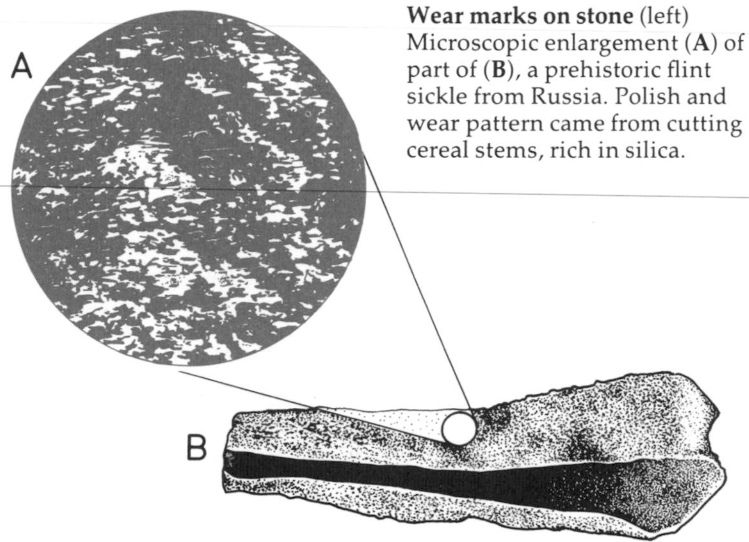

Wear marks on stone (left)
Microscopic enlargement (**A**) of part of (**B**), a prehistoric flint sickle from Russia. Polish and wear pattern came from cutting cereal stems, rich in silica.

and charcoal, with, later, pottery, and post holes. These show how technology advanced, and early man found more efficient ways of getting food, warmth, and shelter.

New study methods also shed fresh light on early tools. Experimenters rediscover how our ancestors made and used stone implements. Lastly, tool and animal remains from camps of modern "Stone Age" hunters like the Eskimos give glimpses into prehistoric social life. Comparing Eskimo and ancient camp remains helps scientists work out the sizes of prehistoric hunting bands, activities their members shared, how long bands spent at individual sites, and at which times of year.

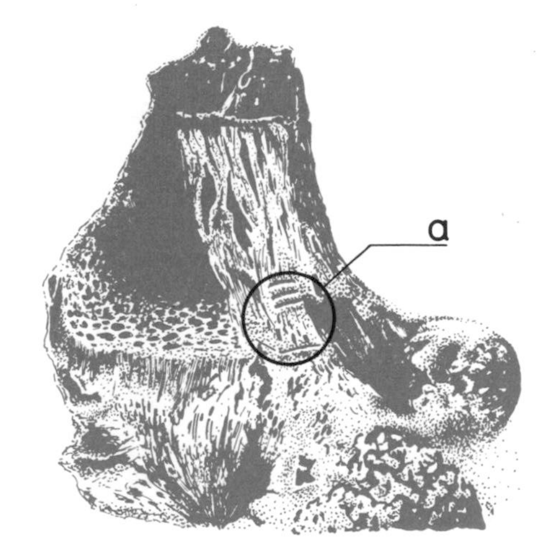

Cut marks on bone (above)
Short cross cuts (**a**) on this antelope bone from Olduvai Gorge show where an ancient hominid sliced off bits of meat with a sharp-edged stone. Cut marks overlying carnivores' tooth marks suggest that early humans were more scavengers than hunters.

Fossil clues to the environment

Fossil landforms, plants, and animals are guides to the climates and countrysides inhabited by early man. Here are just a few examples.

Careful study of layered sediments can unmask old lakes, rivers, or cliffs. Undisturbed sediments of fine particles accumulated on lake beds. Coarse sediments bound by natural cement may represent old river beds. Ripples in old river sediments indicate how fast and which way ancient rivers flowed. Layers of crusty soil sandwiched between grass roots tell of rainy times when sediments were dumped by overflowing lakes or rivers, and of dry intervening periods. Sea-cliff caves above the present level of a beach may mark an ancient shore when the sea stood higher than it stands today.

Fossil leaves, stems, roots, and pollen grains have other tales to tell. Remains of palms, oaks, and larches found respectively in ever higher layers at a single site would show climatic cooling. But glacial lake mud overlain successively by birch, pine, oak, and lime indicate climatic warming following the last Ice Age. A pollen shift from trees to weeds and cereals is a major sign of cultivation after forest clearance.

Clues from pollen (above)
Pollens from a Danish peat bog betray temporary forest clearance by farmers 4300 years ago. In this diagram strips vary in thickness with the abundance of the plants they represent.

A Primeval forest
B Forest clearance
C Regrown forest
a Elm, lime, and ash
b Oak
c Birch
d Herbaceous vegetation
e Wild grasses
f Cereals
g Plantain: a crop weed

River terraces (below)
Paired "benches" on each side of this river are old levels of the valley floor. The river cut deeper each time sea level fell as glaciers advanced.

1 First interglacial
2 Second interglacial
3 Third interglacial
4 Post-glacial

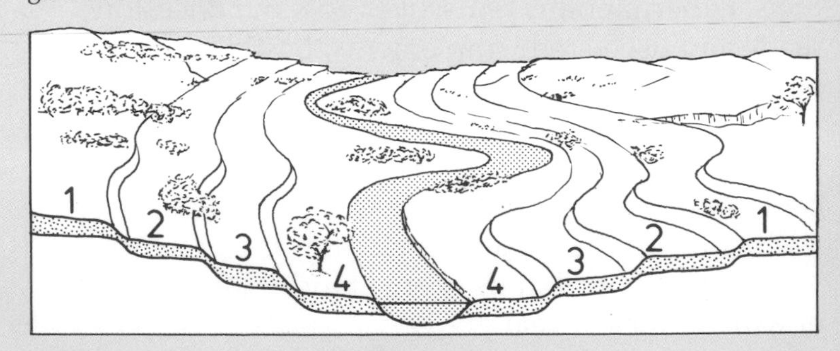

Animal remains are also valuable guides to vanished countrysides. Strands of shells in layered sediments show ancient shorelines. In limy soils, where pollen grains are seldom well preserved, snails reveal past vegetation. Different species flourished in prehistoric tundra, woodland, and grassland, and left their shells at different levels in the soil.

At some sites, a wealth of layered vertebrate remains charts habitat changes. At Olduvai, for instance, moisture-loving rodents at one time gave way to mole rats, which burrow in dry soil – evidence of climatic drying. In Ice Age Europe climatic change replaced warm woodland species like the straight-tusked elephant with musk ox, woolly mammoth and others adapted to cold steppe and tundra.

Fossil rodents
Remains of rodents of known habitats help experts discover how a landscape changed.
A Glacial times: few trees
1 Norway lemming (cold steppe)
2 Birch mouse (birch woodland)
3 Wood mouse (woodland)
4 Garden dormouse (woodland)
5 Ground squirrel (cold steppe)
6 Marmot (scree slopes)
7 Field vole (grassland)
8 Northern water vole (marshes)
9 Arctic lemming (cold steppe)
B Interglacial times: forest
10 Red squirrel (woodland)
11 Fat dormouse (woodland)
12 Wood mouse (woodland)
13 Garden dormouse (woodland)
14 Red-backed vole (woodland – other voles in open spaces)
15 Common dormouse (woodland)

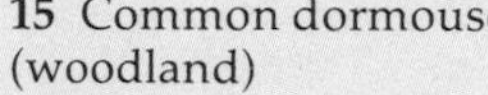

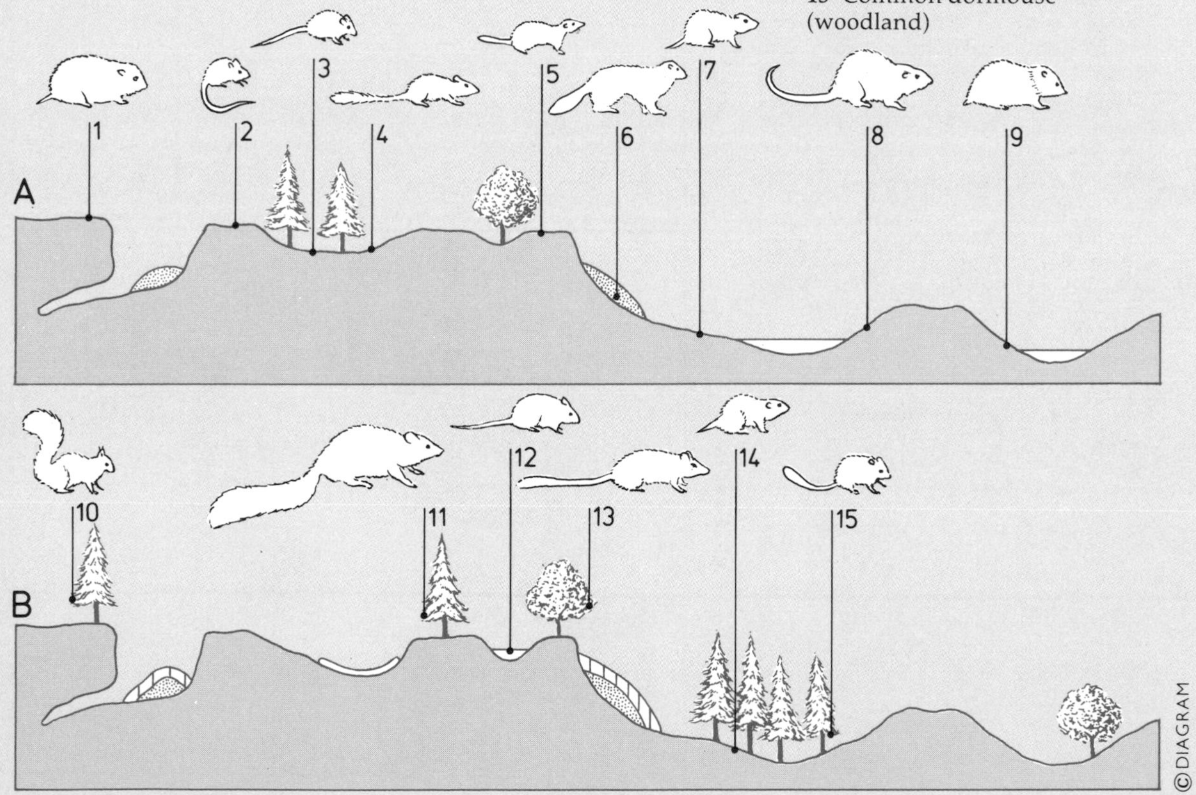

Relative dating

Relative dating is simply finding out if a fossil, artefact, or sedimentary layer is older or younger than (or the same age as) other fossils, artefacts, or layers. This does not show "how old" but scientists can often cross-relate one object to another whose age is more precisely known. Several dating methods may be used.

Biostratigraphy depends on comparing types of fossil animals or plants preserved at different layers in the rocks; usually, older forms lie deeper down than later ones. Studying evolutionary sequences of prehistoric pigs and elephants has helped scientists to cross-relate strata formed at the same time at widely separated sites in East Africa. In turn this helps experts cross-relate the far scarcer finds of fossil man.

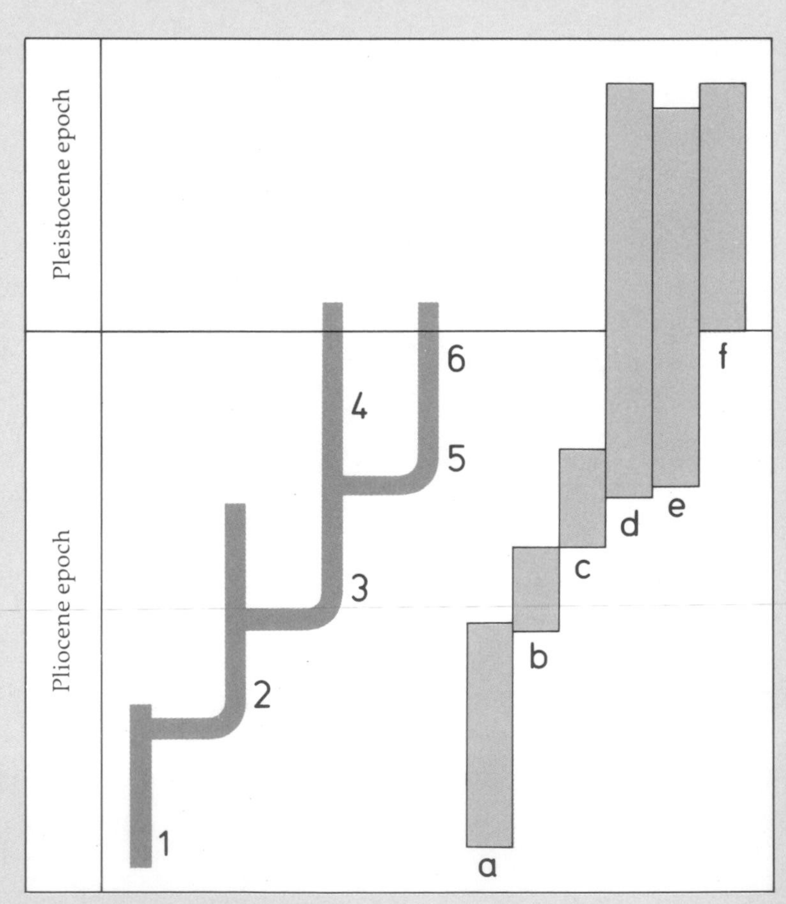

Pigs and places
Finds of evolving sequences of fossil pigs like that above helped scientists to fix the relative ages of early hominids from six East African localities.
Fossil pigs:
1 *Nyanzachoerus tulotos*
2 *Nyanzachoerus kanamensis*
3 *Nyanzachoerus jaegeri*
4 *Notochoerus euilus*
5 *Notochoerus capensis*
6 *Notochoerus scotti*
East African localities:
a Lothagam, Kenya
b Kanapoi, Kenya
c Laetoli, Tanzania
d Omo River, Ethiopia
e Koobi Fora, Kenya
f Olduvai Gorge, Tanzania

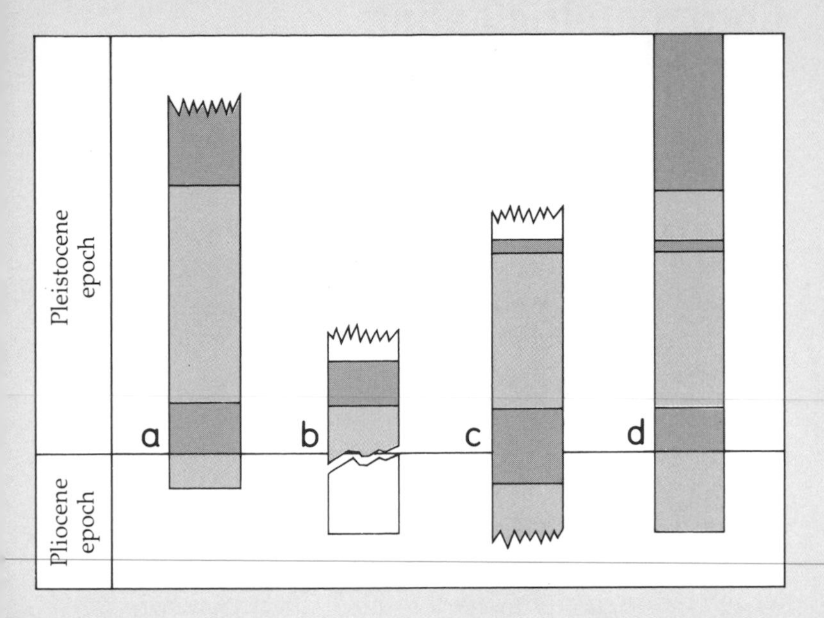

Paleomagnetic sequences (left)
Reversals in Earth's magnetic field recorded in the rocks help scientists to correlate the age of rocks at four East African localities.
a Olduvai Gorge, Tanzania
b Koobi Fora, Kenya
c Omo Delta, Ethiopia
d Omo River, Ethiopia

Normal
Reversed
Not known

Paleomagnetism involves periodic past reversals of the Earth's magnetic poles. Magnetic particles in mineral deposits – particularly lavas – are aligned according to the Earth's magnetic field when these deposits were laid down. Thus the deposits indirectly show relative ages of bones or artefacts buried in them when they formed.

Scientists can often roughly guess the ages of even surface finds. Close study of a bone's morphology (its form and structure) may let them place it in an evolutionary sequence already worked out by stratigraphy. In the same way typology – classifying artefacts by shape – may help archeologists to date a stone artefact by the ways in which they know these altered over time.

Relative bone dating hinges on comparing chemicals in a fossil bone with ones from other bones of known age. Bone absorbs fluorine and/or uranium from groundwater. The longer a bone lies buried, the more fluorine or uranium it accumulates. So measuring a bone's fluorine content or radioactivity level gives a relative idea of age. But this only works for bones from the same area and subject to the same conditions.

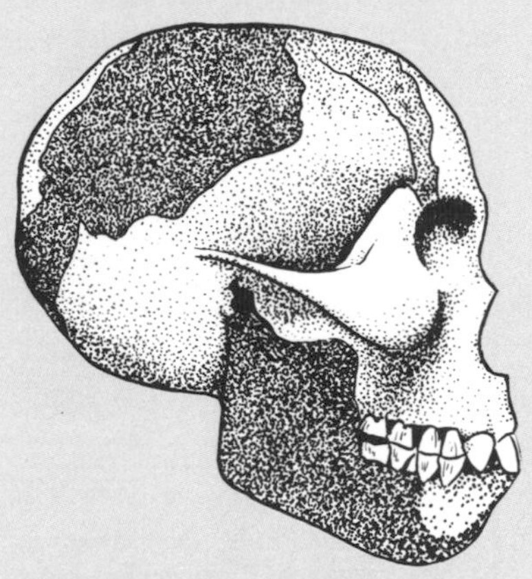

Piltdown forgery (above)
Dug up near Piltdown in South-east England, "Dawn man's" human cranium and ape-like jaw puzzled experts. Then fluorine and other tests revealed the fossil find to be a hoax: part modern man and part orangutan.

Chronometric dating

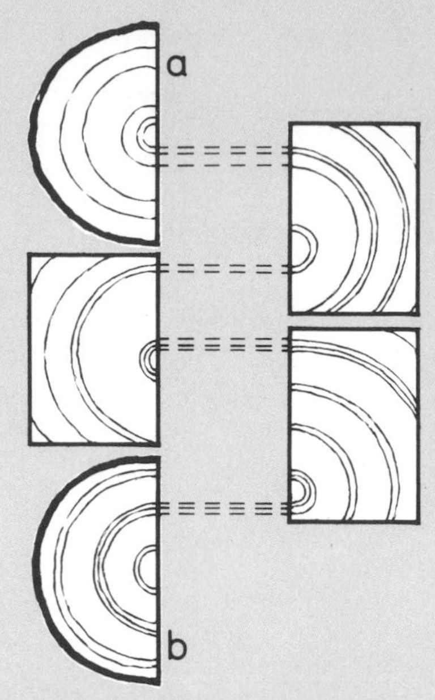

Counting tree rings (above)
Matching growth rings in trees and timbers helps experts date to within a year ancient posts or other wooden structures.
a Modern tree
b Old timber post

Chronometric dating techniques find approximate ages of prehistoric objects. Most of these seven methods measure changes that occur at constant rates in the atomic structures of substances.

1 Dendrochronology is based on counting annual growth rings in timber from trees whose lifespans overlapped. Major applications: certain prehistoric sites up to 6500 years old.

2 Varve analysis involves counting varves (layers of lake sediments deposited seasonally in areas of winter freezing). Application: largely to late Pleistocene chronologies.

3 Oxygen-isotope analysis of deep-sea sediments records changes in amounts of the oxygen isotope 18 which is indirectly geared to global temperature fluctuations. Application: dating mild and cold spells in the Pleistocene and earlier.

4 Radiocarbon dating is based on measuring the amount of the radioactive carbon isotope, carbon-14, in organic matter. In dead animals and plants the amount declines at a constant rate after death. Application: organic substances (eg bones) up to 50,000 years old.

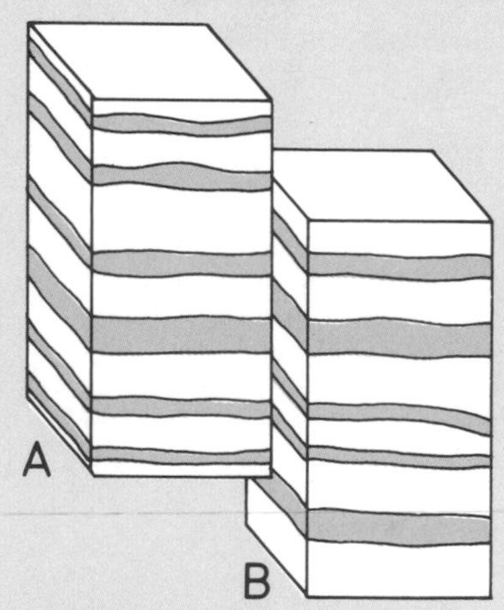

Correlating varves (above)
Layered sediments from different lakes (**A, B**) show matching seasonal deposits. Studying these varves enabled scientists to fix the exact age of some North European objects up to 20,000 years ago.

Radiocarbon dating (right)
A dead body loses carbon-14 by radiation at a measurable rate by which the body's age can be determined. The margin of error increases with the time span.
1 5730 years after death half the carbon-14 remains
2 16,704 years after death one-eighth remains
3 70,000 years after death almost all has disappeared. But sophisticated carbon-14 dating can find the age of even older organic objects.

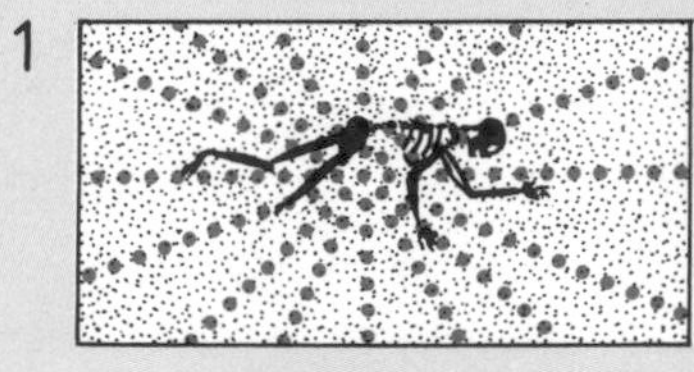

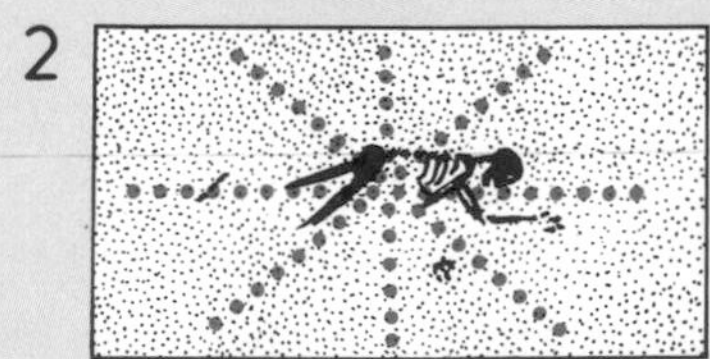

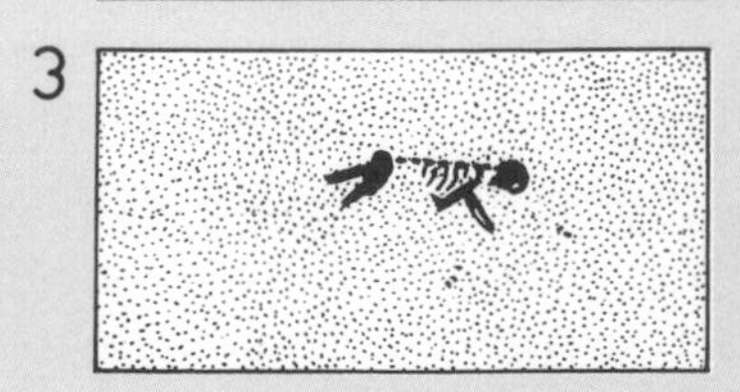

5 Thermoluminescence dating involves measuring light produced by heating certain minerals heated in prehistoric times. The amount of light emitted shows when prehistoric heating had occurred. Application: burnt potsherds, clay figurines, campfire stones, etc, up to 8000 years old.
6 Amino-acid racemization occurs in organisms after death. Optically active protein amino acids (ones deflecting polarized light) slowly racemize, or turn into optically inactive amino acids. Application: bone to 100,000 years old and more.
7 Uranium series dating covers rates of radioactive decay of uranium-234 into protactinium-231 then thorium-230: processes spread over many scores of thousands of years to 350,000 years ago.
8 Fission-track dating measures uranium decay in old stone artefacts as the number of tracks produced by the splitting (fission) of uranium-238 which occurs in nature at a known and constant rate. Application: sites of early man.
9 Potassium-argon dating involves the steady decay of potassium-40 into argon-40. Application: volcanic rocks older than 1 million years.

For paleontologists, items 4, 7, 8, and 9 prove particularly useful. But contamination of tested specimens by other substances may falsify results, so experts cross-check these by using several methods where they can.

Potassium-argon dating
Potassium-40 decays into argon-40 and calcium-40 at the rate shown in this diagram.
1 The mass of potassium-40 in newly formed volcanic rock.
2 The proportion of potassium-40 remaining after one half life (1310 million years)
3 The proportion left after another 1310 million years (two half lives)
4 The proportion left after another 1310 million years (three half lives)

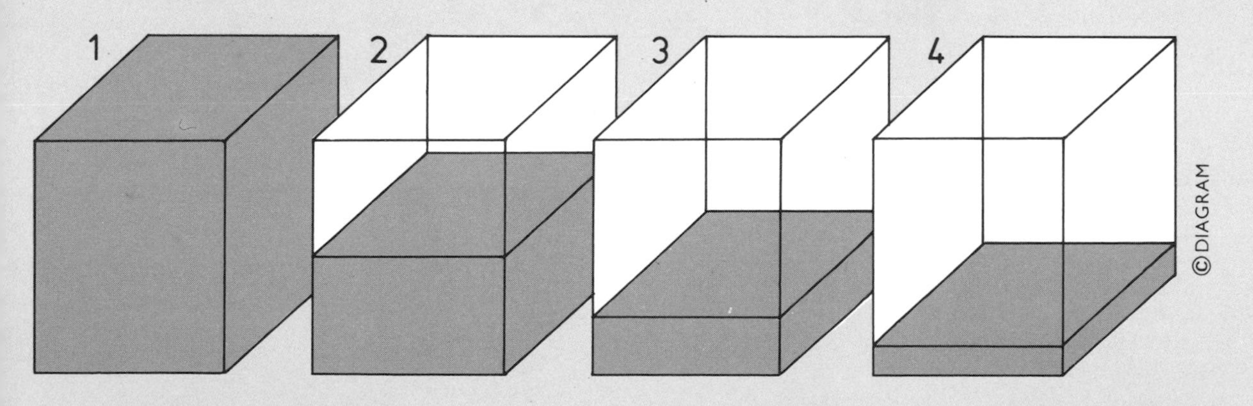

Fossil problems

Despite the methods we have just outlined, tracking the ancestors of *Homo sapiens* means solving puzzles full of missing clues.

Many finds of fossil primates comprise no more than teeth or bits of lower jaw – among a skeleton's more durable remains. Most other bones and of course all soft parts of the body have long disappeared, broken up by carnivores and scavengers, then scattered by animals or water. Few whole early primate skeletons survive.

This dearth of fossil finds forced paleoanthropologists to try identifying early hominids by incomplete remains, some from individuals differing because of sex or age, or bone disease. Not surprisingly, experts disagree about just what some fossils represent.

Even more problematic is how early humans lived, for most behavior leaves no fossil trace, and few old tools of substances as soft as wood survive. We can only guess that early humans – like some modern

a b c

Intrusive burial (above)
This section through the ground shows why bones can seem more ancient than they are. Undisturbed sediments (**a**) increase in age with depth. The skeleton (**b**) lies level with old sediments, but disturbed soil (**c**) betrays this as a recent burial.

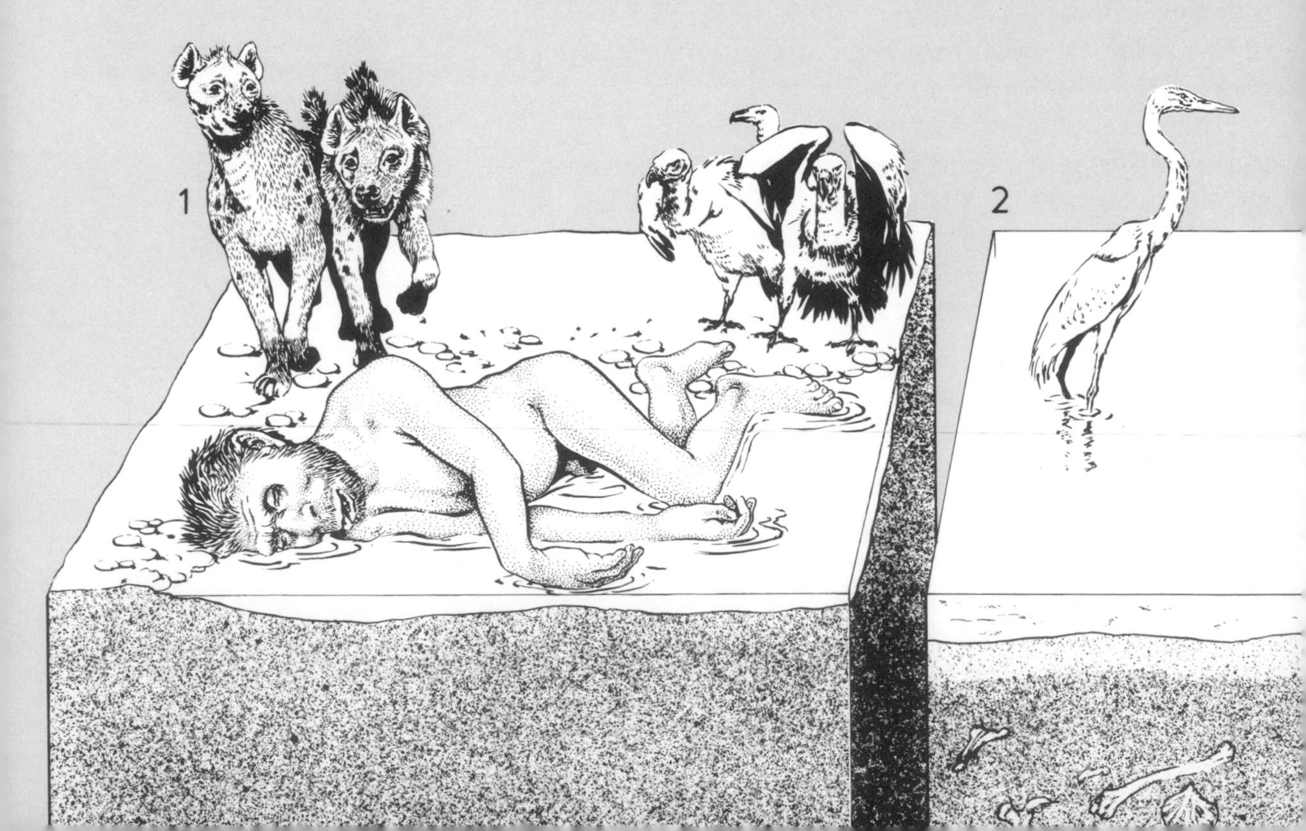

primitives – were foragers, using bark trays to carry gathered roots or fruits back "home."

Then, too, a total lack of finds from certain times and places leaves tantalizing gaps in the fossil primate record, and makes it possible to misinterpret missing links. Even assigning fossils dates is sometimes complicated by contaminated samples, or earth movements that disrupted stratigraphic sequences.

Yet new knowledge of early humans is accumulating fast – from fossils, biochemistry, and comparative studies of living primates including human "primitives." In 1968 all Kenya had yielded a mere three early hominid specimens; by 1984 150 had turned up at Koobi Fora alone. By the year 2000, technical advances and the exploration of still poorly charted sediments promise much more information about exactly when the ape and human lines diverged; how human evolution proceeded in the early Pleistocene; and where and when our own subspecies had its origins.

How evidence is lost
Three pictures show why ancient human fossils are seldom found intact.
1 Hyenas and vultures prepare to dismember a dead hominid lying by a lake or river. Later its broken bones will flake in the sun and passing herds may scatter them.
2 Floods have shed layers of fine sediment upon the now-isolated bits of bone. Meanwhile a river might have carried teeth and tiny bones far downstream before dumping them upon a mudbank.
3 Exposed by a river's back-and-forth migration across its flood plain, this ancient skull endured more weathering before it was discovered.

Tracking early man 1

These four pages list major fossil hunters, archeologists, and others who have traced our species' origins and early cultural development. No list this brief could be complete; ours includes only some of the many scientists involved, and stresses individuals from the Western World.

Jacques Boucher de Perthes

Robert Broom with a cast of *Plesianthropus* (*Australopithecus africanus*)

Bass, George Fletcher American archeologist whose 1960 expedition to a submerged Bronze Age shipwreck off Turkey began scientific underwater archeology by diving archeologists.

Binford, Lewis R. American archeologist who in the 1960s pioneered the "New Archeology" by reexamining old concepts in the light of worldwide comparative archeological and ethnographical research.

Bingham, Hiram (1875–1956) American archeologist who in 1911 rediscovered the lost Inca city Vilcabamba ("Machu Picchu") in the Peruvian Andes – one of the wonders of Precolumbian civilization.

Black, Davidson (1884–1934) Canadian-born anatomist who in 1927 described a tooth of *"Sinanthropus pekinensis" (Homo erectus)* found in China. Later he led major excavations of *Homo erectus* cave sites at Choukoutien near Peking.

Bolk, Louis Dutch anatomist who in 1926 argued that the human species is a neotenous type of ape.

Boucher (de Crèvecoeur) de Perthes, Jacques (1788–1868) French archeologist who discovered hand axes and fossil animal remains in old Somme river gravels, and developed the idea that human prehistory extended back through geological time.

Boule, Marcellin (1861–1942) French paleontologist whose misdescription popularized the notion of Neandertal man as shambling and brutish.

Braidwood, Robert J. American archeologist whose (early 1950s) excavation of Jarmo in Iraq suggested that Middle Eastern agriculture began in the Taurus-Zagros mountain belt, not in the rich valleys of the Fertile Crescent farther south.

Breasted, James Henry (1865–1935) American archeologist and Egyptologist who coined the term "Fertile Crescent" for the rich belt of land from Egypt to Iraq, where Western civilization began.

Breuil, Henri (1877–1961) French priest and archeologist who became the major authority on Paleolithic cave art.

Broom, Robert (1866–1951) Scottish paleontologist who discovered australopithecine remains in South Africa, and in 1938 first described *Australopithecus robustus,* which he called *"Paranthropus robustus."*

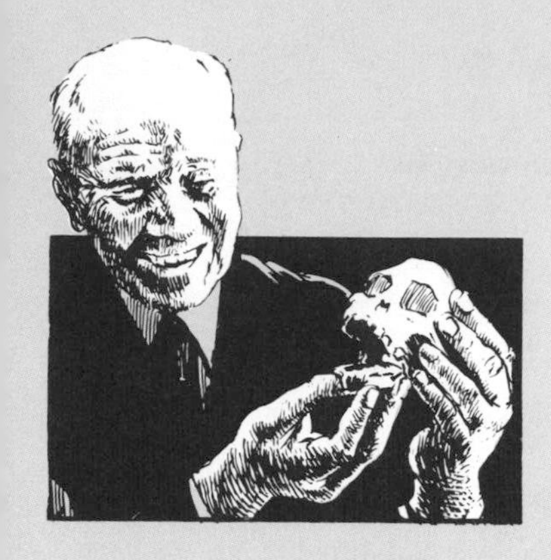

Left: Raymond Dart with the famous Taung skull of a juvenile australopith

Right: Eugene Dubois, discoverer of *Homo erectus*

Dart, Raymond Arthur Australian-born South African anthropologist who in 1925 described *Australopithecus africanus*, the first australopithecine to be discovered.
Darwin, Charles Robert (1809–82) British naturalist whose theory of evolution by natural selection (published 1859) laid the scientific basis for the study of human origins, which he believed had been in Africa.
Dawson, Charles (1864–1916) British amateur geologist whose discovery of *"Eoanthropus dawsonii"* (Piltdown Man) was hailed as an evolutionary missing link in 1912 by paleontologist Arthur Smith Woodward, but proved to be a hoax in 1953 after reappraisals begun by anatomist Joseph S. Weiner.
Douglass, Andrew Ellicott (1867–1962) American astronomer and archeologist who developed dendrochronology (tree-ring dating).
Dubois, (Marie) Eugene (Francois Thomas) (1858–1940) Dutch anatomist who, in Java in 1891, made the first known discovery of *Homo erectus*, which he named *"Pithecanthropus erectus."*
Figgins, Jesse D. American paleontologist whose excavations near Folsom, in New Mexico in 1927, dug up a stone point embedded between ribs of a prehistoric bison – the first conclusive proof of man's antiquity in the Americas.
Frere, John (1740–1807) British antiquary who founded prehistoric archeology. In 1790 he found flint implements and extinct animals in the same strata, at Hoxne in East Anglia.
Geer, Gerard Jakob, Baron de (1858–1943) Swedish geologist who developed a chronometric dating system based on counting varves (seasonally laid-down layers of lake sediments).
Goodall, Jane van Lawick British born ethologist whose intensive studies of chimpanzees at Tanzania's Gombe Stream Reserve in the 1960s and '70s shed light on the origins of human behavior.
Haeckel, Ernst Heinrich (1834–1919) flamboyant German evolutionist who coined the name *"Pithecanthropus"* ("ape man") for a postulated missing link between apes and man.
Jefferson, Thomas (1743–1826) Third president of the United States, who anticipated North American archeology by excavating mounds in Virginia and making stratigraphical observations.
Johanson, Donald American anthropologist who, in Ethiopia in 1974, discovered the oldest known australopith skeleton ("Lucy"), and in 1978 named its species *Australopithecus afarensis*.
King, William Irish anatomist who, in 1864, coined the scientific name *Homo neanderthalensis*.

Jane Goodall with a chimpanzee

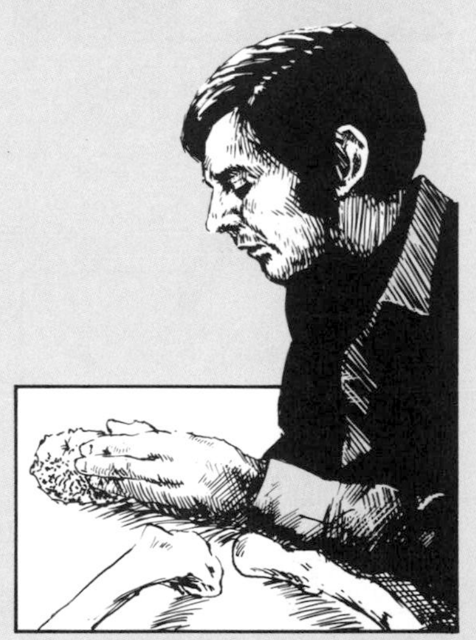

Donald Johanson, discoverer of "Lucy"

Tracking early man 2

Koenigswald, Gustav Heinrich Ralph von German anthropologist who named *Gigantopithecus* in 1935 and discovered *Homo erectus* fossils and part of a massive mandible of "*Meganthropus palaeojavanicus*" in Java (1937–41).

Lartet, Édouard Armand Isidore Hippolyte (1801–71) French geologist, archeologist, and paleontologist. He named *Dryopithecus* and *Pliopithecus*, and excavated many Paleolithic cave sites in the Dordogne.

Lartet, Louis French archeologist and geologist; son of Édouard (qv). He discovered Cro-Magnon Man in 1868 while excavating a rock shelter at Cro-Magnon in the Dordogne.

Leakey, Jonathan Eldest son of Louis (qv); discoverer in 1960 at Tanzania's Olduvai Gorge of the first fossils attributed to *Homo habilis*.

Leakey, Louis Seymour Bazett (1903–72) British anthropologist whose East African expeditions proved human evolution began in Africa. Controversy still surrounds his naming and interpretation of some major fossil finds.

Leakey, Mary (1913–), British-born anthropologist, wife of Louis (qv). She discovered *Australopithecus boisei* (1959), later directed excavations at Tanzania's Olduvai Gorge, and led the expedition that in 1977 found the oldest known human footprints at Laetoli, in Tanzania.

Leakey, Richard E. Director of the National Museums of Kenya; son of Louis (qv). In 1967 his expedition to Ethiopia's Omo region produced possibly the oldest-known skull of fully modern man. Since 1968 his expeditions near Lake Turkana have yielded much new evidence of early hominines.

Lewis, G. Edward Scientist who first identified (1932) and named *Ramapithecus* ("Rama ape") and claimed its (since disputed) position as an ancestor of man.

Libby, Willard Frank (1908–80) American chemist who in 1947 developed radiocarbon dating.

Linnaeus, Carolus (1707–78) Swedish botanist who devised our present system for classifying living things, and in 1758 gave our own species its scientific name: *Homo sapiens*.

Lubbock, Sir John (1834–1913) British amateur archeologist who coined the terms Paleolithic (Old Stone Age) and Neolithic (New Stone Age).

Lyell, Sir Charles (1797–1875) British geologist who helped establish geology as a modern science.

Marshall, Sir John Hubert (1876–1958) British archeologist whose excavations at Harappa (1921) and Mohenjo-daro (1922), now in Pakistan, revealed the prehistoric Indus Valley civilization of 2500–1750 BC.

Marston, A.T. British amateur archeologist who discovered Swanscombe Man near London in 1935.

Mellaart, James British archeologist whose (1961–65) excavations of Çatal Hüyük, Turkey, revealed the largest known Neolithic town in the Middle East.

Mendel, Gregor Johann (1822–84) Austrian monk and botanist, who discovered the principles of heredity, so laying a basis for genetics, which explains the biological mechanisms of evolution.

Morgan, Lewis Henry (1818–81) An American founder of scientific anthropology. His *Ancient Society* (1877) was the first major scientific attempt to trace cultural evolution "*from Savagery through Barbarism to Civilization.*"

Gustav von Koenigswald

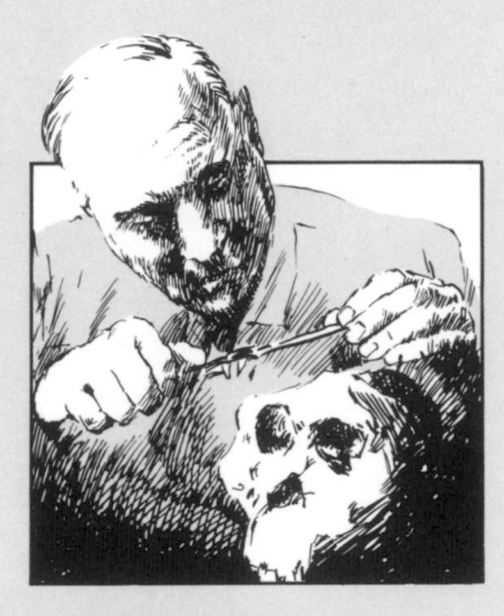

Louis Leakey measuring an *Australopithecus boisei* skull

Mary Leakey, discoverer of *Australopithecus boisei*

Petrie, Sir William Matthew Flinders (1853–1942) British archeologist and Egyptologist who raised excavating standards, stressing detail and the publication of results.
Pitt-Rivers, Augustus Henry (1827–1900) British archeologist who pioneered modern scientific archeological techniques.
Schaaffhausen, Hermann German anthropologist who published an account of Neandertal man after its discovery in 1856.
Schoetensack, Otto German professor who in 1908 published details of the Mauer mandible, alias Heidelberg Man.
Schwalbe, Gustave (1844–1917) German anatomist who argued for *"Pithecanthropus" (Homo erectus)* as an ancestor of modern man.
Semenov, S.A. Russian archeologist whose *Prehistoric Technology* (1957) advanced experimental study of microwear as a clue to use in Stone Age tools.
Simons, Elwyn L. American paleontologist who discovered important early primate fossils in the 1960s in Egypt's Fayum Depression.
Solecki, Ralph American scientist whose (1953–60) excavations in Iraq's Shanidar cave shed new light on Neandertal behavior.
Stephens, John Lloyd (1805–52) American archeologist whose (1839–42) explorations of Maya ruins pioneered the archeology of Middle America.
Susuki, Hisashi Japanese excavator at Amud Cave in Israel in the 1960s where bones and tools suggested transition between Neandertal and modern man, and between Mousterian and Late Paleolithic technology.
Thomsen, Christian Jurgensen (1788–1865) Danish archeologist who divided prehistory into successive Stone, Bronze, and Iron ages.
Weidenreich, Franz (1873–1948) German-born anatomist and anthropologist renowned for reconstructions of "*Sinanthropus*" (Peking Man) which survived the original's destruction in 1941. In 1940 he established the name *Homo erectus.*
Wilson, Alan C. and **Sarich, Vincent M.** American molecular biologists whose "molecular clock" evidence in 1969 showed man's kinship with the apes more recent than fossil finds implied.
Woolley, Sir Charles Leonard (1880–1960) British archeologist whose excavation of the Sumerian city of Ur (1922–34) gave insights into early civilization in Mesopotamia.

Augustus Pitt-Rivers

John Stephens

Franz Weidenreich

Museum displays 1

Thousands of museums show or store tools or bones of prehistoric peoples, many excavated from some of the world's countless ancient burials and other monuments. In some western countries the chief town of every state or county includes a display of prehistoric human life. These six pages simply give a brief selection of museum collections worldwide. Items appear in alphabetical order of country and then city.

Bone flesher in Ottowa's National Museum of Man

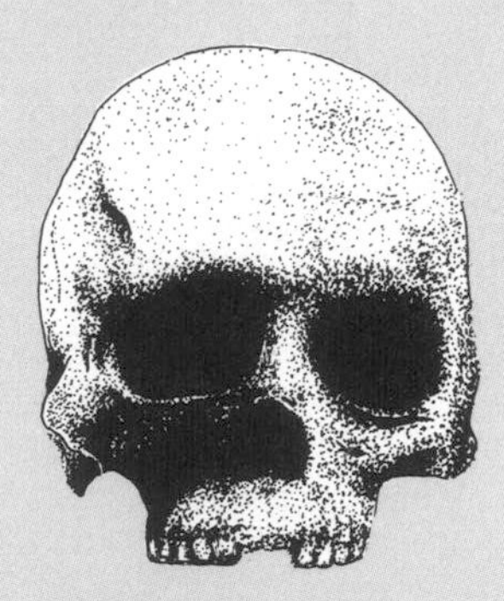

Liujiang Man's skull in a Beijing (Peking) musuem

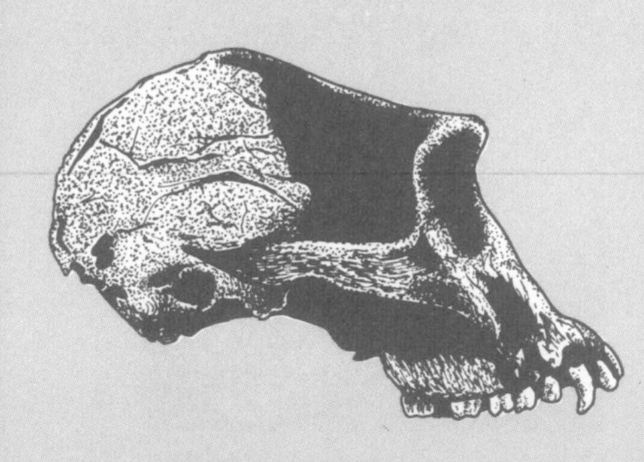

Australopithecus afarensis skull in Addis Ababa

AUSTRIA
Vienna: Natural History Museum The Prehistoric Department has prehistoric material from the former Austro-Hungarian Empire, including Stone Age artefacts, Bronze Age tools and weapons, metal-smelting equipment, and Iron Age implements.

BELGIUM
Brussels: Royal Museum of Art and History Worldwide collections include prehistoric archeological items.

BULGARIA
Sofia: Museum of Sofia's History Traces local developments from the Neolithic onward.

CANADA
Ottawa: National Museum of Man Antiquities include early North American artefacts.

CHINA
Beijing (Peking): Institute of Archeology, Academia Sinica Collections stress Chinese prehistory.
Beijing (Peking): Institute of Vertebrate Paleontology and Paleoanthropology Notable Chinese items include the skull of Liujiang Man.

COLOMBIA
Bogotá: National Museum Includes prehistoric relics.

CYPRUS
Nicosia: Archeological Museum Artefacts illustrate the island's prehistory.

CZECHOSLOVAKIA
Brno: Brno Antropos Museum Includes Late Paleolithic items.
Prague: National Museum Displays include prehistoric artefacts.

DENMARK
Copenhagen: National Museum Prehistoric antiquities include remarkable Bronze and Iron Age artefacts.
Silkeborg: Silkeborg Museum Contains the world-famous 2000-year-old "bog body" Tollund Man.

EGYPT
Cairo: Egyptian Antiquities Museum Features finds from pre-dynastic and Pharaonic sites.
Cairo: Egyptian National Museum Covers prehistoric times through the 6th century AD.

ETHIOPIA
Addis Ababa: National Museum of Ethiopia Has major hominid fossils, including remains of *Australopithecus afarensis*.

FRANCE

Bordeaux: Museum of Aquitaine Prehistoric collections from the Bordelais and Périgord areas feature here.

Carnac, Morbihan: Museum of Prehistory A prehistoric collection (of ceramics, necklaces, etc), largely from nearby burial mounds (but the major local prehistoric landmarks are vast alignments of huge standing stones).

Les Eyzies, Dordogne: National Museum of Prehistory This stresses finds from local Late Paleolithic cave excavations.

Montignac-Lascaux, Dordogne: Lascaux Cave Superb prehistoric cave paintings, no longer open to the public.

Nancy: Museum of Earth Sciences Exhibits trace evolution from early living things to man.

Nice: Museum of Terra Amata This features relics of a camp built on the spot, perhaps 400,000 years ago.

Paris: Louvre Museum Collections include major exhibits from early Western civilizations.

Paris: Museum of Man Contains major Neandertal and Cro-Magnon fossils and artefacts.

Périgueux, Dordogne: Périgord Museum A Neandertal skeleton and Late Paleolithic bison frieze are among its exhibits.

Saint-Germaine-en-Laye, Yvelines: Museum of National Antiquities Has some splendid Late Paleolithic French stone and bone tools and objets d'art.

Talence, Gironde: Institute of the Quaternary, Bordeaux University Holds Late Paleolithic bone and stone implements.

GERMANY (EAST)

Berlin (East): Berlin State Museum of Near Asia Sumerian seals figure among its West Asian antiquities.

GERMANY (WEST)

Berlin (West): Egyptian Museum Stresses Ancient Egyptian sculptures, seals, tools, masks, etc.

Berlin (West): Museum of Prehistory and Early History Deals largely with European and Oriental prehistory.

Bonn: Rhineland Museum Features local prehistory, and includes a Neandertal skull.

Frankfurt: Museum of Prehistory and Early History Stresses the Frankfurt region.

Hamburg: Museum of Ethnology and Prehistory Has worldwide collections.

Stuttgart: Württemberg Museum Features prehistory, early history, etc.

GREECE

Athens: Anthropological and Ethnological Museum Worldwide collections of objects dating back to prehistoric times.

Thessaloniki: Paleontological Museum, University of Thessaloniki Houses the famous Petralona skull (*Homo erectus* or archaic *Homo sapiens*).

GUATEMALA

Guatemala City: Archeological Museum Features Mayan and other artefacts.

HONDURAS

Tegucigalpa: Museum of History and Archeology Contains archeological finds from local excavations.

Venus of Laussel in a Bordeaux Museum

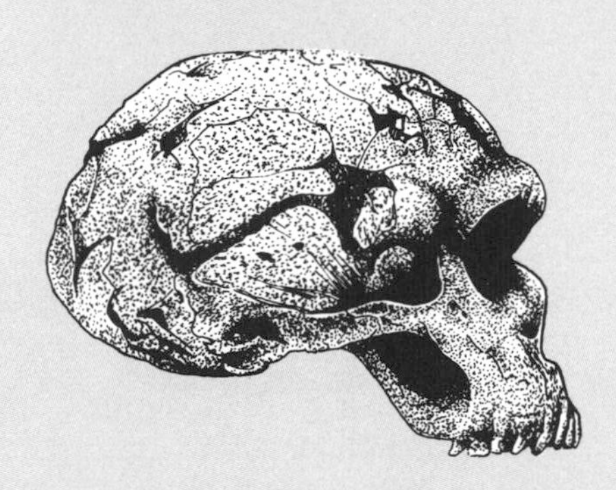

Neandertal skull in the Museum of Man, Paris

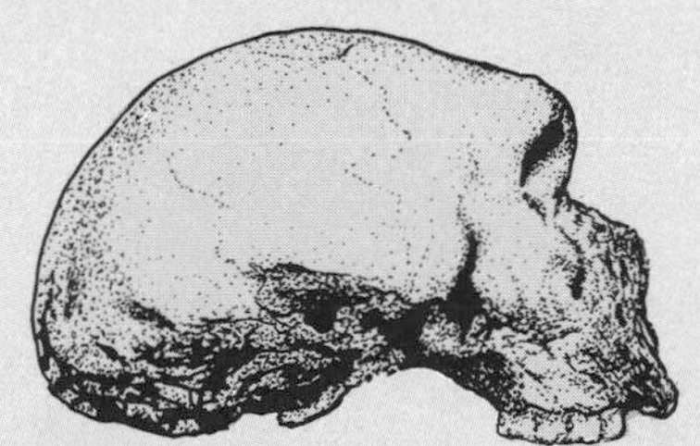

Petralona skull in the Paleontological Museum, Thessaloniki

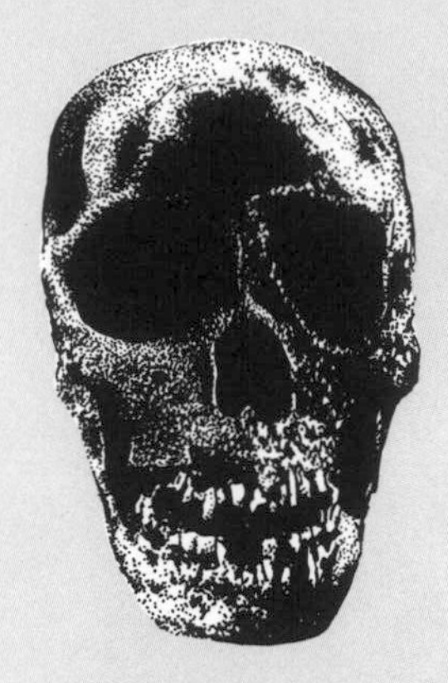

Skull of Shanidar I in Baghdad's Iraq Museum

A Jomon pottery figure in the Tokyo National Museum

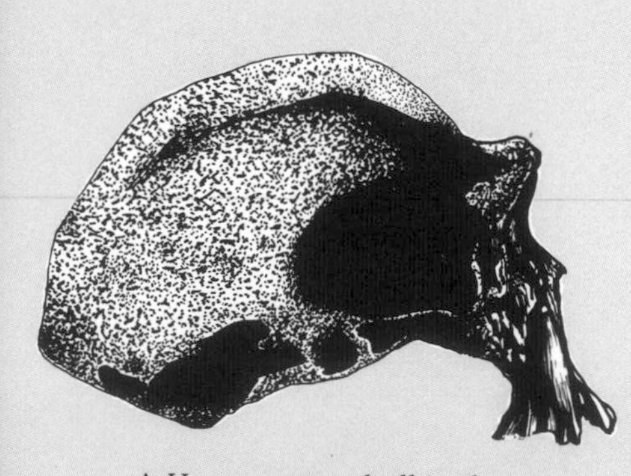

A *Homo erectus* skull in the National Museum, Nairobi

Museum displays 2

HUNGARY
Budapest: Hungarian National Museum Exhibits represent the Old, New, Copper, Bronze, and Iron ages.
INDIA
Calcutta: Indian Museum Has antiquities from prehistoric times onward.
INDONESIA
Gianjar: Museum Gedong Artja Includes Paleolithic, Neolithic, and other prehistoric implements.
IRAN
Tehran: Archeological Museum Holds Paleolithic and later antiquities.
IRAQ
Baghdad: Iraq Museum Has Neandertal remains from Shanidar, Neolithic pottery, and artefacts from early Mesopotamian civilizations.
IRELAND
Dublin: National Museum Contains Mesolithic and Neolithic implements and a wealth of prehistoric gold ornaments.
ISRAEL
Haifa: Museum of Natural History and Prehistory
Jerusalem: Museum of Prehistory Features objects from prehistoric sites in Israel.
ITALY
Milan: Archeological Museum Includes Neolithic and other prehistoric items.
Rome: Collections of the Italian Institute of Human Paleontology
Rome: Museum of Anthropology, the University Has skeletons and skulls including Neandertal finds.
JAPAN
Tokyo: Tokyo National Museum Contains prehistoric Japanese relics.
JORDAN
Amman: Jordan Archeological Museum Covers all periods from Late Paleolithic to medieval.
KENYA
Nairobi: National Museum of Kenya Houses some major fossil primate finds, including *Proconsul*, australopithecines, *Homo habilis*, and *Homo erectus*.
MEXICO
Mérida: Yucatán Museum of Anthropology Deals with Mexican pre-Hispanic cultures.
Mexico City: National Anthropological Museum Exhibits tell the story of Mexico from prehistoric times.
NETHERLANDS
Amsterdam: Archeological Museum of the University of Amsterdam Holds vases, bronzes, etc., from West Asia and Egypt.
Leiden: National Museum of Antiquities Includes West Asian and Egyptian artefacts.
Leiden: National Museum of Ethnology Includes prehistoric Indonesian and Precolumbian items.
PAKISTAN
Karachi: National Museum Features Paleolithic implements from the Soan Valley, and relics of the Indus Valley civilization.
PANAMA
Panama City: Museum of Panama Man Features archeology and ethnography.

PERU
Lima: Museum of Art Illustrates the Peruvian past from Precolumbian times.
POLAND
Warsaw: National Museum of Archeology Covers the prehistoric and later archeology of Poland.
ROMANIA
Bucharest: National Museum of Antiquities Relics illustrate Romanian Stone and Bronze Age prehistory.
SOUTH AFRICA
Pretoria: Transvaal Museum Fossils include remains of *Australopithecus africanus* and *robustus*.
SPAIN
Madrid: National Archeological Museum Includes prehistoric finds from Ambrona.
Santillana del Mar, Santander: Altamira Cave Outstanding Late Paleolithic cave art.
SWEDEN
Stockholm: Historical Museum Includes prehistoric Swedish items.
Stockholm: Museum of Mediterranean and Near Eastern Antiquities One of several that show Asian prehistoric items.
SWITZERLAND
Bern: Bern Historical Museum Has mainly local prehistoric material from the Paleolithic onward.
Zurich: Swiss National Museum Illustrates Swiss prehistory.
SYRIA
Damascus: Damascus National Museum Archeological exhibits represent the country's past civilizations.
TANZANIA
Dar es Salaam: National Museum of Tanzania Items held include the first-found skull of *Australopithecus boisei*.
TURKEY
Ankara: Museum of Anatolian Civilizations Has Paleolithic, Neolithic, Chalcolithic (Copper Age), Bronze Age, and later items.
Istanbul: Archeological Museum of Istanbul Includes Sumerian, Akkadian, Hittite, Assyrian, Egyptian, and other material from early civilizations.
UNITED KINGDOM
Avebury, Wiltshire: Avebury Museum Local late Neolithic finds (Avebury stands inside prehistoric earthworks).
Birmingham: City Museum Includes prehistoric and Egyptian antiquities.
Bristol: City Museum Worldwide collections include items from ancient history.
Cambridge: Museum of Archaeology and Anthropology Features worldwide archeology from the Stone Age onward.
Cardiff: National Museum of Wales Includes an archeological collection.
Cheddar, Somerset: Cheddar Caves Museum contains Paleolithic and later finds.
Doncaster, South Yorkshire: Museum Exhibits include pre-Roman items.
Dorchester, Dorset: Dorset County Museum Has a fine collection of prehistoric implements.

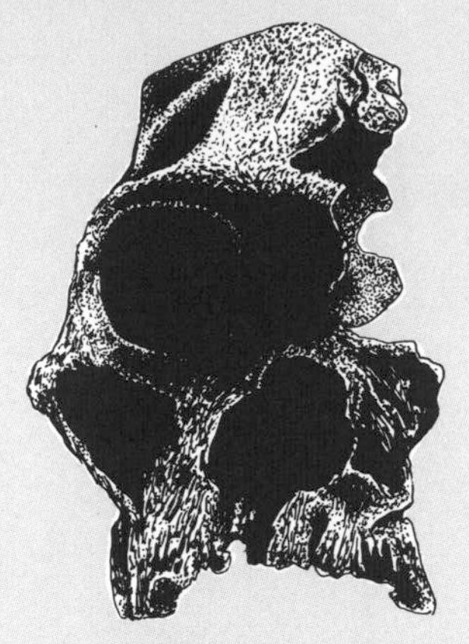

An *Australopithecus africanus* skull in Pretoria

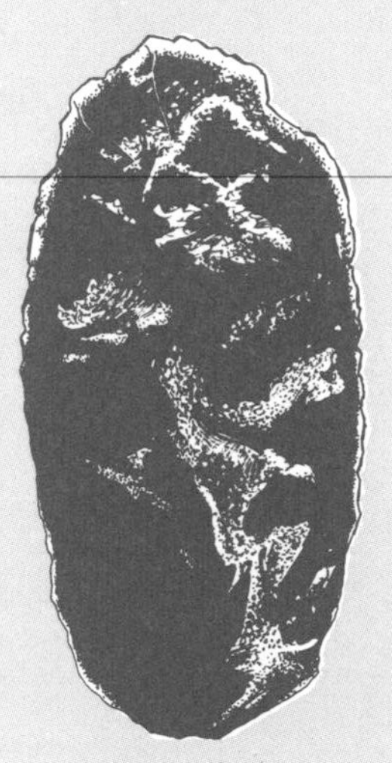

This hand axe is in Madrid

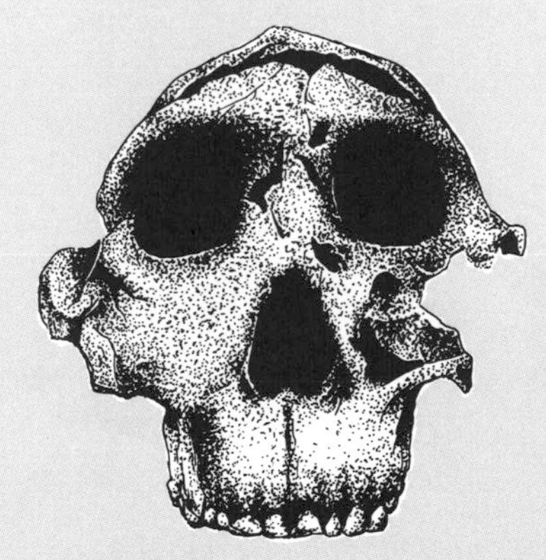

An *Australopithecus boisei* skull in Dar es Salaam

A Paleolithic wooden spearhead in the British Museum (Natural History)

One of the first-known written documents – a 5500-year-old tablet from Kish, Iraq – in Oxford's Ashmolean Museum

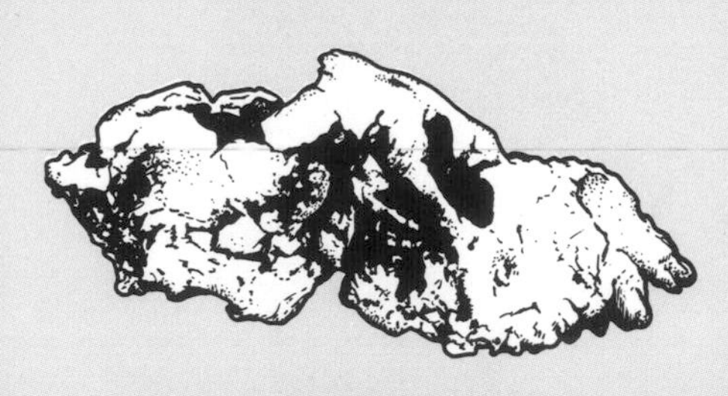

"Aegyptopithecus" (*Propliopithecus*) skull in the Peabody Museum of Natural History, New Haven

Museum displays 3

Edinburgh: National Museum of Antiquities of Scotland Holds extensive collections of prehistoric items.
Glasgow: Glasgow Art Gallery and Museum Includes archeological exhibits.
Glasgow: Hunterian Museum Archeological items are in the university's main building.
Ipswich, Suffolk: The Museum Includes East Anglian prehistoric finds.
London: British Museum A major museum with vast archeological collections of all periods and many lands.
London: British Museum (Natural History) A major museum including exhibits tracing man's place in evolution; has some important fossil human remains.
London: Museum of Mankind A mainly ethnographic museum with archeological collections from Africa and the Americas, most not displayed.
Manchester: Manchester Museum Archeological items include many from Ancient Egypt.
Newcastle upon Tyne: Museum of Antiquities A collection with models and reconstructions.
Oxford: Ashmolean Museum of Art and Archeology Britain's oldest museum with archeological items from Europe and the Near East.
Plymouth: City Museum Includes archeological items.
Salisbury, Wiltshire: Salisbury and South Wiltshire Museum Includes important prehistoric objects from Stonehenge.
Torquay, Devon: Natural History Society Museum Has prehistoric finds from Kent's Cavern.

UNITED STATES

Austin, Texas: Texas Memorial Museum Exhibits show human evolution and American Indian culture.
Berkeley, California: Robert H. Lowie Museum of Anthropology Features anthropological and archeological exhibits from most continents.
Chicago, Illinois: Field Museum of Natural History Features archeological and anthropological items.
Cincinnati, Ohio: Cincinnati Museum of Natural History Archeological material includes American Indian relics.
El Paso, Texas: Wilderness Park Museum Features the archeology and cultures of the South-west US and north Mexico.
Kansas City: Kansas City Museum of History and Science Features Indian history and archeology.
Los Angeles: Los Angeles County Museum of Art Incorporates archeological material.
Madison, Wisconsin: Museum of the State Historical Society of Washington Covers Wisconsin's history from prehistoric times.
Mesa Verde National Park, Colorado Includes hundreds of ruined Indian villages up to 1300 years old.
Miami, Florida: Historical Museum of Southern Florida Covers local anthropology, archeology, etc.
Nashville, Tennessee: Tennessee State Museum Indian artefacts and local history figure here.
New Haven, Connecticut: Peabody Museum of Natural History, Yale University Fossil primates include *"Aegyptopithecus."*
New Orleans, Louisiana: Louisiana State Museum Covers local history and archeology represented by Indian implements.

New York City, New York: (Heye Foundation) Museum of the American Indian Includes a wealth of traditional Indian implements.
Norfolk, Virginia: Chrysler Museum History and archeology of Norfolk.
Oklahoma City, Oklahoma: Historical Society Museum Deals with the state's history from prehistoric times.
Pittsburgh, Pennsylvania: Carnegie Museum of Natural History Includes anthropological exhibits.
Portland, Oregon: Museum of Natural History Has archeological and anthropological Indian material.
Richmond, Virginia: Museum of the Archeological Society of Virginia Stresses Virginian archeology.
Rochester, New York: Museum of the New York State Archeological Association Covers the archeology and anthropology of North-east North America.
Sacramento, California: State Indian Museum The life and artefacts of Californian Indians.
St. Louis, Missouri: Museum of Science and Natural History Includes the anthropology and archeology of Missouri Indians.
San Diego, California: San Diego Museum of Man Includes old Indian skulls.
San Francisco: Asian Art Museum Asian art and culture from Neolithic times onward.
Seattle, Washington: Seattle Art Museum Features Ancient Egyptian and Precolumbian art.
Springfield, Illinois: Illinois State Museum Includes Indian material.
Tucson, Arizona: Arizona State Museum Features Indian culture, notably Hohokam items.
Washington, D.C.: National Museum of Natural History, Smithsonian Institution A major museum including prehistoric exhibits, and stressing American Indians and their ways of life.

USSR

Leningrad: Institute of Archeology Specimens include a prehistoric bone point embedded in a bison's shoulder blade.
Moscow: D.N. Anuchin Anthropology Museum Deals with prehistoric anthropology and archeology.

Gold crown from the Chavín of Peru, in the Museum of the American Indian, New York City

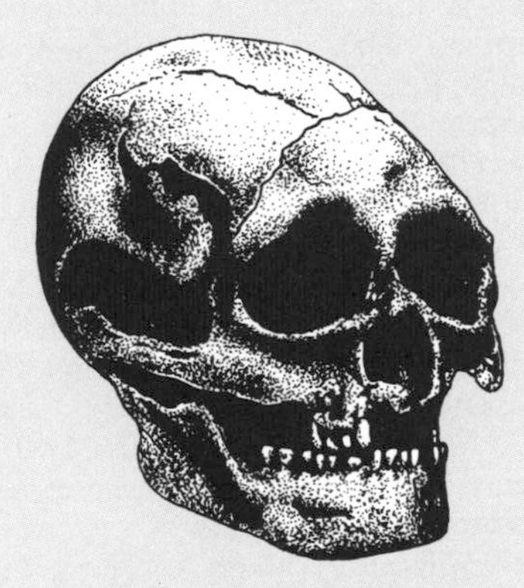

Skull of Del Mar Man in San Diego's Museum of Man

Figures painted on a thousand-year-old Hohokam earthenware vessel, in Tucson's Arizona State Museum

FURTHER READING

General
Aiello, L. *Discovering the Origins of Mankind* Longman, 1982
Brace, C.E. and Montagu, A. *Human Evolution* Macmillan, 1977
Clapham, Frances M. (editor) *The Rise of Man* Sampson Low, 1976
Coles, J.M. and Higgs, E.S. *The Archaeology of Early Man* Faber and Faber, 1969
Jurmain, R., Nelson, H., Kurashina, H., and Turnbaugh, W.A. *Understanding Physical Anthropology and Archeology* West Publishing Co., 1981
Kennedy, G.E. *Paleoanthropology* McGraw Hill, 1980
Leakey, R.E. *The Making of Mankind* Michael Joseph, 1981
Lewin, R. *Human Evolution* Blackwell Scientific Publications, 1984
Oakley, K.P. *Man the Tool-Maker* British Museum (Natural History), 1972
Poirier, F.E. *Fossil Evidence* C.V. Mosby Co., 1981
Tomkins, S. *The Origins of Mankind* Cambridge University Press, 1984
Wood, B. *The Evolution of Early Man* Peter Lowe, 1976
Note: News of major discoveries appears in quality newspapers, *National Geographic,* and science magazines such as *Nature* and *Scientific American.*
Chapter 1: What is Man?
Gribbin, J. and Cherfas, J. *The Monkey Puzzle* Bodley Head, 1982
Romer, A.S. *Man and the Vertebrates* (2 vols) Penguin Books, 1954.
Wood, P., Vaczek, L., Hamblin, D.J., and Leonard, J.N. *Life Before Man* Time Inc, 1972 (with later revisions)
Chapter 2: Primitive Primates and **Chapter 3: Evolving Anthropoids**
Colbert, E.H. *Evolution of the Vertebrates* Wiley, 1980
Napier, P. *Monkeys and Apes* Hamlyn, 1970
Szalay, F.S. and Delson, E. *Evolutionary History of the Primates* Academic Press, 1979
Chapter 4: Apes and Man
Goodall, J. *In the Shadow of Man* Houghton Mifflin, 1971
Gribbin, J. and Cherfas, J.: see above.
See also references for Chapters 2 and 3.
Chapter 5: "Men-Apes" and Early Man
Edey, M.A. *The Missing Link* Time Inc, 1972 (with later revisions)
Leakey, M. *Olduvai Gorge* Collins, 1979
Leakey R.E.: see above.
Reader, J. *Missing Links* Collins, 1981
Szalay, F.S. and Delson, E.: see above.
Chapter 6: Upright Man
Leakey, R. and Walker, A. "Homo Erectus Unearthed." *National Geographic* November 1985
White, E. and Brown, D.M. *The First Men* Time Inc, 1972
Chapter 7: Neandertal Man
Constable, G. *The Neanderthals* Time Inc, 1972
Shackley, M. *Neanderthal Man* Duckworth, 1980
Chapter 8: Modern Man in Europe through **Chapter 10: Since the Ice Age**
Claiborne, Robert *The First Americans* Time Inc, 1973 (with later revisions)
Cole, S. *The Neolithic Revolution* British Museum (Natural History), 1970
Hamblin, D.J. *The First Cities* Time Inc, 1973
Leonard, J.N. *The First Farmers* Time Inc, 1973
McEvedy, C. and Jones, R. *Atlas of World Population History* Allen Lane, 1978
Phillips, P. *The Prehistory of Europe* Penguin Books, 1981
Prideaux, T. *Cro-Magnon Man* Time Inc, 1973
Chapter 11: Discovering Man's Past
Binford, Lewis R. *In Pursuit of the Past* Thames and Hudson, 1983
Brothwell, D.R. *Digging up Bones* British Museum (Natural History)/Oxford University Press, 1981.

See also general reading list.

INDEX